Soil Biology

Volume 58

Series Editor
Ajit Varma, Amity Institute of Microbial Technology,
Amity University, Noida, Uttar Pradesh, India

The Springer series *Soil Biology* publishes topical volumes in the fields of microbiology, environmental sciences, plant sciences, biotechnology, biochemistry, microbial ecology, mycology and agricultural sciences.

Special emphasis is placed on methodological chapters or volumes. This includes coverage of new molecular techniques relevant to soil biology research or to monitoring and assessing soil quality as well as advanced biotechnological applications. Leading international authorities with a background in academia, industry or government will contribute to the series as authors or editors.

Key Topics: microbial-plant interactions; microbial communities; root symbiosis, mycorrhiza; rhizosphere environment; soil fauna, e.g. termites etc.; biochemical processes, soil proteins, enzymes, nucleic acids; degradation of biomaterials, decomposition, nutrient cycles; soil genesis, mineralization; bioremediation of contaminated sites; biotechnological applications of soil microorganisms.

More information about this series at http://www.springer.com/series/5138

Mukesh K. Meghvansi • Ajit Varma
Editors

Biology of Composts

Springer

Editors
Mukesh K. Meghvansi
Bioprocess Technology Division
Defence Research and Development Establishment
Gwalior, Madhya Pradesh, India

Ajit Varma
Amity Institute of Microbial Technology
Amity University
Noida, Uttar Pradesh, India

ISSN 1613-3382 ISSN 2196-4831 (electronic)
Soil Biology
ISBN 978-3-030-39172-0 ISBN 978-3-030-39173-7 (eBook)
https://doi.org/10.1007/978-3-030-39173-7

This Springer imprint is published by the registered company Springer Nature Switzerland AG.
The registered company address is: Gewerbestrasse 11, 6330 Cham, Switzerland

Preface

Composting is a process of decomposition of organic waste mediated by microorganisms. The end product is called "*compost*" that can be used as an amendment for improving the soil fertility. When this process relies on earthworms, it is termed as "*vermicomposting*" and the finished product as "*vermicompost*." Since the municipal waste generated in most of the developing countries contains a substantial amount of organic matter suitable for composting, this technology offers a win-win opportunity for the stakeholders in terms of getting rid of organic waste and providing organic fertilizer for agriculture. Soil amendment with compost reduces the dependency on chemical fertilizers required in agriculture, thereby indirectly contributing to reduced carbon footprint. Hence, soil amendment with compost is a sustainable and environment-friendly alternative to harmful chemical fertilizers.

This volume summarizes and updates the information about the fundamental aspects of composting, the interaction of various microorganisms, and the underlying mechanisms during composting and vermicomposting. In total, fourteen chapters contributed by the international subject experts have been organized into three parts. Part I deals with paradigms and mechanisms of composting. Part II discusses modern tools and techniques for composting research, whereas Part III is dedicated to composting applications/case studies. It is believed that this new volume on biology of composts will be of immense interest to students, researchers, and scholars who are working in the field of soil fertilization, plant nutrient management, and organic waste management. In addition, this book will facilitate the policy makers and administrative authorities across the world in making strategies for the management of organic waste through composting and vermicomposting.

The editors thank all the authors for their valuable contributions. It was a wonderful experience reading through the exciting knowledge synthesized by the authors in the form of book chapters. We take this opportunity to thank all the staff members of Springer, especially Mr. Anand Venkatachalam, Project Coordinator (Books), and Dr. Sabine Schwarz, Executive Editor (Springer, Biomedicine; Genetics & Microbiology) for their critical evaluation, constant support, and encouragement.

Dr. Mukesh K. Meghvansi wishes to express gratitude to Dr. Vijay Veer, ex-Director, Defence Research Laboratory, Tezpur, India, for his constant support and encouragement. Thanks are also due to my nematologist friend Dr. K.K. Chaudhary for his valuable suggestions and help in editing some of the chapters. Also, thanks to Mrs. Manju Meghvansi (wife) and two wonderful kids (Miss Lakshita Meghvansi and Miss Parnika Meghvansi) for their love, patience, support, and understanding.

Gwalior, India — Mukesh K. Meghvansi
Noida, India — Ajit Varma

Contents

Contributors

S. A. Alamri College of Science, Biology Department, King Khalid University, Abha, Saudi Arabia

Mohammed A. Al-Kahtani College of Science, Biology Department, King Khalid University, Abha, Saudi Arabia

S. A. Alrumman College of Science, Biology Department, King Khalid University, Abha, Saudi Arabia

Sanjeev Kumar Awasthi College of Natural Resources and Environment, Yangling, Shaanxi Province, People's Republic of China

Amel Ayari-Akkari Biology Department, College of Science, King Khalid University, Abha, Saudi Arabia

B. Basant Kumar Pillai National Institute of Technology, Warangal, Telangana, India

Laura Buzón-Durán Agriculture and Forestry Engineering Department, ETSIIAA, Universidad de Valladolid, Palencia, Spain

K. K. Chaudhary Department of Plant Protection, Hamelmalo Agricultural College, Hamelmalo, Eritrea

Lelde Grantina-Ievina Animal Disease Diagnostic Laboratory, Institute of Food Safety, Animal Health and Environment "BIOR", Riga, Latvia

Gederts Ievinsh Department of Plant Physiology, Faculty of Biology, University of Latvia, Riga, Latvia

Mouna Jeridi Biology Department, College of Science, King Khalid University, Abha, Saudi Arabia

Mohammad Haneef Khan Professor Joy Michelle Bergelson Lab, Department of Ecology and Evolution, University of Chicago, Chicago, IL, USA

N. Manoj Kumar Vellore Institute of Technology, Vellore, India

Jesús Martín-Gil Agriculture and Forestry Engineering Department, ETSIIAA, Universidad de Valladolid, Palencia, Spain

Pablo Martín-Ramos Department of Agricultural and Environmental Sciences, EPS, Instituto Universitario de Investigación en Ciencias Ambientales de Aragón (IUCA), Universidad de Zaragoza, Huesca, Spain

P. W. Mashela Green Biotechnologies Research Centre of Excellence, School of Agricultural and Environmental Sciences, University of Limpopo, Sovenga, South Africa

Mukesh K. Meghvansi Defence Research Laboratory, Tezpur, Assam, India
Bioprocess Technology Division, Defence Research and Development Establishment, Gwalior, Madhya Pradesh, India

M. Moustafa College of Science, Biology Department, King Khalid University, Abha, Saudi Arabia

Eduardo Pérez-Lebeña Agriculture and Forestry Engineering Department, ETSIIAA, Universidad de Valladolid, Palencia, Spain

K. M. Pofu Agricultural Research Council, Pretoria, South Africa

Richard Rampersaud Inter-American Institute for Cooperation on Agriculture, Couva, Trinidad and Tobago

Ieva Rodze Animal Disease Diagnostic Laboratory, Institute of Food Safety, Animal Health and Environment "BIOR,", Riga, Latvia

J. Rouse-Miller Department of Life Science, The University of the West Indies, St. Augustine, Trinidad and Tobago

Mercedes Sánchez-Báscones Department of Agroforestry Sciences, ETSIIAA, University of Valladolid, Palencia, Spain

E. Shokoohi Green Biotechnologies Research Centre of Excellence, School of Agricultural and Environmental Sciences, University of Limpopo, Sovenga, South Africa

T. Shumail College of Science, Biology Department, King Khalid University, Abha, Saudi Arabia

Sazada Siddiqui Department of Biology, College of Science, King Khalid University, Abha, Saudi Arabia

J. Siva Rani EID Parry, Chennai, India

Vijay Kumar Srivastava Sankalchand Patel Vidyadham, Sankalchand Patel University, Visnagar, Gujarat, India

C. C. G. St. Martin Inter-American Institute for Cooperation on Agriculture, Couva, Trinidad and Tobago

M. Chaitanya Sudha Department of Environmental Science, S.V University, Tirupati, India

G. Thomas-Barry Department of Life Science, The University of the West Indies, St. Augustine, Trinidad and Tobago

Sunita Varjani Gujarat Pollution Control Board, Sector-10A, Gandhinagar, Gujarat, India

Ajit Varma Amity Institute of Microbial Technology, Amity University Uttar Pradesh, Noida, Delhi-NCR, India

P. Vilpigue Department of Food Production, The University of the West Indies, St. Augustine, Trinidad and Tobago

Part I
Composting: Paradigms and Mechanisms

Chapter 1
Compost and Compost Tea Microbiology: The "-Omics" Era

Chaney C. G. St. Martin, Judy Rouse-Miller, Gem Thomas Barry, and Piterson Vilpigue

Abstract Composting is largely driven and mediated by microorganisms interacting with abiotic factors. However, until recently our knowledge of compost microbes has been heavily informed by culture-dependent methods that capture <1% of microorganisms involved in composting. This suggests that challenges related to optimizing the process of composting and the effectiveness of its products may be due to a partial understanding of microbial community structure, diversity, and function. Recent advances in molecular biology, bioinformatics, and sequencing technologies have presented opportunities to gain unprecedented insights into the microbiology of compost and compost tea by using "-omics" approaches. This chapter summarizes research aimed at better understanding the microbiology and effect of compost and compost tea using -omics approaches (genomics, metagenomics, metaproteomics, metaprotegenomics, metatranscriptomics, and metabolomics). Reference to findings from metaprofiling work done using genetic fingerprinting and culture-dependent techniques are made when necessary. To this end, a systematic framework that facilitates data integration and analysis from multi-omics and culture-dependent approaches are recommended to continue improving our knowledge of compost microbes.

Keywords Microbial ecology · Genomics, DNA sequencing · Microbial communities · Functional genes · Biomass degradation

---

C. C. G. St. Martin (✉)
Inter-American Institute for Cooperation on Agriculture, Couva, Trinidad and Tobago
e-mail: chaney.stmartin@iica.int

J. Rouse-Miller · G. T. Barry
Department of Life Science, The University of the West Indies, St. Augustine, Trinidad and Tobago
e-mail: judy.rouse-miller@sta.uwi.edu; gem.thomas@my.uwi.edu

P. Vilpigue
Department of Food Production, The University of the West Indies, St. Augustine, Trinidad and Tobago

M. K. Meghvansi, A. Varma (eds.), *Biology of Composts*, Soil Biology 58,
https://doi.org/10.1007/978-3-030-39173-7_1

1.1 Introduction

Microbial detection, identification, characterization, and quantification often pose several challenges during composting or in compost products. These challenges result from difficulty in extracting DNA and cells from complex compost matrices, which contains humic acids and other compounds that bind to DNA (Ogram et al. 1987; Pfaller et al. 1994; LaMontagne et al. 2002). Further challenges are posed by diverse microbial communities with complex interaction among biotic and abiotic factors. Despite these challenges, traditional culture-based methods have used to study the compost microbiology. Though useful, in soils, <1% of the genetic diversity of prokaryotes is captured using culture-based methods, with an unknown percentage captured in compost products (Torsvik et al. 1990; Hugenholtz 2002). This suggests that the three fundamental questions related to microbial ecology of composting: (1) What microbial types and community structures are present during composting and in compost products? (2) What are the roles and functions of these microbial types? and (3) What is the relationship between the activities of these microbial types and predictable results (disease suppression and plant growth enhancement)? have been incompletely informed by traditional culture-based methods. Such information is important since mechanism of action for compost-based products have been attributed in part or in full to the activities of diverse microbial communities or specific microbial species.

In this context, microbial studies on composting and compost-based products have advanced considerably since the seminal International Conference (Innsbruck, Austria in 2000) and publication on "Microbiology of Composting" (Insam et al. 2002). Research in this publication mainly used culture-plate methods alone or complemented with culture-independent approaches to provide "up-to-date" insight into processes and microorganisms associated with compost production. These include techniques, which allowed partial microbial community analysis (metaprofiling) such as, denaturing gradient gel electrophoresis (DGGE), phospholipid analysis (PLFA'S), community level physiological profiles (CLPPs), terminal restriction fragment length polymorphism (TRFLP), clone libraries, and amplified ribosomal DNA restriction analysis (ANDRA). Limited work was presented on effects of compost and compost tea on soil, rhizosphere, or phyllosphere microbial communities. Moreover, studies on the characterization of microbes during the compost tea brewing processes and the resulting end-products were outside the scope of this publication.

Notwithstanding these scope limits, the use of metaprofiling approaches, which includes polymerase chain reaction (PCR)-based methods with sequencing and phylogenetic analysis of 16S or 18S rRNA, have added much to the literature on compost and compost tea microbiology (Peters et al. 2000; Tiquia et al. 2005; Danon et al. 2008; de Gannes et al. 2013; Larkin and Tavantzis 2013). This includes a "general" consensus that: (1) aerobic composting is characterized as a microbially driven, self-heating process, which results in temperatures >50 °C, with subsequent and sustained temperatures of 60–80 °C, followed by a steady cooling of the compost heap (Ryckeboer et al. 2003; Kumar 2011); (2) mesophilic and thermophilic microorganisms with different physiological requirements and tolerance levels, decompose the organic matter as is consistent with continuous environmental fluxes during composting (Alfreider et al. 2002; Partanen et al. 2010; Jurado et al. 2014); (3) the level of abundance of routinely found bacterial phyla such as Actinobacteria, Bacteroidetes, Firmicutes, and Proteobacteria is dependent on the characteristics of the starting materials and the type of composting procedure used (Ryckeboer et al. 2003; de Gannes et al. 2013); and (4) fungi generally seem to be most important in the cooling and curing stages of composting (Neher et al. 2015) since they are not readily or abundantly detected at temperatures >65 °C (thermophilic stage) (Langarica-Fuentes et al. 2014a, b). This suggests that relative to bacteria, their degradative activities are minor during the thermophilic phase (Langarica-Fuentes et al. 2014a, b).

However, as approaches informing the biology of composting, there are three main limitations of PCR-based analyses of 16S and 18S rRNA amplicons (Zhou et al. 2010): (1) obtaining information to sequence between primers is limited with the use of PCR. As such, the amount of functional information captured using PCR is limited. (2) most PCR-based measurements provide mainly relative abundance information since PCR-based analysis is only somewhat quantitative. (3) the probability of entirely missing some lineages due to PCR-primer mismatches is of concern, particularly with complex environmental samples. Furthermore, 16S or 18S rRNA amplicon is a highly conserved molecule, as such, they do not provide sufficient species and strain-level resolution as it targets single or few genes (Konstantinidis et al. 2006).

Most of these challenges have been addressed with advances in molecular biology, bioinformatics, and sequencing technologies (Handelsman 2005; Tringe et al. 2005). These advances have allowed deeper study ("-omics") of biomolecules along the central dogma framework of molecular biology. These include the study of total: DNA/genome (genomics), the mRNA/transcripts (transcriptomics), proteins (proteomics), and metabolites (metabolomics) of an organism. When the total complement of these respective biomolecules is examined for entire communities of organisms, the prefix "meta" (meaning beyond) is added to the root word that indicates the type of molecule being studied. For example, "metaproteomics" studies the entire protein complement of microbial communities from environmental samples. Collectively, these "-omics" studies have advanced an era in microbial ecology, which has allowed unprecedented discovery of new taxa, genes, and functions. Specifically, the combination of DNA (genomic)-, mRNA-, protein-, and metabolite-based (postgenomic) analyses of microbial communities from distinct environments has allowed for in-depth elucidation of the structure, diversity, functions, and interactions of microbial communities, which are linked to various environmental processes (Simon and Daniel 2011).

This chapter aims to summarize research findings aimed at better understanding the microbiology and effect of compost and compost tea using -omics approaches (genomics, metagenomics, metaproteomics, metaprotegenomics, metatranscriptomics, and metametabolomics). The technical definition applied to metagenomics in this chapter does not include studies that use PCR to amplify gene cassettes (Holmes et al. 2003) or random PCR primers to access genes of interest (Eschenfeldt et al. 2001; Brzostowicz et al. 2003), since these methods provide limited genomic information beyond the amplified genes. It also excludes the broader definitions, which refer to metagenomics as any type of analysis of DNA acquired directly from environmental samples (Handelsman et al. 1998). Instead, the definition is more process oriented, involving either the direct analysis of total community DNA or vector cloning before analysis (whole metagenome shotgun sequencing).

Owing to the limited -omics studies on compost tea and aspects of compost and composting, findings obtained using metaprofiling approaches are included when relevant. Details on definitions, standards, uses, disease suppression, and challenges with composting, compost products are not presented in this chapter since these have been extensively reviewed by Litterick et al. (2004), Scheuerell et al. (2005), St. Martin (2014), and St. Martin and Ramsubhag (2015). Due to the many variations in composting methods, feedstocks, and abiotic factors, a serious attempt is made not to suggest a "typical" microbiological profile, process, or ecology for compost and compost tea, particularly at a genus or species level. Instead, greater emphasis is placed on detailing unique findings and highlighting emerging research trends on compost microbiology. Some specific limitations of -omics approaches are briefly stated in the conclusions or at the end of some subsections.

1.2 Genomic Approaches

1.2.1 Genomics

Numerous microorganisms involved in composting have been identified and extracted to evaluate their potential roles in various agricultural, environmental, and industrial applications. Until recently, most of these microorganisms have been identified using phenotypic characterization techniques (morphology, biochemical profiles, diagnostic staining, and media) and 16S or 18S rRNA gene sequencing and/or analysis of phospholipid profiles (Insam et al. 2002; Ryckeboer et al. 2003). To date, a major application focus of many studies has been single-species microbial isolation from compost or compost tea to increase the understanding and predictability of plant disease suppression or growth enhancement. To this end, the suppressiveness of compost-based products have been attributed to bacterial species mainly from the genera *Bacillus*, *Serratia*, *Pseudomonas*, *Stenotrophomonas*, *Flavobacterium*, *Streptomyces*, and *Enterobacter* (Kwok et al. 1987; Hoitink 1990; Phae et al. 1990; Inbar et al. 2005; Kerkeni et al. 2007; Ryan et al. 2009; Kouki et al. 2012; Khaldi et al. 2015). Whereas, fungal species from the genera *Trichoderma*, *Penicillium*, *Aspergillus*, and *Gliocladium* and *Fusarium* (non-pathogenic) have been reported as the main taxa related to the disease suppressive effect of compost and compost tea (Hoitink and Fahy 1986; Kwok et al. 1987; Malandraki et al. 2008; Daami-remadi et al. 2012).

Though useful, the identification of some of these taxa from compost microbiomes is represented by draft or incomplete genomes in various gene banks (INSDC 2018). This is partly due to the previously high cost and processing speed limitations of second-generation sequencing technologies, which limited more extensive and in-depth examination of microbial species (Ku and Roukos 2013). This means that incomplete genomes only provide genomic information on the genes that are amplified. Furthermore, owing to the genome sequencing of microbial species being severely skewed toward a few phyla that contain model organisms (Land et al. 2015), many microorganisms present in compost have not been fully sequenced. This highlights the tremendous scope for genomic information that can significantly impact our understanding of microbiology of composting and compost products.

Genomics, which refers to the study of the complete genetic complement of a species, rather than the study of only single genes, provides tremendous opportunities for more in-depth insights into the structural and functional characteristics of microorganisms in composting. Specifically, structural genomics offers the opportunity for sequencing the complete DNA of an organism (genome) and determining the complete set and three-dimensional structure of proteins produced by an organism. Whereas, functional genomics focuses on gene transcription, translation, and protein–protein interactions. More specifically, it involves the study of mRNA (transcriptomics), proteins (proteomics), and metabolites (metabolomics) in a biological sample.

Most of the genomics work on compost have been done using cultivated "bulk" cell populations of single microbial species, which falls more aptly in the domain of isolation genomics. Though useful, particularly for comparative genomics, information from such studies is limited to microbial species that can be cultured. More so, with exception of rare instances where cells can be accurately synchronized, bulk measurements destroy important biological information such as cell phenotypes, metabolic states, and transition between states and cellular functions by averaging individual cell signals (Trapnell 2015). In contrast, single-cell genomics, which refers to the sequencing of a genome of a single cell selected from a population of mixed cells, makes possible the study of genomes of uncultivated microorganisms, particularly from complex communities such as compost (Rinke et al. 2013). As such, single-cell genomics provides a critical link between isolate genomics and metagenomics. Such a link is important to gain insights into growing formerly uncultivable microorganisms and reconstructing genomes of dominant microbial species in environmental samples.

Progress in this direction is already evident with the advent of culturomics, a highly diverse culture conditions-rapid microbial identification approach, which has resulted in the first-time cultivation of many bacteria (Lagier et al. 2015). Culturomics also addresses a limitation of metagenomics methods, which is the inability to detect minority microbial populations (species $<10^5$ per gram) such as, *Salmonella enterica* serovar (pathogen), in environmental samples (Lagier et al.

2012). Therefore, single-cell genomics complemented with culturomics has the potential to transform our understanding of gene regulation during plant disease development and suppression with compost or compost tea. Also, a combined approach of stable isotope-labeled substrates and single-cell analyses could provide insights into the in situ function of uncultivated microbes during composting or soils treated with compost-based products (Eichorst et al. 2015).

To this end, Matteoli et al. (2018) full genome sequenced a *Serratia marcescens* strain from vermicompost and assessed its plant growth-promoting properties. They reported that *S. marcescens* solubilized P and Zn, produced indole compounds, colonized hyphae and countered the growth of phytopathogenic fungi *Fusarium oxysporum* and *F. solani* in vitro. Using a genome-centric analysis of a group of thermophilic and cellulolytic bacteria isolated from compost, Lemos et al. (2017) discovered four novel genomes. The novel genomes encoded several glycoside hydrolases and possessed genes related to lignocellulose breakdown. Likewise, Akita et al. (2017) isolated and full genome sequenced *Ureibacillus thermosphaericus* A1, a thermophilic *Bacillus* from compost. *U. thermosphaericus* A1 produced several enzymes of industrial importance including catalase, amino acid dehydrogenase, and esterase. Additionally, *U. thermosphaericus* A1, which grew at temperature ranging from 37 °C to 55 °C, was used as a biocatalyst for degrading lignocellulosic biomass. Brumm et al. (2016) also full genome sequenced *Geobacillus* sp., which was one of several thermophilic microorganisms isolated from wood compost. According to the researchers, this genus is known for thriving in varied harsh environments, which suggests that this species may contain enzymes suitable for industrial applications.

Other interesting microbial isolates with the ability to decaffeinate coffee, tea, and chocolate (Divine 2014), decolorize dyes in effluents (bioremediation) (Abd El-Kader et al. 2019) and degrade plastics (Dang et al. 2018) have been extracted from compost/compost tea and identified using 16S rDNA sequencing. Limited published work on the extraction of viruses from compost for beneficial agricultural use has been done. However, as indicated by the work of Heringa et al. (2010), viruses extracted from sewage effluent may have a potential role in disease suppression with compost-based products. Heringa et al. (2010) reported that within 4 h of applying an effluent extracted mixture of five strains of bacteriophages to dairy manure compost, a >2-log reduction in *Salmonella enterica* was observed across moisture levels compared with controls. It is possible that strains of bacteriophages with similar human and plant pathogen and disease suppressive effects may be in compost tea. Though the study of single-species isolates is useful, it is often activities and interactions of different types of microorganisms that have been attributed to the increased efficacy of compost products (St. Martin and Brathwaite 2012; Cook and Baker 1983). Therefore, the analysis of the microbial communities of compost and compost tea is equally important.

1.2.2 Metagenomics

In principle, the basis of metagenomics is that the entire genetic complement of microbial communities from environmental samples could be sequenced and analyzed in a like manner as whole genome sequencing a single microbial isolate. As such, metagenomics refers to the sequence (computational) and function-based (experimental) analysis of the collective microbial genomes contained in environmental samples. Isolation and lab cultivation of individual species are not necessary for such analysis and prior knowledge of the microbial communities is not required (Riesenfeld et al. 2004). A detailed description of the process of metagenomics is provided by Sabree et al. (2009) and can be summarized as extraction of DNA directly from the microbial community, followed by cloning of DNA into a surrogate host then analysis of metagenomic DNA (sequence- or function-driven). Conceptually, the sequence-driven analysis identifies the genes and "metabolic pathways" by comparing metagenomic DNA with genes found in other samples with known functions. Whereas, functional-driven analysis screens for expression of activities (enzymes or antibiotic production) of interest conferred by the metagenomic DNA.

Though crucial for relating microbial ecology to the efficacy of processes and compost products, metagenomics studies on compost and compost tea have been limited. Previous studies have focused on metaprofiling composting phases (Insam et al. 2002; Klammer et al. 2005; Danon et al. 2008) and elucidating the mechanisms of plant disease suppression using compost-based products (Scheuerell and Mahaffee 2004). Resulting from this research trend is a preponderance of work on soil-borne pathogens and compost-induced changes in the rhizosphere/soil (St. Martin 2015) with less work on aerial pathogens and induced changes in the phyllosphere. Therefore, compared to the rhizosphere, our knowledge of the microbiology of phyllosphere as affected by compost tea or compost is lagging. Furthermore, Vorholt (2012) noted that for the most part, basic questions related to which microbial types are present in the phyllosphere and their functions, remain unanswered.

In one of the first reports of metagenomic studies on composting, Martins et al. (2013) presented findings that contrasted results from previous works done using culture-dependent (Golueke et al. 1954) and -independent studies (Peters et al. 2000; Ishii et al. 2000; Alfreider et al. 2002; Schloss et al. 2003; Partanen et al. 2010). They reported that *Lactobacillus* genus (particularly *L. brevis*) had a clear dominance in the older (thermophilic) compost sample whereas mesophilic compost sample was dominated by members of *Acinetobacter* and *Stenotrophomones* genera. Traditionally, composting literature has shown that the initial stage of composting is dominated by mesophilic organic acid-producing bacteria such as *Lactobacillus* spp. and *Acetobacter* spp., which degrade readily degradable compounds (e.g., sugars). This results in lower pH levels (Golueke et al. 1954; Yu 2014), growth inhibition of other microbes (Yu 2014) and high odor emission, particularly when *Clostridia* is present (Sundberg et al. 2011, 2013). The authors contended that the dominance of *Lactobacillus* spp. in older compost may be due to the competitive advantage of the species, achieved partly by the production of bacteriocins. Peters et al. (2000) noted that *Lactobacilli* were typically the dominant microorganisms under oxygen limitation degrading relatively wet plant material or substrate. Though scarcely reported in compost, studies have identified thermophilic Lactobacilli in traditional yogurt and cheeses (Randazzo et al. 2002; Azadnia et al. 2011). These studies may further support the findings of Martins et al. (2013) and highlight the usefulness of metagenomics in advancing knowledge in compost microbiology.

Martins et al. (2013) further reported that bacterial enzymes, possibly from Clostridiales and Actinomycetales were fully responsible for degrading recalcitrant lignocellulose (Allgaier et al. 2010; Bugg et al. 2011). This finding fits well with the current understanding of the degradation of recalcitrant lignocellulose during composting as reviewed by Bugg et al. (2011). However, traditionally, in composting literature, the degradation of recalcitrant lignocellulose has been mainly attributed to fungi (Tuomela et al. 2000; Sánchez 2009). Martins et al. (2013) explained that the relatively frequent anaerobic and thermophilic conditions during composting possibly diminish the degradation role of fungi as it relates to recalcitrant lignocellulose.

In a more recent metagenomic study, Antunes et al. (2016) explored the microbial community structure of large-scale thermophilic composting using shotgun DNA and 16S rRNA gene sequencing techniques. They reported that at the phylum and order level, results of the shotgun DNA and 16S amplicon analyzes generally agreed with each other. However, at the genus level, 16S results on microbial composition structure starkly contrasted shotgun DNA findings. That is, none of the five most abundant OTUs in 16S analysis seemed to correspond to the species *Rhodothermus marinus*, *Thermobispora bispora*, *Symbiobacterium thermophilum*, *Sphaerobacter thermophilus*, and *Thermobifida fusca* classified using MyTaxa (Chengwei et al. 2014) through shotgun DNA data. The authors attributed this discrepancy to the unavailability of complete "reference" genomes of microorganisms present during composting, which precluded identification during the analyses of the shotgun DNA metagenomics dataset. The unavailability of complete reference genomes poses a serious bottleneck challenge in metagenomic works on composting and compost products. This challenge will persist until taxonomic databases are more comprehensively populated with full genomic entries and more novel classification schemes are developed.

Nonetheless, Antunes et al. (2016) noted that the most abundant orders for shotgun DNA and 16S amplicon were Clostridiales, Bacillales, and Actinomycetales. These orders along with Enterobacteriales and Thermoanaerobacterales were proposed as the bacterial core group mainly responsible for degrading lignocellulosic biomass at different stages of composting. Actinomycetales played a primary role in lignocellulosic degradation throughout composting; Bacillales at the start and middle, and Clostridiales and Enterobacteriales at the start and end, respectively. Interestingly, the relatively high abundance of Clostridiales, which include micro-aerophilic or anaerobic species, in the initial stages of composting suggests quasi-static conditions that favored the fluctuations between anaerobic and aerobic micro-environments (Ryckeboer et al. 2003; Jurado et al. 2014). Hemsworth et al. (2015) reported that anaerobic microorganisms play a major role in degrading biomass. Moreover, members of the Clostridiales and Bacillales orders have been reported to possess genes that encode enzymes, which degrade hemicellulose and cellulose (Kanokratana et al. 2011; Ventorino et al. 2015).

In contrast to the findings of Antunes et al. (2016), the relative abundance of *Enterobacter* spp. are generally described as highest during early composting phases (Chandna et al. 2013). Nonetheless, *Enterobacter* spp. have been associated with lower temperatures (<60 °C) (Gbolagade 2006; Chandna et al. 2013), which may explain their relative abundance and lignocellulosic degrading activity during the end phase of composting. This was particularly evident for *Klebsiella pneumoniae*, the predominant species of the Enterobacteriales order, known to perform cellulose and hemicellulose degradation in composting ecosystems (Droffner et al. 1995) and wood termite guts (Doolittle et al. 2008). Antunes et al. (2016) also reported the almost complete genome construction of a novel biodegrading bacterial species (order Bacillales) capable of bioconverting all components in plant biomass.

Results from such studies provided the foundation or link to emerging trends in compost metagenomic studies, which focus on: (1) compost microbiomes as rich and diverse sources for discovering biochemical catalysts and pathways for advanced biofuel production or other industrial and bioremediation (soil, wetlands, and plastic polluted spaces) applications (bioprospecting) (Dougherty et al. 2012; Yi-fang et al. 2013; Wang et al. 2016); (2) composting as a strategy to eliminate antibiotic resistance genes (ARG)/resistome and residues from animal manure (Wang et al. 2017; Chen et al. 2018; Gou et al. 2018) that pose potential global health risks, particularly for the antibiotics tetracycline, sulfonamide, and fluoroquinolone, commonly used by humans and in livestock production; and (3) compost as a carrier medium or enhancer for biocontrol (entomopathogenic nematodes) (Herren et al. 2018) and plant growth promoting (Rhizobacteria and Mycorrhiza) agents (Yang et al. 2018).

In this context, in one of the first study to use a functional metagenomics approach, Yeh et al. (2013) identified and cloned a novel endogluconase gene (RS-EG1) from the metagenome of rice straw compost. The RS-EG1 shared approx. 70% similarity with its closest known bacterial cellulase from *Micromonospora aurantiaca* and *Thermobispora* sp. and was stable over wide ranges of temperature and pH. The authors concluded that the novel endogluconase was potentially useful in the production of cellulosic biofuel. Allgaier et al. (2010) and Dougherty et al. (2012) who successfully used targeted metagenomics to identify several active enzymes with differing hemicellulose degrading activities reported similar results.

As it relates to ARG studies, Wang et al. (2017) integrated metagenomics and time-series metatranscriptomics data to determine if changes in resistome expression were related to the evolution of active microbiome profiles during composting. They found that the principal determinant that defined the diverse transcriptional response of the resistome was the microbial phylogeny during composting. During the mesophilic and thermophilic stages, the most prevalent phylum that harbored ARG families were Firmicutes, whereas Actinobacteria and Ascomycota were primary source in the matured compost. Moreover, amending composting with biochar significantly reduced the aggregated level of ARG expression by an additional 38% compared non-amended compost treatment. The researchers suggested that the quicker microbial succession observed in biochar-amended composting, which was mainly due to enhanced composting kinetics, may account for this difference.

Further to the composting process, Gou et al. (2018) investigated the effect of compost-treated soils on the temporal succession of ARGs. They found that diversity and abundance of ARGs in compost-treated soils were significantly lower compared to manure-treated soils. Therefore, it is concluded that composting was an effective method to limit ARG dispersal, which was linked with land application of organic wastes. In a related study, Chen et al. (2018) found that sewage sludge and manure applications to the soil over 10 years increased the incidence and abundance of ARGs in phyllosphere. Therefore, they concluded that soil may serve as an antibiotic resistome reservoir for phyllosphere. Furthermore, ARGs profiles were strongly correlated with bacterial communities, which were dominated by Proteobacteria, Bacteroidetes, Actinobacteria, and Firmicutes.

As it relates to studies on compost as a carrier medium or enhancer for biocontrol, Yang et al. (2018) used PCR-based method with sequencing of 18S gene to investigate the impact of compost applications rates (0, 11.25, 22.5, and 45 Mg/ha) on the composition and abundance of arbuscular mycorrhizal fungi (AMF) communities at the seedling, flowering, and mature stage of soybean. They found that moderate and high compost applications rates significantly increased AMF root colonization and extraradical hyphal density and the abundance of *Paraglomus* sp. generally decreased along the compost application gradient, while *Rhizophagus fasciculatum* showed an opposite trend.

Using Nematode Indicator of Compost Maturity as proposed by Steel et al. (2018), Herren et al. (2018) reported that *Steinernema feltiae*, an entomopathogenic nematode, which was applied to soil via mature compost, had the highest survival and virulence against *Galleria mellonella* compared to when it was applied without compost or via immature compost. They suggested that higher survival rate of EFN in mature compost was due to reduced predation pressure (by mites and collembolans) on the EPN, in favor of other nematodes. These other nematodes generally had higher and more diverse populations in mature compared to immature compost. Nitrate-N concentration was the only characteristic differentiating mature from immature composts and was significantly higher in mature compost. Pan et al. (2015) and Griffiths et al. (1992) reported that with the application of nitrate-N nematode diversity and evenness increased and dominance decreased, which may support the predation pressure inference made by Herren et al. (2018).

In a recent study, Blaya et al. (2016) assessed the microbial consortia of composts made of different agro–industrial waste and with varying levels of suppressiveness against *P. nicotianae.* They found that Ascomycota phylum had a higher relative abundance in suppressive composts compared to non-suppressive media and was negatively correlated with Phytophthora root rot incidence. The researchers postulated that the high proportions (67–75%) vineyard pruning waste promoted higher relative abundance of Ascomycota (particularly of Sordariales and Hypocreales taxa) and microbial activity, which were essential for controlling the disease. No unique fungi or bacteria that would be suggestive of suppressive or conducive compost were identified. However, conducive compost contained relatively higher abundance of Actinobacteria and Gemmatimonadetes compared to suppressive composts.

To date and there is limited published articles on the metagenome of compost tea and how it may differ to that of compost. Currently, a research team from University of Alabama, Huntsville, USA is undertaking a project on the metagenomic analysis of aerated compost tea (Cseke 2016). To this end, metaprofiling studies have shown that bacterial phyla Actinobacteria, Bacteroidetes, Firmicutes, Proteobacteria, Verrucomicrobia, Chloroflexi, Planctomycetes, and Acidobacteria dominated non-aerated compost teas (NCTs) made from different compost sources (vermicompost and agricultural and municipal wastes) (Mengesha et al. 2017). Ascomycota was the predominant fungal phylum detected in NCTs. However, results revealed dissimilarities at genera level with *Thiobacillus*, *Malikia*, *Hydrogenophaga*, *Desulfomicrobium*, and *Prolixibacter* observed only in NCTs made from agricultural waste. Whereas, genera *Oligosphaera*, *Paracoccus*, *Synergistes*, and *Anoxybacillus* were only observed in NCTs made from solid municipal waste. Although compost source affected microbial structure of the NCTs, it is uncertain how the compost tea brewing process affected characteristics such as relative abundance of microbial species. This is definitely an area that requires further work in most compost–compost tea studies.

To this end, based on a high number of DGGE bands, Diánez et al. (2018) suggested regardless of the incubation conditions (aeration versus not aeration) or compost source (spent mushroom and grape marc compost, and crop residues vermicompost), all compost teas contained a high diversity of species. It is unclear what can be attributed to the high population of microbial species in all compost tea given that compost source and incubation conditions were not suggested as significant factors. Nevertheless, they reported that aerated compost teas (ACT) provided higher bacterial richness, diversity, and evenness values compared to NCT. That is, during incubation the effect of aeration on bacterial richness and diversity was compost type-specific. Bacteroidetes and Proteobacteria were reported as the main phyla dominating the ACTs and NCTs in this study. Kim et al. (2015) found that the dominant bacterial genera were *Bacillus*, *Ochrobactrum*, and *Spingomonas* reported similar results. However, the density of fungal populations were significantly lower than those of bacteria and decreased significantly in all compost teas after 2 days incubation.

As it relates to soil application, the efficacy of ACT in introducing beneficial microorganisms and its effect on soil microbes and disease dynamics under crop rotation systems was investigated by Larkin (2008) using various techniques including soil dilution plating and profiles of fatty acid methyl ester and substrate utilization. It was reported that ACT successfully delivered microorganisms into the soil, significantly increased soil microbial population and activity, reduced soil-borne disease, and improved yield under the barley/rye crop rotation, but not under other rotations or plots continuously cultivated with potato. Due to these results, Larkin (2008) suggested that to be effective, ACT required a minimum level of support from the soil microbial environment. It was, however, evident that the "minimum" soil support level required was mediated by crop rotation, which had a more dominant, positive, and distinct effect on soil microbial community characteristics than ACT. Fritz et al. (2012) reported a similar positive response to plant growth, which was associated with vermicompost teas with the highest microbial population and diversity. They reported that the addition of different carbon substrates during brewing significantly affected the richness and diversity of microbial communities. Furthermore, the microbial communities of the solid composts were distinct from that of the compost teas produced from them. DGGE profiles showed that compost tea stored beyond 1 week at ≥ 10 °C resulted in a significant change in microbial communities, which was probably related to a loss in quality. The authors postulated that the difference between the microbial communities of solid composts and that of the corresponding compost teas was that compost bacteria were not being extracted proportionally into the tea.

Using COMPOCHIP microarray analyses, the authors found that the main discriminatory microbial species/group across compost types were Acinetobacter, specifically, *Acinetobacter lwoffii* and *Acinetobacter calcoaceticus*. Microbial species reported to be involved in nitrogen cycling (*Nitrosovibrio* spp. and *Nitrosospira* spp.) (Kowalchuk et al. 1999; Innerebner et al. 2006; Danon et al. 2008), degradation of biopolymers (Alfreider et al. 2002), plant disease suppression, and human pathogens (*Xanthomonas* spp. and *Stenotrophomonas* spp.) (Franke-Whittle et al. 2009) were also detected in compost teas. Of concern to the authors was the presence of *A. calcoaceticus*, which has been linked to bovine spongiform encephalopathy (Cayuela et al. 2008) and the seemingly inconclusive results on the reproducibility of compost tea. Both concerns have serious implications, which can limit the marketability of compost teas. In contrast, Ottesen et al. (2009) reported that in the clone library of compost tea, neither of the bacterial genera most commonly associated with produce-related illness outbreaks (*Salmonella* and *Escherichia*) were observed.

Findings from metagenomic studies on microbiology of composting process were generally in agreement with cultivable and partial microbial community analysis at the domain and phyla levels, which showed bacteria was the dominant domain through all phases of composting with the abundance of the dominant phyla Firmicutes decreasing after thermophilic, while that of Proteobacteria, Bacteroidetes, and Actinobacteria increased. However, stark differences in findings across studies were more apparent at the genus and species levels as dictated by contrasting feedstock characteristics and nutrients, oxygen, temperature, and to some extent pH levels. This ability of differing microbial taxon to carry out processes/composting at similar rate despite differences in microbial community composition speaks strongly to the extent of functional similarity and/or redundancy in compost microbial consortia. Although much work is still needed to better access and represent compost metagenome by investigating DNA extraction biases, obtaining representative samples and providing better tools to study microbial diversity, current metagenomic libraries are proving to be a great resource to postgenomic or function-related studies such as metatranscriptomics, metaproteomics, and metametabolomics.

1.3 Postgenomic Approaches

1.3.1 Metatranscriptomics

While metagenomic DNA-based analyses can provide information on the functional and metabolic capacity of microbial communities, it cannot differentiate between expressed and non-expressed genes. Thus, it does not reflect the actual metabolic activity of microbial communities (Sorek and Cossart 2010). By focusing on genes expressed by the entire microbial community, metatranscriptomics is used to examine the *active* functional profile of a microbial community. With this approach and the use of point or time sampling, transcriptional profiles associated with specific microbial populations within a community can be produced (Carvalhais et al. 2012). Such profiles are useful in obtaining a more in-depth understanding of the potential activities and regulatory mechanisms of microbial communities in compost-based products.

In contrast to qPCR and microarrays, respectively, metatranscriptomics does not use primers or probes. Rather, as with metagenomics, it consists of the random sequencing of mRNA of the microbial community. As such, according to Moran (2009) and Carvalhais et al. (2012), the constraints of preselecting which and how many genes should be surveyed in a study when using qPCR and microarrays is overcome with the use of metatranscriptomics. For this reason, employing metatranscriptomics results in the sequencing of transcripts from microbial communities with less bias compared with the use of microarray or qPCR techniques. Moreover, Moran (2009) reported that using metatranscriptomics, it is possible to distinguish paralogous sequences that might cross-hybridize on a microarray.

In this light, to obtain an overall profile of gene function during composting, Antunes et al. (2016) used Cluster of Orthologous Groups (COGs) (Galperin et al. 2015) to classify coding sequences from two composts assembled shotgun reads. They found the most abundant functions in the composts were related to cell maintenance and proliferation, signal transduction, and defense mechanisms. Hierarchical clustering of the metatranscriptome coding sequences (CDSs) revealed distinctions among the beginning/turning, middle, and end phases of composting as evident by their corresponding COG category differential abundance levels. The main functional groupings of metabolism, cellular processes and signaling, and information storage and processing seemed more equally distributed with higher COGs abundance levels in the end phase of composting. Whereas, beginning/turning of composting phase was more skewed to functions related to metabolism (energy production and transport/coenzyme metabolism). Among the composting phases, the middle phase of composting had the lowest in the abundance of COGs related to metabolism but had noticeably high COGs related to replication/recombination/repair activities. The authors inferred that functions related to the microbial metabolism in the beginning phase or after turning compost were related to the degradation and utilization of easily degradable organic nutrients. Whereas, the higher expression levels of genes in the replication/recombination/repair category may be explained by changes in microbial composition between the beginning/turning and middle groups as was evident in the taxonomic analysis.

A clear and solid inference of the functional profiles obtained at the end of composting was not provided by the authors. However, the general inferences made about results trend in beginning and middle of composting are consistent with that made for a typical aerobic composting process (Ryckeboer et al. 2003; Kumar 2011). Antunes et al. (2016) also confirmed that most coding sequences (CDSs) related to plant biomass degradation belonged to members of the Bacillales, Clostridiales, Actinomycetales, and Thermoanaerobacterales. Members of these orders were associated with five CAZyme classes: glycoside hydrolases, carbohydrate-binding modules, glycosyl transferases, carbohydrate esterases, and polysaccharide lyases.

Similar results were reported by Wang et al. (2016). They found that the dominant phylum Actinobacteria in rice straw-adapted (RSA) microbial consortia enriched from compost, contained about 46.1% of Carbohydrate-active enzyme (CAZyme) genes, which harbored an extensive catalog of the cellobiohydrolase, β-glucosidase, acetyl xylan esterase, arabinofuranosidase, pectin lyase, and ligninase genes. Both studies showed degradation of hemicellulose, cellulose, pectin, and lignin occurred throughout composting. However, Antunes et al. (2016) specifically noted that turning compost temporarily slowdown the degradation process. This "slowdown" was probably linked to restoring a microbial population profile similar to what was seen at the beginning of composting. Neither Antunes et al. (2016) nor Wang et al. (2016) found any CDSs that could be annotated as fungal ligninolytic enzymes during thermophilic composting. Hence, they concluded that the degradation of lignocellulose during thermophilic composting is mainly or even exclusively the result of bacterial enzymatic activity.

As it relates to ARGs, Wang et al. (2017) reported that during the active phase of composting, the relative abundance of expressed resistome significantly increased but decreased during the cooling and maturing phases. This trend was particularly evident for the three most prevalent resistance mechanisms (ATP-binding cassette antibiotic efflux pumps, tetracycline resistance, and vancomycin resistance proteins). The expression of tetracycline resistance genes (*tetM-tetW-tetO-tetS*) declined as composting progressed and composting had no effect on the expression of sulfonamide and fluoroquinolone resistance genes.

Emerging trends and interest in plant nutritional genomics are expected to further fuel metatranscriptomics works involving compost and compost tea. That is, metatranscriptomics works focused on linking plant nutritional status under stressed conditions to plant–microbe interactions as facilitated or affected by compost or compost tea. In this regard, research works by Carvalhais et al. (2013) have set a strong foundation in this direction. They found that the nutritional status of maize affected the transcriptome of a beneficial root colonizing bacterium, *Bacillus amyloliquefaciens,* due to compositional changes in root exudates. Exudates from nitrogen-deprived maize triggered a general stress response in *B. amyloliquefaciens* and downregulated genes associated with chemotaxis and motility in the exponential growth phase. These results have serious implications for beneficial bacteria such as *Pseudomonas fluorescens* (plant disease control) and *Azospirillum brasilense* (nitrogen fixing) that depend on chemotactic motility to colonize roots/rhizosphere. Notwithstanding these results, much is needed to extend such works from transcriptome to metatranscriptome level for PGPR focusing on using compost or compost tea to enhance resident rhizosphere microbial populations or as a carrier for PGPR with diverse modes of action.

As noted by Borek et al. (1958) and Schut et al. (1993), mRNA is more sensitive to environmental fluxes than proteins due in part to its shorter half-life and lower inventory in cells. As such, the metatranscriptome may be better able to reflect more near real-time regulatory reactions of cells to environmental changes compared to the metaproteome (Moran 2009; Moran et al. 2013). A review of the limitations and biases of metatranscriptomics as it relates to difficult protocols for synthesizing and amplifying cDNA and isolating mRNA is provided by Carvalhais et al. (2012) and Moran (2009). Moran (2009) noted that most of these limitations are associated with the instability and impurity of mRNA during isolation or storage (Redon et al. 2005; Deutscher 2006; Opel et al. 2010) and its relatively low quantum in microbial communities from environmental samples (He et al. 2010). Another limitation related to specifically to prokaryotic microorganisms is the lack of 3′-poly-A tails (Moran 2009).

1.3.2 Metaproteomics and Metaproteogenomics

Metatranscriptomics analyses do not give any insight into whether transcripts are translated to proteins, or if constitutively expressed genes are differentially post-translationally modified. As such, metatranscriptomics is arguably a less suitable and direct way of profiling microbial community function compared to metaproteomics, which refers to the characterization of the protein composition of microbiota from environmental samples (Maron et al. 2007; Carvalhais et al. 2012). This is so because ultimately, in a cell, specific functions are carried out by proteins, more specifically, by enzymes (Chistoserdova 2013). Interestingly, Maier et al. (2009) found that there was a very weak correlation between protein synthesis and the abundance of transcripts, which mediated the synthesis process. Moreover, extracting, separating, and identifying proteins pose several technical challenges which include removing contaminants (organic and inorganic). In light of these challenges, Moran (2009) noted that metaproteomics may be more arduous than metatranscriptomics, particularly when dealing with environmental samples with high levels of microbial diversity. In such cases, due to the dilution of each protein in a complex sample, it is likely that only proteins that are most abundant will be identified (Schneider and Riedel 2010), which often results in specific microorganisms being under- or overrepresented. Furthermore, microbial species that have never been genome sequenced or in vitro studied may be present in environmental samples. This means that the protein sequences for such microorganisms, which are required for identification using mass spectrometry, are not present in public databases. Banfield et al. (2005) noted that a combined metaproteomics–metagenomic approach, which is termed metaproteogenomics can be used to address limitations of the two approaches. More specifically, metaproteogenomics allows for a more systematic way of linking phylogenetic identities or diversity of microorganisms with their biological functions since total DNA and proteins are extracted from the same environmental sample (Rastogi and Sani 2011).

In this light, Liu et al. (2015) used compost metaproteomes to evaluate microbial succession during composting phases and infer the predominant metabolic processes by bacteria and fungi. Results showed that the diversity of fungi was lower compared to bacteria, and the abundance of Actinobacteria and *Saccharomyces* increased significantly with composting time. Fungi (*Fusarium oxysporum*, *Neurospora crassa,* and *Phanerochaete chrysosporium*) were the main producers of cellulase in earlier phase of composting, bacterial communities (*Bacillus subtilis* and *Thermobifida fusca*) replaced the cellulolytic fungal communities during active phase and decomposition of cellulose in the curing phase required the synergy between bacteria and fungi. Interestingly, thermophilic fungi are not active through the thermophilic phase and Saccharomycetes, Schizosaccharomycetes, Sordariomycetes, Eurotiomycetes, and Basidiomycota were the most abundant fungal classes.

In a study, more akin to cold rather than active composting, Schneider et al. (2012) used metaproteomics to study the influence of environmental factors and nutrients on the structure and function of the decomposer during beech litter decomposition. In contrast to Liu et al. (2015), Schneider et al. (2012) found that the primary producers of extracellular hydrolytic enzymes were fungi, with no bacterial hydrolases detected. This may be due to the lack of sustained high-temperature phase (thermophilic) during litter decomposition, which tends to preclude fungal microorganisms and activity. Schneider et al. (2012) noted that the stoichiometry of C:N:P affected the decomposer community structure. Moreover, microbial activity was stimulated at higher nutrient contents through higher abundance and activity of extracellular enzymes. In an in vitro metaproteogenomics study, Ros et al. (2018) reported differences in phylogenetic structure and functional levels between compost suppressive and not suppressive to *Phytophthora nicotianae*. The authors concluded that *Proteobacteria* could be indicators of *P. nicotianae* suppression and annotated proteins using COGs, to the carbohydrate process, cell wall structure, and inorganic ion transport and metabolism.

Metaproteogenomic studies focused on characterizing microbiota in compost tea-treated phyllospheres have been limited. Even more limited is research on compost tea that focused on profiling the phyllosphere and rhizosphere in a single study, as was done by Knief et al. (2012) to characterize microbiota in rice cultivars. Such research is important to gain a greater understanding of disease suppression mechanisms since the foliar application of compost tea has been reported to result in a phyllosphere-mediated disease suppressive effect (Weltzien 1991). More specifically, as it relates to the contrasting environments of NCT and ACT, metaproteogenomic approaches should prove useful in the identification of specific stress protein production linkages and new functional genes and metabolic pathway tracking (Maron et al. 2007). All of which speaks to reevaluating microbial ecology concepts from a more functional perspective or "lens" (Maron et al. 2007).

1.3.3 Metametabolomics

Microbial communities release metabolites (naturally occurring low-molecular weight organic molecules), often as final responses to environmental fluxes, toxic compounds, disease, or genetic alterations into their immediate environment (Khanfir et al. 2009). As such, metabolomics refers to the study of these molecules and their relationship with microbial communities and the environment (Dunn and Ellis 2005). Metabolomic profiles per se do not reflect microbial functionality directly or in totality, however, they may indicate a level of dependency between microbiome and environmental factors such as climatic stresses and available nutrients. Thus, providing valuable information not just about the characteristics of the microbiome but about the interactions of the microbial community with the host environment. In this way, metametabolomics complements the information provided by the other omics and is considered the most direct indicator of the health of an environment or of the alterations in homeostasis (Bernini et al. 2009). That is, metametabolomics aims to improve our understanding of the role of the microbiome in the transformation of nutrients and pollutants, and other abiotic factors that may affect the homeostasis of the host environment (Aguiar-Pulido et al. 2016).

In this light, metametabolomics is also regarded as the end point of the "omics" cascade (Dettmer and Hammock 2004) since the metabolome is most predictive of phenotype. Metametabolomics also represents an approach to pathway analysis since variation in the production of signature metabolites is related to changes in the activity of metabolic routes (Manor et al. 2014). In turn, the combination of metabolomic and pathway information can lead to new hypotheses. Metabolome analysis can also provide information on the signaling processes which characterize communication between bacteria such as in quorum sensing Bassler (2002). Such information is important in advancing or improving strategies for disease control with compost and compost tea. More so, it is critical for optimizing application efficacy of compost and compost tea, particularly in sustainably intensified farming systems.

In light of this, Cronin et al. (1996) and Scheuerell (2002) reported that the disease suppressive capacity of compost-based products was related to secondary metabolite production by native microorganisms. However, for the most part, the identification and purification of such metabolites have not been very successful (Cronin et al. 1996; Sang and Kim 2011). As such, there is limited information on the metabolic profiles and function of compost-based products. In one of the first compost–metametabalome disease suppression studies, Blaya et al. (2016) reported that 54 major metabolites from compost and peat extracts clustered in a way that was associated with their suppressive ability of Phytophthora root rot (*P. nicotianae*). In suppressive compost, most of these metabolites (identified by values, not name) were found in lower relative abundance. Whereas, several mass compounds predominantly contributed to the separation of peat from the composts and others to the separation of suppressive from conductive composts. Blaya et al. (2016), however, concluded that results obtained were preliminary and there was a need to further investigate the relationship between metabolites and the capacity of composts to suppress *P. nicotianae*.

Other studies have focused on the effect of compost on the metabolome of crops or the effect of specific metabolites from microbial isolates in compost products on the suppression pathogens. For example, Vinci et al. (2018) reported that a synergic effect on plant growth, phosphorus uptake, and plant metabolite expression was observed with composts and *Trichoderma harzianum* (strain OMG-08) inoculation. That is, plant growth and phosphorus uptake were enhanced, and a relationship between the expression of different metabolites and improved photosynthetic activity was observed. Conversely, plant metabolome revealed compounds typical of biotic or abiotic stresses when *T. harzianum* (strain OMG-08) was applied with inorganic fertilizers. The authors attributed the stress profile to a reduced capacity of inorganic fertilizers to provide sufficient phosphorus availability during plant growth. Vinci et al. (2018) reported similar results with *Bacillus amyloliquefaciens*–composts treatments, which resulted in significant increases in glucose, fructose, alanine, and GABA metabolites in maize leaves. De Juras (2008) also reported a 32% increase in flavonoid content in sambong (*Blumea balsamifera*) plants treated with compost. Likewise, Neugart et al. (2018) demonstrated that the solid biological waste composts induced specific changes in the metabolite profiles. Furthermore, the changes were dependent on the type of organic residues and its concentration in soil. Targeted analysis of selected plant metabolites revealed a 3.2-fold (maxima) increase in the concentrations of carotenoids and 4.7-fold and 1.5-fold decrease in glucosinolates and phenolic compounds, respectively (Neugart et al. 2018).

Using a genomics-guided discovery process, Yang et al. (2016) isolated three antibacterial active metabolites (Aurantinins B, C, and D) from compost-associated *B. subtilis* fmb60. Aurantinins C and D were identified as new antimicrobial compounds and all three metabolites showed significant bactericidal activity against multidrug-resistant *Staphylococcus aureus* and *Clostridium sporogenes*. Cell membrane disruption was reported as the main bactericidal activity of the metabolites. On the premise that secondary metabolite production is dependent on the nutrient state of microorganisms, Li et al. (2004) investigated geosmin concentration as a possible indicator of compost stability. Results demonstrated that geosmin correlated with C/N ratio and could be used as an index for the compost stability assessment across different composting processes with various organic solid wastes.

Results from these studies show the potential applications of metametabolomics, which is not without limitations. One of the major challenges of metametabolomics is difficulty in mapping metabolites to individual species of the microbiome. Furthermore, if conclusions are to be made about which genes, enzymes, or pathways are associated with a specific metabolite, the results obtained from a metabolomic study must be combined with other -omics data (Aguiar-Pulido et al. 2016). Moreover, as noted by Miller (2007), databases with sufficient information for the identification of metabolites is lacking. So is software for automating the process of identifying and quantifying metabolites. However, Putri et al. (2013) noted that the increasing use of metabolomics techniques to study microbiology and in related fields have resulted in the development of more comprehensive public metabolomics databases. This has concurrently occurred with more advanced statistical and bioinformatics approaches to process and manage big data.

1.4 Conclusions and Future Work

High-throughput *-omics* techniques have revolutionized molecular biology and advanced our understanding of compost and compost tea microbiology. However, there is still a need to better understand and relate microbial dynamics to the effectiveness of the composting process and compost products. Such an understanding is important in optimizing production and use protocols for compost and compost tea without risking human health. More so, it is critical in reducing the variability in efficacy associated with using compost products. To achieve this understanding, scientific consensus on a theoretical framework that examines the interactions between the physical, chemical, and biological parameters of compost and abiotic factors such as temperature and moisture are needed. Embedded in this framework, -omics techniques can be used as a "toolbox" for systematic analyses and integration of information on the diversity, function, and ecology of microorganisms in compost products. Multi-omics approaches are also important in supporting emerging or next frontier trends such as metaphenomics (Nesme et al. 2018). Metaphenomics will allow for mapping physiological states of microorganisms to available resources, outputs, and ultimately effects of products such as compost and compost tea. Therefore, it has the potential to provide deeper insights into predictability *problematique* of compost products than metagenomics, which analyzes DNA from microbes with vastly varying physiological states.

In this light, although extremely useful, -omics tools are not without challenges, particularly the "meta" approaches. The correct annotation of only a small fraction of a very large number of ecologically important genes presents a cross-cutting challenge for all "meta" approaches (Moran 2009). More so, the rapid expansion of the sequence database is still representative of the most abundant genes in the environment (Moran 2009), with a relatively limited contribution from compost-based samples. To further complex this issue, many of these sequences cannot be confidently assigned to a function because there is no close matching in public sequence databases (Poretsky et al. 2005). However, with recent surge of compost studies using -omics approaches, the severity of this challenge is gradually decreasing.

Developing more comprehensive public sequencing databases through integrative visualization of multiple biological datasets and gene expression, biomarkers and pathway analysis are now possible through online multi-omics platforms such as PaintOmics (Hernández-de-Diego et al. 2018). However, there is still a need for contributions to better reflect a balance of entries from compost rhizosphere and phyllosphere studies, particularly for fungal species. Moreover, as more metagenomic datasets are generated, it becomes increasingly important to have standardized procedures and shared data storage and analysis to ensure that outputs of individual projects can be assessed and compared (Thomas et al. 2012). This becomes even more critical with the impending use of real-time metagenomic next-generation sequencing in studying species diversity during the composting and compost tea brewing process.

Other technical challenges persist with -omics approaches including DNA, RNA, and protein extraction from environmental samples like compost, mRNA instability, and low abundance of certain gene transcripts in total RNA. Biases associated with nucleic isolation and PCR result in perhaps the greatest challenge, which is the quantitative assessment of microbial communities. According to researchers, more specific challenges related to metagenomics is sequence assembly, particularly from complex microbial communities (Pell et al. 2012; Nagarajan and Pop 2013). Although metagenome-specific assembly algorithms and methods for "binning" genomes from metagenome data have resulted in successes, this remains one of the biggest challenges in bioinformatics (Pell et al. 2012; Verberkmoes et al. 2012).

In advancing the knowledge on compost microbiology, it should be noted that culture-based and culture-independent molecular techniques are neither contradictory nor excluding and should be considered complementary. Moreover, none of the molecular approaches provides complete access to the genetic and functional diversity of complex microbial communities. As such, an -omics approach should be selected based on the biological questions and objectives of the study. To this end, metaprofiling remains a cost-effective and viable tool for exploratory analysis of microbial community to inform the direction of subsequent "omics" studies. This is particularly relevant for large-scale characterization of complex or differing environmental samples.

References

Abd El-Kader SF, El-Chaghaby GA, Khalafalla GM et al (2019) A novel microbial consortium from sheep compost for decolorization and degradation of Congo red. Glob J Environ Sci Manag 5:61–70. https://doi.org/10.22034/gjesm.2019.01.05

Aguiar-Pulido V, Huang W, Suarez-Ulloa V et al (2016) Metagenomics, metatranscriptomics, and metabolomics approaches for microbiome analysis. Evol Bioinforma 12:5–16

Akita H, Kimura Z-I, Matsushika A (2017) Complete genome sequence of *Ureibacillus thermosphaericus* A1, a thermophilic *Bacillus* isolated from compost. Genome Announc 5: e00910–e00917. https://doi.org/10.1128/genomeA.00910-17

Alfreider A, Peters S, Tebbe CC et al (2002) Microbial community dynamics during composting of organic matter as determined by 16S ribosomal DNA analysis. Compost Sci Util 10:303–312. https://doi.org/10.1080/1065657X.2002.10702094

Allgaier M, Reddy A, Park JI et al (2010) Targeted discovery of glycoside hydrolases from a switchgrass-adapted compost community. PLoS One 5:e8812. https://doi.org/10.1371/journal.pone.0008812

Antunes LP, Martins LF, Pereira RV, et al (2016) Microbial community structure and dynamics in thermophilic composting viewed through metagenomics and metatranscriptomics. Nat Publ Gr 1–13. doi: https://doi.org/10.1038/srep38915

Azadnia P, Zamam MH, Ghasemi SA et al (2011) Isolation and identification of thermophilic lactobacilli from traditional yoghurts of tribes of Kazerun. J Anim Vet Adv 10:774–776. https://doi.org/10.3923/javaa.2011.774.776

Banfield JF, Verberkmoes NC, Hettich RL, Thelen MP (2005) Proteogenomic approaches for the Molecular Characterization of Natural Microbial Communities. 9:301–333

Bassler BL (2002) Small talk. Cell-to-cell communication in bacteria. Cell 109:421–424

Bernini P, Bertini I, Luchinat C et al (2009) Individual human phenotypes in metabolic space and time. J Proteome Res 8:4264–4271. https://doi.org/10.1021/pr900344m

Blaya J, Marhuenda FC, Pascual JA, Ros M (2016) Microbiota characterization of compost using omics approaches opens new perspectives for phytophthora root rot control. PLoS One 11: e0158048. https://doi.org/10.1371/journal.pone.0158048

Borek E, Ponticorvo L, Rittenberg D (1958) Protein turnover in micro-organisms. Proc Natl Acad Sci U S A 44:369–374

Brumm PJ, Land ML, Mead DA (2016) Complete genome sequences of *Geobacillus* sp. WCH70, a thermophilic strain isolated from wood compost. Stand Genomic Sci 11:33. https://doi.org/10.1186/s40793-016-0153-y

Brzostowicz PC, Walters DM, Thomas SM et al (2003) mRNA differential display in a microbial enrichment culture: simultaneous identification of three cyclohexanone monooxygenases from three species. Appl Environ Microbiol 69:334–342. https://doi.org/10.1128/AEM.69.1.334-342.2003

Bugg TD, Ahmad M, Hardiman EM, Singh R (2011) The emerging role for bacteria in lignin degradation and bio-product formation. Curr Opin Biotechnol 22:394–400. https://doi.org/10.1016/J.COPBIO.2010.10.009

Carvalhais LC, Dennis PG, Tyson GW, Schenk PM (2012) Application of metatranscriptomics to soil environments. J Microbiol Methods 91:246–251

Carvalhais LC, Dennis PG, Fan B et al (2013) Linking plant nutritional status to plant-microbe interactions. PLoS One 8:e68555. https://doi.org/10.1371/journal.pone.0068555

Cayuela ML, Millner PD, Meyer SLF, Roig A (2008) Potential of olive mill waste and compost as biobased pesticides against weeds, fungi, and nematodes. Sci Total Environ 399:11–8. doi: https://doi.org/10.1016/j.scitotenv.2008.03.031

Chandna P, Nain L, Singh S, Kuhad RC (2013) Assessment of bacterial diversity during composting of agricultural byproducts. BMC Microbiol 13:99. https://doi.org/10.1186/1471-2180-13-99

Chen QL, An XL, Zheng BX et al (2018) Long-term organic fertilization increased antibiotic resistome in phyllosphere of maize. Sci Total Environ 645:1230–1237. https://doi.org/10.1016/j.scitotenv.2018.07.260

Chengwei L, Rodriguez LM, Konstantinidis KT (2014) MyTaxa: an advanced taxonomic classifier for genomic and metagenomic sequences. Nucleic Acids Res 42:e73. https://doi.org/10.1093/nar/gku169

Chistoserdova L (2013) Biotechnology and Genetic Engineering Reviews Functional Metagenomics : Recent Advances and Future Challenges. 37–41. doi: https://doi.org/10.5661/bger-26-335

Cook RJ, Baker KF (1983) The nature and practice of biological control of plant pathogens. American Phytopathological Society, St. Paul, MN

Cronin MJ, Yohalem DS, Harris RF, Andrews JH (1996) Putative mechanism and dynamics of inhibition of the apple scab pathogen *Venturia inaequalis* by compost extracts. Soil Biol Biochem 28:1241–1249. https://doi.org/10.1016/0038-0717(96)00131-9

Cseke L (2016) Metagenomic analysis of aerated compost tea. In: Metagenomic analysis of aerated compost tea. https://www.researchgate.net/project/Metagenomic-Analysis-of-Aerated-Compost-Tea.

Daami-remadi M, Dkhili I, Jabnoun-Khiareddine H, El Mahjoub M (2012) Biological control of potato leak with antagonistic fungi isolated from compost teas and solarized and non-solarized soils. Pest Technol 6:32–40

Dang TCH, Nguyen DT, Thai H et al (2018) Plastic degradation by thermophilic *Bacillus* sp. BCBT21 isolated from composting agricultural residual in Vietnam. Adv Nat Sci Nanosci Nanotechnol 9:015014. https://doi.org/10.1088/2043-6254/aaabaf

Danon M, Franke-whittle IH, Insam H, et al (2008) Molecular analysis of bacterial community succession during prolonged compost curing. doi: https://doi.org/10.1111/j.1574-6941.2008.00506.x

de Gannes V, Eudoxie G, Hickey WJ (2013) Prokaryotic successions and diversity in composts as revealed by 454-pyrosequencing. Bioresour Technol 133:573–580. https://doi.org/10.1016/j.biortech.2013.01.138

De Juras RJ (2008) Growth of and secondary metabolite production in sambong (*Blumea balsamifera*) as influenced by compost and nitrogen. University of the Philippines, Los Baños

Dettmer K, Hammock BD (2004) Metabolomics--a new exciting field within the "omics" sciences. Environ Health Perspect 112:A396–A397. https://doi.org/10.1007/s11060-008-9572-y

Deutscher MP (2006) Degradation of RNA in bacteria: comparison of mRNA and stable RNA. Nucleic Acids Res 34:659–666. https://doi.org/10.1093/nar/gkj472

Diánez F, Marín F, Santos M et al (2018) Genetic analysis and in vitro enzymatic determination of bacterial community in compost teas from different sources genetic analysis and in vitro enzymatic determination of bacterial community in compost teas from different sources. Compost Sci Util 26:1–15. https://doi.org/10.1080/1065657X.2018.1496045

Divine R (2014) Isolation of bacteria from compost for potential use in biodecaffeination. Cornell University

Doolittle M, Raina A, Lax A, Boopathy R (2008) Presence of nitrogen fixing *Klebsiella pneumoniae* in the gut of the Formosan subterranean termite (*Coptotermes formosanus*). Bioresour Technol 99:3297–3300. https://doi.org/10.1016/j.biortech.2007.07.013

Dougherty MJ, Patrik D, Hazen TC et al (2012) Glycoside hydrolases from a targeted compost metagenome, activity-screening and functional characterization. BMC Biotechnol 12(1):38. https://doi.org/10.1186/1472-6750-12-38

Droffner ML, Brinton WF, Evans E (1995) Evidence for the prominence of well characterized mesophilic bacteria in thermophilic (50-70°C) composting environments. Biomass Bioenergy 8:191–195. https://doi.org/10.1016/0961-9534(95)00002-O

Dunn WB, Ellis DI (2005) Metabolomics: current analytical platforms and methodologies. TrAC Trends Anal Chem 24:285–294. https://doi.org/10.1016/j.trac.2004.11.021

Eichorst SA, Strasser F, Woyke T et al (2015) Advancements in the application of NanoSIMS and Raman microspectroscopy to investigate the activity of microbial cells in soils. FEMS Microbiol Ecol 91:fiv106. https://doi.org/10.1093/femsec/fiv106

Eschenfeldt W, Stols L, Rosenbaum H et al (2001) DNA from uncultured organisms as a source of 2,5-Diketo-d-Gluconic acid reductases. Appl Environ Microbiol 67:4206–4214. https://doi.org/10.1128/AEM.67.9.4206-4214.2001

Franke-Whittle IH, B a K, Fuchs J et al (2009) Application of COMPOCHIP microarray to investigate the bacterial communities of different composts. Microb Ecol 57:510–521. https://doi.org/10.1007/s00248-008-9435-2

Fritz JI, Franke-Whittle IH, Haindl S et al (2012) Microbiological community analysis of vermicompost tea and its influence on the growth of vegetables and cereals. Can J Microbiol 58:836–847. https://doi.org/10.1139/w2012-061

Galperin MY, Makarova KS, Wolf YI, Koonin EV (2015) Expanded microbial genome coverage and improved protein family annotation in the COG database. Nucleic Acids Res 43:D261–D269. https://doi.org/10.1093/nar/gku1223

Gbolagade JS (2006) Bacteria associated with compost used for cultivation of Nigerian edible mushrooms *Pleurotus tuber-regium* (Fr.) singer, and *Lentinus squarrosulus* (Berk.). Afr J Biotechnol 5:338–342. https://doi.org/10.5897/AJB05.408

Golueke CG, Card BJ, McGauhey PH (1954) A critical evaluation of inoculums in composting. Appl Microbiol 2:45–53

Gou M, Hu H, Zhang Y, Wang JT (2018) Aerobic composting reduces antibiotic resistance genes in cattle manure and the resistome dissemination in agricultural soils. Sci Total Environ 612:1300–1310

Griffiths BS, Welschen R, van Arendonk JJCM, Lambers H (1992) The effect of nitrate-nitrogen supply on bacteria and bacterial-feeding fauna in the rhizosphere of different grass species. Oecologia 91:253–259. https://doi.org/10.1007/BF00317793

Handelsman J (2005) Metagenomics: application of genomics to uncultured microorganisms. Microbiol Mol Biol Rev 69:195–195. https://doi.org/10.1128/MMBR.69.1.195.2005

Handelsman J, Rondon MR, Brady SF et al (1998) Molecular biological access to the chemistry of unknown soil microbes: a new frontier for natural products. Chem Biol 5:R245–R249. https://doi.org/10.1016/S1074-5521(98)90108-9

He S, Wurtzel O, Singh K et al (2010) Validation of two ribosomal RNA removal methods for microbial metatranscriptomics. Nat Methods 7:807–812. https://doi.org/10.1038/nmeth.1507

Hemsworth GR, Johnston EM, Davies GJ, Walton PH (2015) Lytic polysaccharide monooxygenases in biomass conversion. Trends Biotechnol 33:747–761. https://doi.org/10.1016/j.tibtech.2015.09.006

Heringa SD, Kim JK, Jiang X et al (2010) Use of a mixture of bacteriophages for biological control of *Salmonella enterica* strains in compose. Appl Environ Microbiol 76:5327–5332. https://doi.org/10.1128/AEM.00075-10

Hernández-de-Diego R, Tarazona S, Martínez-Mira C et al (2018) PaintOmics 3: a web resource for the pathway analysis and visualization of multi-omics data. Nucleic Acids Res 46:W503–W509. https://doi.org/10.1093/nar/gky466

Herren GL, Binnemans I, Joos L et al (2018) Compost as a carrier medium for entomopathogenic nematodes – the influence of compost maturity on their virulence and survival. Biol Control 125:29–38. https://doi.org/10.1016/j.biocontrol.2018.06.007

Hoitink HAJ (1990) Production of disease suppressive compost and container media, and microorganism culture for use therein. US Patent 4:900,348

Hoitink HAJ, Fahy P (1986) Basis for the control of soilborne plant pathogens with composts. 93–114

Holmes AJ, Gillings MR, Nield BS et al (2003) The gene cassette metagenome is a basic resource for bacterial genome evolution. Environ Microbiol 5:383–394. https://doi.org/10.1046/j.1462-2920.2003.00429.x

Hugenholtz P (2002) Exploring prokaryotic diversity in the genomic era. Genome Biol 3:1–8. https://doi.org/10.1186/gb-2002-3-2-reviews0003

Inbar E, Green SJ, Hadar Y, Minz D (2005) Competing factors of compost concentration and proximity to root affect the distribution of streptomycetes. Microb Ecol 50:73–81. https://doi.org/10.1007/s00248-004-0111-x

Innerebner G, Knapp B, Vasara T et al (2006) Traceability of ammonia-oxidizing bacteria in compost-treated soils. Soil Biol Biochem 38:1092–1100. https://doi.org/10.1016/j.soilbio.2005.09.008

Insam H, Riddech N, Klammer S (2002) Microbiology of composting. Springer, Berlin

INSDC (2018) No title. In: Int. Nucleotide Seq. Database Collab. http://www.insdc.org/about. Accessed 14 Nov 2018

Ishii K, Fukui M, Takii S (2000) Microbial succession during a composting process as evaluated by denaturing gradient gel electrophoresis analysis. J Appl Microbiol 89:768–777. https://doi.org/10.1046/j.1365-2672.2000.01177.x

Jurado M, López MJ, Suárez-Estrella F et al (2014) Exploiting composting biodiversity: study of the persistent and biotechnologically relevant microorganisms from lignocellulose-based composting. Bioresour Technol 162:283–293. https://doi.org/10.1016/j.biortech.2014.03.145

Kanokratana P, Uengwetwanit T, Rattanachomsri U et al (2011) Insights into the phylogeny and metabolic potential of a primary tropical peat swamp forest microbial community by metagenomic analysis. Microb Ecol 61:518–528

Kerkeni A, Daami-Remadi M, Tarchoun N, Ben KM (2007) In vitro assessment of the antifungal activity of several compost extracts obtained from composted animal manure mixtures. Int J Agric Res 2:786–794. https://doi.org/10.3923/ijar.2007.786.794

Khaldi R El, Daami-remadi M, Hamada W et al (2015) Plant pathology & microbiology: the potential of *Serratia marcescens*: an indigenous strain isolated from date palm compost as biocontrol agent of *Rhizoctonia solani* on potato. doi: https://doi.org/10.4172/2157-7471.S3-006

Khanfir R, Jenana B, Haouala R et al (2009) Composts, compost extracts and bacterial suppressive action on *Pythium aphanidermatum* in tomato. Pak J Bot 41:315–327

Kim MJ, Shim CK, Kim YK et al (2015) Effect of aerated compost tea on the growth promotion of lettuce, soybean, and sweet corn in organic cultivation. Plant Pathol J 31:259–268. https://doi.org/10.5423/PPJ.OA.02.2015.0024

Klammer S, Mondini C, Insam H (2005) Microbial community fingerprints of composts stored under different conditions. Ann Microbiol 55:299–305

Knief C, Delmotte N, Chaffron S et al (2012) Metaproteogenomic analysis of microbial communities in the phyllosphere and rhizosphere of rice. ISME J 6:1378–1390. https://doi.org/10.1038/ismej.2011.192

Konstantinidis KT, Ramette A, Tiedje JM (2006) The bacterial species definition in the genomic era. Philos Trans R Soc B Biol Sci USA 361:1929–1940

Kouki S, Saidi N, Rajeb A Ben, et al (2012) Control of Fusarium wilt of tomato caused by *Fusarium oxysporum* F. Sp. Radicis-Lycopersici using mixture of vegetable and *Posidonia oceanica* compost. https://doi.org/10.1155/2012/239639

Kowalchuk GA, Naoumenko ZS, Derikx PJL et al (1999) Molecular analysis of ammonia-oxidizing bacteria of the subdivision of the class Proteobacteria in compost and composted materials. Appl Environ Microbiol 65:396–403

Ku C, Roukos DH (2013) From next-generation sequencing to nanopore sequencing technology : paving the way to personalized. Genom Med:1–6

Kumar S (2011) Composting of municipal solid waste. Crit Rev Biotechnol 31:112–136. https://doi.org/10.3109/07388551.2010.492207

Kwok O, Fahy P, Hoitink H, Kuter G (1987) Interactions between bacteria and *Trichoderma hamatum* in suppression of Rhizoctonia damping-off in bark compost media. Phytopathology 77:1206. https://doi.org/10.1094/Phyto-77-1206

Lagier J, Armougom F, Million M et al (2012) Microbial culturomics: paradigm shift in the human gut microbiome study. Clin Microbiol Infect 18:1185–1193. https://doi.org/10.1111/1469-0691.12023

Lagier J, Hugon P, Khelaifia S et al (2015) The rebirth of culture in microbiology through the example of culturomics to study human gut microbiota. Clin Microbiol Rev 28:237–264. https://doi.org/10.1128/CMR.00014-14

LaMontagne MG, Michel FC, P a H, C a R (2002) Evaluation of extraction and purification methods for obtaining PCR-amplifiable DNA from compost for microbial community analysis. J Microbiol Methods 49:255–264

Land M, Hauser L, Jun S-R et al (2015) Insights from 20 years of bacterial genome sequencing. Funct Integr Genomics 15:141–161. https://doi.org/10.1007/s10142-015-0433-4

Langarica-Fuentes A, Handley PS, Houlden A et al (2014a) An investigation of the biodiversity of thermophilic and thermotolerant fungal species in composts using culture-based and molecular techniques. Fungal Ecol 11:132–144. https://doi.org/10.1016/j.funeco.2014.05.007

Langarica-Fuentes A, Zafar U, Heyworth A et al (2014b) Fungal succession in an in-vessel composting system characterized using 454 pyrosequencing. FEMS Microbiol Ecol 88:296–308. https://doi.org/10.1111/1574-6941.12293

Larkin RP (2008) Relative effects of biological amendments and crop rotations on soil microbial communities and soilborne diseases of potato. Soil Biol Biochem 40:1341–1351. https://doi.org/10.1016/j.soilbio.2007.03.005

Larkin RP, Tavantzis S (2013) Use of biocontrol organisms and compost amendments for improved control of soilborne diseases and increased potato production. Am J Potato Res 90(3):261–270. https://doi.org/10.1007/s12230-013-9301-8

Lemos LN, Pereira RV, Quaggio RB et al (2017) Genome-centric analysis of a thermophilic and cellulolytic bacterial consortium derived from composting. Front Microbiol 8:1–16. https://doi.org/10.3389/fmicb.2017.00644

Li HF, Imai T, Ukita M et al (2004) Compost stability assessment using a secondary metabolite: geosmin. Environ Technol 25:1305–1312. https://doi.org/10.1080/09593332508618374

Litterick a M, Harrier L, Wallace P et al (2004) The role of uncomposted materials, composts, manures, and compost extracts in reducing pest and disease incidence and severity in sustainable temperate agricultural and horticultural crop production—a review. CRC Crit Rev Plant Sci 23:453–479. https://doi.org/10.1080/07352680490886815

Liu D, Li M, Xi B et al (2015) Metaproteomics reveals major microbial players and their biodegradation functions in a large-scale aerobic composting plant. Microb Biotechnol 8:950–960. https://doi.org/10.1111/1751-7915.12290

Maier T, Güell M, Serrano L (2009) Correlation of mRNA and protein in complex biological samples. FEBS Lett 583:3966–3973. https://doi.org/10.1016/j.febslet.2009.10.036

Malandraki I, Tjamos SE, Pantelides IS, Paplomatas EJ (2008) Thermal inactivation of compost suppressiveness implicates possible biological factors in disease management. 44:180–187. doi: https://doi.org/10.1016/j.biocontrol.2007.10.006

Manor O, Levy R, Borenstein E (2014) Mapping the inner workings of the microbiome: genomic- and metagenomic-based study of metabolism and metabolic interactions in the human microbiome. Cell Metab 20:742–752. https://doi.org/10.1016/j.cmet.2014.07.021

Maron P-A, Ranjard L, Mougel C, Lemanceau P (2007) Metaproteomics: a new approach for studying functional microbial ecology. Microb Ecol 53:486–493. https://doi.org/10.1007/s00248-006-9196-8

Martins LF, Antunes LP, Pascon RC et al (2013) Metagenomic analysis of a tropical composting operation at the São Paulo Zoo Park reveals diversity of biomass degradation functions and organisms. PLoS One 8:e61928. https://doi.org/10.1371/journal.pone.0061928

Matteoli FP, Passarelli-Araujo H, Reis RJA et al (2018) Genome sequencing and assessment of plant growth-promoting properties of a *Serratia marcescens* strain isolated from vermicompost. BMC Genomics 19:750. https://doi.org/10.1186/s12864-018-5130-y

Mengesha WK, Powell SM, Evans KJ, Barry KM (2017) Diverse microbial communities in non-aerated compost teas suppress bacterial wilt. World J Microbiol Biotechnol 33:1–14. https://doi.org/10.1007/s11274-017-2212-y

Miller MG (2007) Environmental metabolomics: a SWOT analysis (strengths, weaknesses, opportunities, and threats). J Proteome Res 6:540–545. https://doi.org/10.1021/pr060623x

Moran MA (2009) Metatranscriptomics: eavesdropping on complex microbial communities. Microbe Mag 4:329–335. https://doi.org/10.1128/microbe.4.329.1

Moran MA, Satinsky B, Gifford SM et al (2013) Sizing up metatranscriptomics. ISME J 7:237–243

Nagarajan N, Pop M (2013) Sequence assembly demystified. Nat Rev Genet 7:157–167

Neher DA, Weicht TR, Dunseith P (2015) Compost for management of weed seeds, pathogen, and early blight on brassicas in organic farmer fields. Agroecol Sustain Food Syst 39:3–18. https://doi.org/10.1080/21683565.2014.884516

Nesme J, Bailey M, Wagner M (2018) The soil microbiome—from metagenomics to metaphenomics. Curr Opin Microbiol 43:162–168
Neugart S, Wiesner-Reinhold M, Frede K et al (2018) Effect of solid biological waste compost on the metabolite profile of *Brassica rapa* ssp. chinensis. Front Plant Sci 9:305. https://doi.org/10.3389/fpls.2018.00305
Ogram A, Sayler GS, Barkay T (1987) The extraction and purification of microbial DNA from sediments. J Microbiol Methods 7:57–66. https://doi.org/10.1016/0167-7012(87)90025-X
Opel KL, Chung D, McCord BR (2010) A study of PCR inhibition mechanisms using real time PCR. J Forensic Sci 55:25–33
Ottesen AR, White JR, Skaltsas DN et al (2009) Impact of organic and conventional management on the phyllosphere microbial ecology of an apple crop. J Food Prot 72:2321–2325
Pan K, Gong P, Wang J et al (2015) Applications of nitrate and ammonium fertilizers alter soil nematode food webs in a continuous cucumber cropping system in southwestern Sichuan, China. Eurasian J Soil Sci 4:287. https://doi.org/10.18393/ejss.2015.4.287-300
Partanen P, Hultman J, Paulin L et al (2010) Bacterial diversity at different stages of the composting process. BMC Microbiol 10:94. https://doi.org/10.1186/1471-2180-10-94
Pell J, Hintze A, Canino-Koning R et al (2012) Scaling metagenome sequence assembly with probabilistic de Bruijn graphs. Proc Natl Acad Sci 109:13272–13277. https://doi.org/10.1073/pnas.1121464109
Peters S, Koschinsky S, Schwieger F, Tebbe CC (2000) Succession of microbial communities during hot composting as detected by PCR-single-strand-conformation polymorphism-based genetic profiles of small-subunit rRNA genes. Appl Environ Microbiol 66:930–936
Pfaller SL, Vesper SJ, Moreno H (1994) The use of PCR to detect a pathogen in compost. Compost Sci Util 2:48–54. https://doi.org/10.1080/1065657X.1994.10771139
Phae CGUN, Shoda M, Kubota AYDH (1990) Suppressive effect of *Bacillus subtilis* & and it's products on phytopathogenic microorganisms J Ferment Bioeng 69:1–7
Poretsky RS, Bano N, Buchan A et al (2005) Analysis of microbial gene transcripts in environmental samples. Appl Environ Microbiol 71:4121–4126. https://doi.org/10.1128/AEM.71.7.4121-4126.2005
Putri SP, Nakayama Y, Matsuda F et al (2013) Current metabolomics: practical applications. J Biosci Bioeng 115:579–589. https://doi.org/10.1016/j.jbiosc.2012.12.007
Randazzo CL, Torriani S, Akkermans ADL et al (2002) Diversity, dynamics, and activity of bacterial communities during production of an artisanal Sicilian cheese as evaluated by 16S rRNA analysis. Appl Environ Microbiol 68:1882–1892
Rastogi G, Sani RK (2011) Molecular techniques to assess microbial community structure, function, and dynamics in the environment. In: Microbes and microbial technology. Springer, New York, NY, pp 29–57
Redon E, Loubière P, Cocaign-Bousquet M (2005) Role of mRNA stability during genome-wide adaptation of *Lactococcus lactis* to carbon starvation. J Biol Chem 280:36380–36385. https://doi.org/10.1074/jbc.M506006200
Riesenfeld CS, Schloss PD, Handelsman J (2004) Metagenomics: genomic analysis of microbial communities. Annu Rev Genet 38:525–552. https://doi.org/10.1146/annurev.genet.38.072902.091216
Rinke C, Schwientek P, Sczyrba A et al (2013) Insights into the phylogeny and coding potential of microbial dark matter. Nature 499:431–437. https://doi.org/10.1038/nature12352
Ros M, Blaya J, Baldrian P et al (2018) In vitro elucidation of suppression effects of composts to soil-borne pathogen *Phytophthora nicotianae* on pepper plants using 16S amplicon sequencing and. Renew Agric Food Syst:1–9. https://doi.org/10.1017/S1742170518000467
Ryan RP, Monchy S, Cardinale M et al (2009) The versatility and adaptation of bacteria from the genus Stenotrophomonas. Nat Rev Microbiol 7:514–525. https://doi.org/10.1038/nrmicro2163
Ryckeboer J, Mergaert J, Vaes K, Klammer S (2003) A survey of bacteria and fungi occurring during composting and self-heating processes. Ann Microbiol 53:349–410

Sabree Z, Rondon M, Handelsman J (2009) Metagenomics. In: Schaechter M (ed) Encyclopedia of microbiology, 1st edn. Elsevier, Amsterdam, pp 622–632
Sánchez C (2009) Lignocellulosic residues: biodegradation and bioconversion by fungi. Biotechnol Adv 27:185–194. https://doi.org/10.1016/J.BIOTECHADV.2008.11.001
Sang MK, Kim KD (2011) Biocontrol activity and primed systemic resistance by compost water extracts against anthracnoses of pepper and cucumber. Phytopathology 101:732–740. https://doi.org/10.1094/PHYTO-10-10-0287
Scheuerell SJ (2002) Compost teas and compost amended container media for plant disease control. Oregon State University
Scheuerell SJ, Mahaffee WF (2004) Compost tea as a container medium drench for suppressing seedling damping-off caused by *Pythium ultimum*. Phytopathology 94:1156–1163. https://doi.org/10.1094/PHYTO.2004.94.11.1156
Scheuerell SJ, Sullivan DM, Mahaffee WF (2005) Suppression of seedling damping-off caused by *Pythium ultimum*, *P. irregulare*, and *Rhizoctonia solani* in container media amended with a diverse range of Pacific Northwest compost sources. Phytopathology 95:306–315. https://doi.org/10.1094/PHYTO-95-0306
Schloss PD, Hay AG, Wilson DB, Walker LP (2003) Tracking temporal changes of bacterial community fingerprints during the initial stages of composting. FEMS Microbiol Ecol 46:1–9. https://doi.org/10.1016/S0168-6496(03)00153-3
Schneider T, Riedel K (2010) Environmental proteomics: analysis of structure and function of microbial communities. Proteomics 10:785–798. https://doi.org/10.1002/pmic.200900450
Schneider T, Keiblinger KM, Schmid E et al (2012) Who is who in litter decomposition? Metaproteomics reveals major microbial players and their biogeochemical functions. ISME J 6:1749–1762. https://doi.org/10.1038/ismej.2012.11
Schut F, De Vries EJ, Gottschal JC et al (1993) Isolation of typical marine bacteria by dilution culture: growth, maintenance, and characteristics of isolates under laboratory conditions. Appl Environ Microbiol 59:2150–2160. https://doi.org/10.1007/s000120300002
Simon C, Daniel R (2011) Metagenomic analyses: past and future trends. Appl Environ Microbiol 77:1153–1161. https://doi.org/10.1128/AEM.02345-10
Sorek R, Cossart P (2010) Prokaryotic transcriptomics: a new view on regulation, physiology and pathogenicity. Nat Rev Genet 11:9–16. https://doi.org/10.1038/nrg2695
St. Martin CCG (2014) Potential of compost tea for suppressing plant diseases. CAB Rev Perspect Agric Vet Sci Nutr Nat Resour 9:1–38. https://doi.org/10.1079/PAVSNNR20149032
St. Martin CCG (2015) Enhancing soil suppressiveness using compost and compost tea. In: Meghvansi MK, Varma A (eds) Organic amendments and soil Suppressiveness in plant disease management, 46th edn. Springer, London, pp 25–49
St. Martin CCG, Brathwaite RAI (2012) Compost and compost tea: principles and prospects as substrates and soil-borne disease management strategies in soil-less vegetable production. Biol Agric Hortic 28:1–33. https://doi.org/10.1080/01448765.2012.671516
St. Martin CCG, Ramsubhag A (2015) Potential of compost for suppressing plant diseases. In: Ganesan S, Kurucheve Vadivel JJ (eds) Sustainable crop disease management using natural products. CABI, pp 345–388
Steel H, Moens T, Vandecasteele B, Hendrickx F, De Neve S, Neher DA, Bert W (2018) Factors influencing the nematode community during composting and nematode based criteria for compost maturity. Ecol Indic 85:409–421
Sundberg C, Franke-Whittle IH, Kauppi S et al (2011) Characterisation of source-separated household waste intended for composting. Bioresour Technol 102:2859–2867. https://doi.org/10.1016/j.biortech.2010.10.075
Sundberg C, Yu D, Franke-Whittle I et al (2013) Effects of pH and microbial composition on odour in food waste composting. Waste Manag 33:204–211. https://doi.org/10.1016/j.wasman.2012.09.017
Thomas T, Gilbert J, Meyer F (2012) Metagenomics - a guide from sampling to data analysis. Microb Inform Exp 2:3. https://doi.org/10.1186/2042-5783-2-3

Tiquia SM, Ichida JM, Keener HM et al (2005) Bacterial community profiles on feathers during composting as determined by terminal restriction fragment length polymorphism analysis of 16S rDNA genes. Appl Microbiol Biotechnol 67:412–419. https://doi.org/10.1007/s00253-004-1788-y

Torsvik V, Goksøyr J, Daae FL (1990) High diversity in DNA of soil bacteria. Appl Environ Microbiol 56:782–787

Trapnell C (2015) Defining cell types and states with single-cell genomics. Genome Res 25:1491–1498

Tringe SG, von Mering C, Kobayashi A et al (2005) Comparative metagenomics of microbial communities. Science 308:554–557. https://doi.org/10.1126/science.1107851

Tuomela M, Vikman M, Hatakka A, It M (2000) Biodegradation of lignin in a compost environment: a review. Bioresour Technol 72:169–183

Ventorino V, Aliberti A, Faraco V et al (2015) Exploring the microbiota dynamics related to vegetable biomasses degradation and study of lignocellulose-degrading bacteria for industrial biotechnological application. Sci Rep 5:8161. https://doi.org/10.1038/srep08161

Verberkmoes NC, Wilkins MJ, Hettich RL, Lipton MS (2012) Fermentation, hydrogen, and sulfur metabolism in multiple uncultivated bacterial phyla. Science (80-) 337:1661–1665

Vinci G, Cozzolino V, Mazzei P et al (2018) An alternative to mineral phosphorus fertilizers: the combined effects of *Trichoderma harzianum* and compost on *Zea mays*, as revealed by 1H NMR and GC-MS metabolomics. PLoS One 13:e0209664. https://doi.org/10.1371/journal.pone.0209664

Vorholt JA (2012) Microbial life in the phyllosphere. Nat Publ Gr 10:828–840. https://doi.org/10.1038/nrmicro2910

Wang C, Dong D, Wang H et al (2016) Metagenomic analysis of microbial consortia enriched from compost: new insights into the role of Actinobacteria in lignocellulose decomposition. Biotechnol Biofuels 9:22. https://doi.org/10.1186/s13068-016-0440-2

Wang C, Dong D, Strong PJ et al (2017) Microbial phylogeny determines transcriptional response of resistome to dynamic composting processes. Microbiome 5:103. https://doi.org/10.1186/s40168-017-0324-0

Weltzien HC (1991) Biocontrol of foliar fungal diseases with compost extracts. In: Andrews J, Hirano S (eds) Microbiology of leaves. Springer, New York, NY, pp 430–450

Yang J, Zhu X, Cao M et al (2016) Genomics-inspired discovery of three antibacterial active metabolites, aurantinins B, C, and D from compost-associated *Bacillus subtilis* fmb60. J Agric Food Chem 64:8811–8820. https://doi.org/10.1021/acs.jafc.6b04455

Yang W, Gu S, Xin Y et al (2018) Compost addition enhanced hyphal growth and sporulation of arbuscular mycorrhizal Fungi without affecting their community composition in the soil. Front Microbiol 9. https://doi.org/10.3389/fmicb.2018.00169

Yeh Y, Chang S, Kuo H et al (2013) A metagenomic approach for the identification and cloning of an endoglucanase from rice straw compost. Gene 519:360–366

Yi-fang Y, Chia-yu CS, Hsion-wen K et al (2013) A metagenomic approach for the identification and cloning of an endoglucanase from rice straw compost. Gene 519:360–366. https://doi.org/10.1016/j.gene.2012.07.076

Yu D (2014) Microbial community profiling of biodegradable municipal solid waste treatments – aerobic composting and anaerobic digestion. University of Helsinki, Helsinki

Zhou J, Deng Y, He Z et al (2010) Applying GeoChip analysis to disparate microbial communities. Microbe 5:60–65. https://doi.org/10.1128/microbe.5.60.1

Chapter 2
Biological Sterilisation, Detoxification and Stimulation of Cucurbitacin-Containing Manure

Phatu William Mashela, Kgabo Martha Pofu, and Ebrahim Shokoohi

Abstract Wild watermelon (*Cucumis africanus*) contains cucurbitacin B in all organs—a highly stable chemical compound, with both thermophilic and bioactivity attributes, which are essential in manure. The manufacture of Nemafric-manure from this plant is a two-step process. The first step comprises fermenting ground fruit dried at 50 °C using effective microorganisms (EM) to produce concentrated solutions with pH of 3.7. The second step to spray either dried or fresh biomass stacks of whole plants with diluted solutions to produce a special manure that undergoes fermentation process, with EM components that have the ability to sterilise, detoxify and pasteurise the product. On the other hand, the product has attributes of stimulating other EM components. Nemafric-manure, through its active ingredient cucurbitacin B, has the potential to suppress plant–parasitic nematodes and soil-borne pathogens like *Fusarium* species, while at the same time stimulating plant growth and productivity.

Keywords *Cucumis* species · Organic manure · Nematodes · Thermophilic microbes · Toxic gases

2.1 Introduction

Manure is an integral component of climate-smart agriculture (CSA) in context of contributing towards adaptation and mitigation strategies (FAO 2018). The CSA aims to enhance the capacity of agricultural systems to enhance food security, through incorporating adaptation and mitigation activities that support innovative

P. W. Mashela (✉) · E. Shokoohi
Green Biotechnologies Research Centre of Excellence, School of Agricultural and Environmental Sciences, University of Limpopo, Sovenga, South Africa
e-mail: phatu.mashela@ul.ac.za; ebrahim.shokoohi@ul.ac.za

K. M. Pofu
Agricultural Research Council, Pretoria, South Africa
e-mail: MarthaP@arc.agric.za

M. K. Meghvansi, A. Varma (eds.), *Biology of Composts*, Soil Biology 58,
https://doi.org/10.1007/978-3-030-39173-7_2

and sustainable development (FAO 2018). In crop production systems, adaptation and mitigation activities range from improving soil health, thereby enhancing physical properties, beneficial chemicals and effective microorganisms (EM), with the overall purpose being the reduction of agriculture-induced carbon fingerprint (Sinha et al. 2016). Consequently, manure could play a central role in adaptation and mitigation strategies. Notwithstanding, manure farming could, like any other innovation, have unintended consequences by introducing unfit-for-purpose drawbacks, which may include but not limited to emitting greenhouse gases such as CO_2, SO_2 and methane (CH_4) (Singh 2016).

Fit-for-purpose is a concept adopted by the registration authorities of agricultural inputs, where zero tolerance is being meted for unfit-for-purpose products (OECD 2017). The unfit-for-purpose concept could imply various things to different end users and therefore, the requirement for legally binding label claims. In CSA, unfit-for-purpose for agricultural inputs could be intrinsic or extrinsic attributes, which could even limit commercialisation of inputs. Intrinsic unfit-for-purpose in manure could include phytotoxicity to crops, which is viewed as being intolerable at all cost (OECD 2017). Intrinsic properties of manure, particularly during the fermentation process, could reduce soil pH to below 5.5, resulting in Al^{3+}, H^+ and Mn^{2+} toxicities and/or P, Ca^{2+}, Mg^{2+} and Mo deficiencies (Läuchli and Grattan 2012). Extrinsic unfit-for-purpose includes but not limited to unintentionally introducing seeds of aggressive invasive weeds that are difficult to control such as wild cucumber (*Cucumis myriocarpus* Naude.) in Australia and quarantined pathogens, nematodes or insects (Shaik et al. 2016; Singh 2016). In some instances, where the aim was to take advantage of suppressive attributes of allelochemicals in manure on soil-borne diseases and nematodes, the active ingredients could be degraded by high fermentation-induced temperatures. Consequently, EM used in fermentation should have thermophilic attributes that could allow for gradual adaptation to increasing temperatures during fermentation (Stetter 2006). Successful uses of manure in context of adaptation and mitigation strategies in CSA depend on the intended focus, degree of intrinsic fit-for-purpose or extrinsic unfit-for-purpose and thermophile potential of EM.

Increasing soil temperature in various regions coincided with the withdrawal of most synthetic chemical nematicides from the agrochemical markets (Mashela et al. 2017). High soil temperatures had been associated with unmanageable incidents of increasing soil-borne diseases such as fusarium (*Fusarium oxysporum* f.sp. *cubense*) and lesion nematode (*Pratylenchus* spp.) in various farming systems. Relative to 3 years prior to the withdrawal of methyl bromide as a biocide fumigant in 2005 from the agrochemical markets (Chitwood 2003), 13 years after the withdrawal, global crop failures ascribed to nematode damage had increased to as high as 37% (Mashela et al. 2016). Exclusive reliance on plant resistance in soils with increasing temperatures is vulnerable since most resistant genes in trusted genotypes failed to express themselves in soils with high temperatures, as observed in genes for resistance against root-knot (*Meloidogyne* species) nematodes (Dropkin 1969; Fuller et al. 2008). Such failures are critical in crop rotation systems intended to manage soil-borne diseases in crops without genotypes with resistance to the related pathogens or nematodes (Thies and Levi 2006). In soils with increasing soil temperature, most soil-borne pathogens and nematodes have increasing shorter life cycles (Sikora et al. 2018), others becoming highly aggressive than ever recorded previously (Fuller et al. 2008). Another challenge is that other important crops in food security clusters such as potato (*Solanum tuberosum* L.) and watermelon (*Citrullus lanatus* Thunb), do not have genotypes with resistance to the notorious root-knot (*Meloidogyne* species) nematodes (Pofu et al. 2011; Thies and Levi 2006). Root-knot nematodes could reduce plant growth in such crops from as high as 50% to complete crop failure (Thies and Levi 2006). Most crops with pre-infectional nematode resistance, due to the unique attribute of cryptobiosis in nematode juveniles, appear to have auto-nematode resistance, where susceptible successor crops are not protected against nematode juveniles that emerged from cryptobiosis (Pofu et al. 2019).

All listed challenges could enhance the importance of manure as a multipurpose input in the context of CSA, for example in (a) managing soil-borne diseases and nematodes, (b) providing nutrient elements and (c) improving soil health. Such a multipurpose input requires empirically based information for its disease suppressing capabilities, active ingredients that could withstand thermophilic conditions pre- and post-fermentation. Similarly, since manure relies on microbial degradation, the EM used in such processes should also have some degree to withstand heat during fermentation. Additionally, such a manure should not have allelopathic effects on a wide range of crops. Cucurbitacin-containing manure derived from shoots and fruit of wild watermelon (*Cucumis africanus* L.F.) had been identified to contain a wide range of attributes, as shown in its uses as a phytonematicide (Mashela et al. 2017; Tseke and Mashela 2017). In view of the above, the objective of the present chapter was to provide some empirically based insight on biological sterilisation, detoxification of gases and stimulation of plant growth through cucurbitacin-containing manure using locally available resources, with attributes for suppression of soil-borne diseases and nematodes in context of CSA.

2.2 Biosynthesis and Bioactivities of Cucurbitacins

Allelochemicals, stored in various organs, comprise a wide range of chemical compounds, with extraordinary high oxidative potential (Van Wyk and Wink 2004). Basically, allelochemicals are excess hydrocarbons that are detoured from the anaerobic glycolysis biosynthesis pathway during cellular respiration. Three main biosynthetic pathways for allelochemicals, namely, malonic acid pathway (MaAP), shikimic acid pathway (ShAP) and mevalonic acid pathway (MeAP), had been identified as such detours (Dudareva et al. 2013). MaAP and ShAP occur during glycolysis in the cytoplasm of cells, whereas MeAP occurs inside the mitochondrion just prior to acetyl coA entering the Krebs cycle. Allelochemicals are therefore products removed along the respiration pathway to ensure that the high-energy acetyl coA molecules do not accumulate at the entry side into the Krebs cycle. The MaAP pathway outside the mitochondria plays a role in the biosynthesis of flavonoid compounds, whereas ShAP is responsible for the biosynthesis of aromatic compounds that form tryptophan, tyrosine and phenylalanine (Van Wyk and Wink 2004). MeAP and MEP/DOXP biosynthetic pathways, with a common exit pathway for acetyl coA from the respiratory pathway in the mitochondrion, biosynthesise triterpenoids, comprising cucurbitacins and steroids (Chen et al. 2014; Van Wyk and Wink 2004). MeAP and MEP/DOXP biosynthetic pathways comprise eight and six precursors, respectively (Chen et al. 2014), which have been shown to emulate thermophilic attributes (Shadung et al. 2016a).

Cucurbitacins, a class of highly oxidised tetracyclic triterpenoids have a cucurbitane skeleton (C30), with a wide range of bioactivities against pathogenic bacteria and fungi, along with nematodes and diseases (Chen et al. 2005; Kaushik et al. 2015; Mashela et al. 2017; Mphahlele et al. 2012; Van Wyk and Wink 2004). Cucurbitacin B ($C_{32}H_{48}O_8$) and D ($C_{30}H_{44}O_7$) are the most abundant and with the highest concentration in tissues of most plants (Guo et al. 2014). In the current study, the focus was on cucurbitacin B, which occurs in all organs of *C. africanus* plants (Shadung and Mashela 2016), with its centre of biodiversity being in Limpopo Province, South Africa (Kristkova et al. 2003). This perennial plant species has one of the shortest erect stems, with succulent vines that dieback every winter, and an excessively deep root system that confers the plant some drought tolerance (Pofu 2012). *Cucumis africanus* is a vigorous prostrate growing plant, with resistance to tropical *Meloidogyne* species, *M. incognita* race 2, *M. incognita* race 4 and *M. javanica* (Pofu 2012; Pofu et al. 2011). Resistance to *M. incognita* on *C. africanus* was confirmed in China, in addition to showing that the plant species were resistant to *F. oxysporum* (Liu et al. 2015). Incidentally, *C. lanatus* cultivars, renowned for unknown genotypes without resistance to *Meloidogyne* species (Thies and Levi 2006), have also no genotypes with resistance to *F. oxysporum* (Thies and Levi 2006). In most cases, rootstocks with resistance to *Meloidogyne* species and *F. oxysporum* are used in genotype cultivars without resistance (Liu et al. 2015; Pofu 2012) (Fig. 2.1).

Cucurbitacin B in Nemafric-BL phytonematicide (B = cucurbitacin B, L = liquid formulation), prepared as fermented crude extracts of *C. africanus* fruit, consistently suppress nematode population densities on various crops (Mashela 2002; Mashela et al. 2011, 2017). The product is compatible with biocontrol agents such as *Steinernema feltiae* and *Trichoderma harzianum* (Madaure et al. 2017). Briefly, mature fruit at 93 days after transplanting are collected from cultivated fields (Shadung et al. 2015), cut into pieces and oven dried at 52 °C for 72 h, followed by grinding in a Wiley mill to increase the surface area during fermentation (Mashela et al. 2017). In contrast, plant parts for medicinal purposes are dried from 30 to 50 °C to preserve the active ingredients from thermal degradation (Müller and Heindl 2006). However, when *C. africanus* fruit was oven dried at less than 50 °C, contamination with *Penicillium simplicissimum* (Oudem.) Thom consistently occurred (Mphahlele et al. 2012). Most *Cucumis* species are host to this plant growth-promoting fungus (Mphahlele et al. 2012).

During drying *C. africanus* fruit from 52 to 100 °C, cucurbitacin B was decreased linearly from 6.8 to 1.0 $\mu g.ml^{-1}$ (Shadung et al. 2016a). Relative to 52 °C oven

Fig. 2.1 A field of *Cucumis africanus* (**a**) and exposed fruit (**b**): Photo by authors

drying, freeze drying at −18 °C and 100% RH reduced cucurbitacin B by 60% (Shadung et al. 2016a). At sea level, the boiling point for cucurbitacin B is approximately 699 °C (Krieger 2001). Also, cucurbitacin B is nonpolar and therefore insoluble in water (Chen et al. 2005). Storage of *C. africanus* fruit dried at 52 °C in open or airtight containers for 6 months, or Nemafric-BL phytonematicide under similar conditions and duration, was characterised by cucurbitacin B and storage time exhibiting positive linear relations (Shadung et al. 2016b). At 6 months after storage, cucurbitacin B concentration was still in the stimulation phase of the dose-response growth patterns (Shadung and Mashela 2017). All cited studies (Krieger 2001; Shadung and Mashela 2017; Shadung et al. 2016a, b) suggest that cucurbitacin B is thermostable, an important attribute for achieving successful fermentation in the overall preparation of manure. Moreover, due to its large molecular size, products containing cucurbitacin B as active ingredient hardly leave chemical residues in plant tissues (Shadung et al. 2017).

2.3 Preparation of Cucurbitacin-Containing Manure

2.3.1 Cultivation of Cucumis africanus

Seeds of *C. africanus* exhibited the attributes of both chemical and physical dormancies, which could not be broken using either stratification or scarification or both (Maila et al. 2016). Subjecting seeds to pre-sowing temperatures from 40 to 60 °C resulted in 100% germination and then emergence (Mafeo 2014). Seeds are sown in 120-hole polystyrene seedling trays, containing Hygromix-T (Hygrotech, Pretoria North, South Africa) growing mixture, held in a growth chamber at 50 °C for 5 days and transferred to the greenhouse. At four-leaf stage, seedlings are hardened-off outside the greenhouse using intermittent withholding of irrigation and irrigating to full capacity for 10 days. Thereafter, seedlings can be transplanted to the field at 1.0 × 1.0 m spacing with irrigation and fertilisation being as described previously (Shadung 2016).

2.3.2 Preparation of Nemafric-BL Stock Solution

At 92 days after transplanting, whole plants (shoots, fruit) have an optimum concentration of cucurbitacin B (Shadung et al. 2015). Whole plant can be removed and chopped into pieces and dried at 52 °C for 72 h to enable storage of inventories prior to processing. The latter is necessary to protect the produce from *P. simplicissimum*-induced post-harvest rots. Nemafric-BL solution is then prepared by mixing approximately 500 g powdered *C. africanus* fruit with 1250 g brown sugar, 3750 ml molasses and 3750 ml EM in a hermetically sealed half-filled 400-L-aluminium fermenter (Fig. 2.2a) at 35 °C (Mashela et al. 2017). A small tube is hermetically inserted in a hole and dangled into a bottle half-filled with water to trap any escaping toxic gases. Fermentation is allowed for 14 days until pH is at 3.7 (Higa and Parr 1994). The EM used consists of lactic acid bacteria, yeast, actinomycetes, photosynthetic bacteria and some unknown fungal species. Since EM can survive high temperature (up to 70 °C) during fermentation, constituents of EM are classified as thermophiles (Stetter 2006; Von Wolferen et al. 2013).

2.3.3 Preparation of Nemafric-Manure

Approximately 250 L dried whole plant material is moistened using a 1:5 diluted solution from 200 L Nemafric-stock solution (200 L/5 = 40 L stock solution, mixed with 160 L chlorine-free tap water) and placed in 250 L heavy duty plastic fermenters (Fig. 2.2b). Each fermenter has a thermometer with a 1-m long probe and an inbuilt tube for releasing gases into a litre bottle half-filled with water for trapping

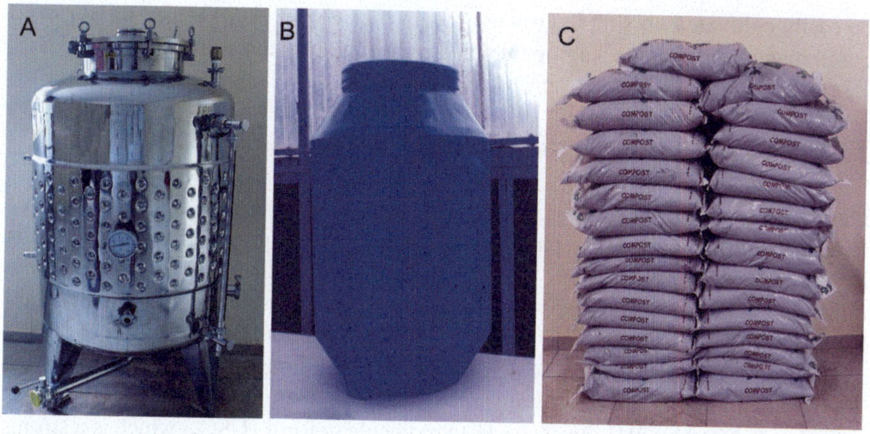

Fig. 2.2 Pilot fermentation plant: (**a**) 400 aluminium fermenter, (**b**) 200 L fermenter and (**c**) 25 kg sample bags of cucurbitacin-containing manure

greenhouse gases during fermentation, should any be released. At T_0 (initial day), T_7 (7 days) and T_{21} (21 days) during fermentation, leachate is sampled, with pH averaging 6.7, 3.9 and 3.5, respectively, whereas temperature of manure usually averages 27, 56 and 59 °C, respectively. At 23 days, the manure is spread onto open plastic containers for a three-week curing period and then packaged in 25 kg opaque plastic bags and sealed (Fig. 2.2c).

2.3.3.1 Potential Sterilisation

Lactic acid bacteria are responsible for breaking cellulose and lignin (Madigan and Martino 2006), which are the toughest materials in cell walls of plants. The end product of cellulose and lignin degradation is lactic acid, which further reduces pH of organic manure (Higa and Parr 1994). Yeast bacteria are also responsible for lowering pH to acidic medium by breaking glucose during an anaerobic glycolysis to pyruvic acid (Stetter 2006), which cannot proceed to Krebs cycle without oxygen that serves as the final electron acceptor. The pH of manure is gradually decreased, and most fungal pathogens cannot grow under acidic media (Madigan and Martino 2006), whereas such conditions favour the flourishing of bacteria. Yeast also releases antimicrobial chemicals that add to the sterilisation of organic manure by preventing fungal growth that include the *Cucumis*-resident *P. simplicissimum*. Actinomycete bacteria have capabilities to release chitinase, which is an enzyme with capabilities to hydrolyse chitin in exoskeleton of insects, insect eggs, nematode cuticles, nematode eggs and mycelia of various fungi (Madigan and Martino 2006). Actinomycetes are naturally aerobic bacteria and continue their bioactivities in the early stages of fermentation and post application. The Gram-positive actinomycete bacteria can be easily cultivated on simple growing media (Prescott et al. 2002) for mixing with cured manure prior to packaging, which sterilise the product completely.

2.3.3.2 Potential Detoxification and Pasteurisation

During decomposition, plant organs release large quantities of sulphur in the form of hydrogen sulphide (H_2S). In photosynthetic plants, during Photosystem II, H_2O is hydrolysed to release $2H^+$ ions, with one being used to reduce nicotinamide adenine dinucleotide (NAD) to energy-rich NADH molecule. In contrast, in photosynthetic bacteria, NAD is reduced by H^+ ion from H_2S during Photosystem II (Stetter 2006), whereas S is oxidised to SO_2 which is released as a potentially lethal gas (Stetter 2006), which is further oxidised to form a strong oxidising agent, SO_4^{2-}. The latter is further reduced by $2H^+$ ions from H_2O molecule to form H_2SO_4, which is a strong acid. The produced NADH, when oxidised, increases temperature of manure to as high as over 65 °C, thereby further pasteurising the manure. Due to the lithotrophic nature of photosynthetic bacteria that oxidise hydrogen sulphide to H_2SO_4, the toxic SO_2 gas is readily removed during the fermentation process (Stetter 2006). Additionally, due to the end product being a strong acid (H_2SO_4), EM fermentation should be carried out in plastic containers (Fig. 2.2b) and thermometers should be removed soon after taking the temperature measurement, with the probes temporarily dipped in a basic solution prior to storage.

2.3.3.3 Potential Stimulation

Among many other attributes, allelochemicals have the attribute to induce dose-response growth (DRG) patterns in various entities, as shown previously (Liu et al. 2003; Mashela et al. 2016, 2017). In one instance such growth was viewed as 'fertiliser effects' (Mashela 2002). The Curve-fitting Allelochemical Response Dosage (CARD) computer-based algorithm model, developed in Australia (Liu et al. 2003), quantifies the DRG phases, namely, stimulation, neutral and inhibition phases, using seven biological indices. The first two biological indices, the threshold stimulation point (D_m) and the saturation stimulation point (R_h), initiates and terminates the stimulation phase, respectively (Liu et al. 2003; Mashela et al. 2017). The two biological indices were empirically used to develop the Mean Concentration Stimulation Point [MCSP = (D_m + (R_h/2)] (Mashela et al. 2017). The MCSP is the quantity of a phytonematicide that could be applied to suppress nematode population densities without inducing phytotoxicity on crops protected against nematode damage. Technically, MCSP is referred to as the first law of phytonematicides (Mashela et al. 2017). Due to the nature of biological indices as proportions (Liu et al. 2003), the law could also be used to establish the quantity of manure to be applied, as shown later. Additionally, the CARD model provides two other biological indices, the sensitivity index (k) and the coefficient of determination (R^2). The k index provides information on the sensitivity of entities measured in the manure to allelochemicals such as cucurbitacin B, R^2 gives the CARD model an attribute of being a statistical tool (Mashela et al. 2017). Generally, when k approaches zero, the susceptibility of the test entity to the manure increased to the point of susceptibility (Liu et al. 2003; Mashela et al. 2017). The overall sensitivity (Σk) is the sum of k values of different parts of the same entity that was exposed to increasing concentration of the test allelochemical (Liu et al. 2003). For instance, when tomato plants are exposed to Nemafric-manure, dry shoot mass, plant height, fruit mass, stem diameter and photosynthesis (Mashela et al. 2017) are variables that can be used to contribute their individual k values to Σk. The Σk value at zero shows that the plant is highly sensitive to the test manure, and therefore, the product is not suitable for use in the test crop.

2.4 Application and Effects of Nemafric-Manure

A study was conducted on an open field at the Green Biotechnologies Research Centre of Excellence, University of Limpopo, South Africa (23°53″10′S, 29°44″15′E). The location has summer (October–December) rainfall with a mean annual rainfall of less than 500 mm, with hot and dry summers having maximum temperatures ranging from 28 to 38 °C. The site contained Hutton soil (65% sand, 30% clay, 5% silt) with 1.6% organic C. A field was sampled for initial nematode population (Pi) densities, soil pH (H_2O) = 6.5, electrical conductivity (EC = 0.148 dS/m) and soil nutrient elements. A tomato cv. 'Floradade' experiment, with manure at 0, 2, 4, 8, 16, 32 and 64 kg per 2 m^2, arranged in a randomised complete block design, replicated 20 times, was conducted during autumn (January–March) 2016 and repeated in 2017. Cultural practices were as in commercial tomato production systems, but without fertilisation. At harvest, 110 days after transplanting, dry shoot mass and fresh fruit mass were determined (Shadung 2016). Mature healthy leaves were collected, prepared, dried and quantified for Ca, Mg, K, Fe and Zn (Shadung 2016). Nematodes were extracted from soil and root samples using standard procedures (Marais et al. 2017). Soil samples were collected, for soil bulk density, organic matter, nutrient elements (Ca, Mg, Na K) and soil pH measurements using standard procedures. Prior to analysis, treatments were $\log_2 2^x$-transformed (0, 1, 2, 3, 4, 5 and 6 kg/2 m^2) to homogenise the x-axis interval (Causton 1977), thereby avoiding skewness to the right (Liu et al. 2003). Data were subjected to the CARD model to generate D_m and R_h for computing MCSP for Nemafric-manure (Mashela et al. 2017).

2.4.1 Effects on Plant Variables

2.4.1.1 Dry Shoot Mass and Fresh Fruit Yield

For the purpose of explaining the concept using Nemafric-manure in tomato production, dry shoot mass and fresh fruit yield (Table 2.1) were subjected to the CARD model to generate the biological indices (Table 2.2).

The variables (Table 2.1) were subjected to the CARD model to generate the improved version of the standard curves (Fig. 2.3) (Liu and An 2005) and the biological indices (Table 2.2). The biological indices are proportions and therefore unit less (Mashela et al. 2017), thereby providing the opportunity to use the first two, namely, D_m and R_h, in the calculation of MCSP, with the unit being that of the concentration used to generate the indices. The latter is outside the scope of the current study and would not be described further.

During 2016 growing season, the models for dry shoot mass and fresh fruit mass were explained by 93 and 0.92%, respectively, whereas during 2017 the models were explained by 89 and 78%, respectively (Table 2.2). In 2016 and 2017, the MCSP [$= D_m + (R_h/2)$] values were 33.687 kg/2 m^2 (16.844 t.ha^{-1}) and 16.678 kg/

Table 2.1 Influence of Nemafric-manure on dry shoot mass and fresh fruit mass of tomato cultivar 'Floradade'

	2016		2017	
Manure (kg/2 m^2)	Dry shoot mass (kg)	Fresh fruit mass (kg)	Dry shoot mass (kg)	Fresh fruit mass (kg)
0	40.93	24.92	55.35	30.18
1	55.38	32.44	90.24	32.41
2	98.93	49.11	128.13	58.31
3	93.17	48.11	126.17	56.11
4	99.03	47.29	101.03	45.75
5	78.48	30.28	48.48	49.34
6	62.18	19.89	35.62	19.93

Table 2.2 Biological indices generated by CARD computer-based model on tomato plants during two seasons

	2016		2017	
Biological indices	DSM	FFM	DSM	FFM
Threshold stimulation (D_m)	3.366	2.654	2.896	2.950
Saturation point (R_h)	60.833	61.875	25.371	29.646
0% inhibition (D_0)	6.732	5.309	5.753	5.901
50% inhibition (D_{50})	7.208	5.87	6.343	6.483
100% inhibition (D_{100})	7.6	6.3	6.8	7.0
R^2	0.93	0.92	0.89	0.78
Sensitivity (k)	3	1	4	2
Total sensitivity ($\sum k$)	4		6	

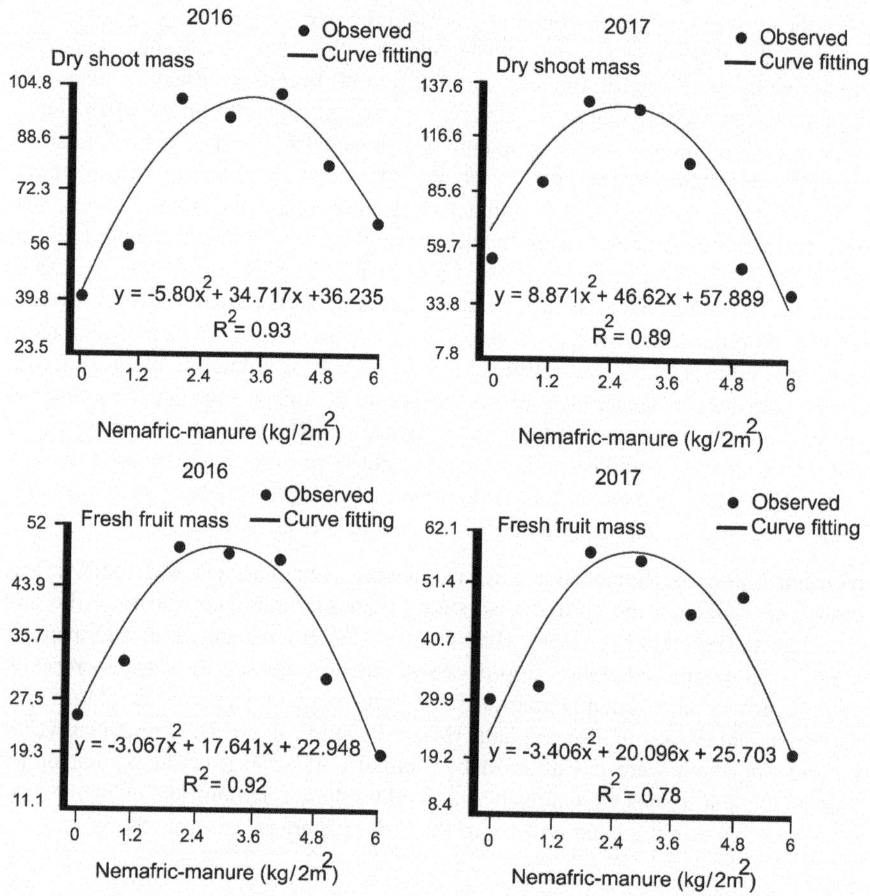

Fig. 2.3 CARD model-generated curves for tomato dry shoot mass and fresh fruit mass

2 m^2 (8.339 $t.ha^{-1}$), respectively, averaging to 13 $t.ha^{-1}$. In the Eastern Cape Province, South Africa, smallholder farmers use cattle manure at 0.3 and 182 t. ha^{-1} (Yoganathan et al. 1998), whereas empirically based recommendations stand at 25–100 $t.ha^{-1}$ cattle manure (Mkile 2001). Tester (1990), without empirical support, suggested that adding 100 $t.ha^{-1}$ compost per annum could be enough to induce physical changes in the soil and increase yield. However, some empirically based studies in South Africa, suggested manure application rates from 5.5 to 11 $t.ha^{-1}$ for field crops, whereas doubling the rates were recommended for crops with higher nutrient requirements such as vegetables (Malherbe 1964). In most cases, the quantity of manure to be applied depends on the objective that needed to be achieved, which could include improved plant growth and yield or improved soil health. In addition to targeted benefits, drawbacks such as phytotoxicity should always be borne in mind, since registration authorities have zero tolerance on agricultural inputs with unfit-for-purpose attributes (OECD 2017).

In the current case study, tomato plants had the overall sensitivity ($\sum k$) of 4 and 6 during the respective seasons, suggesting that the crop was highly tolerant to phytotoxicity of Nemafric-manure was far remote because estimating mean 0% inhibition phase occurred at $D_0 = 267$ t.ha^{-1} in comparison with MCSP of 13 t.ha^{-1}.

As shown in tomato–Nemafric-manure relations, DRG patterns with stimulation, neutral and inhibition phases (Liu et al. 2003) are characterised by quadratic relations (Mashela et al. 2017). Numerous hypotheses, with contradictions, had been advanced to explain observed increases in plant growth under manure (Arshad 2006; Arshad and Frankenberger 1991; Fujita et al. 1997; Kato et al. 1997; Xu 2000; Yamada et al. 1996). Consistent with the first law of phytonematicides (MCSP), the CARD computer-based model was able to generate the quantity of Nemafric-manure to be applied as 13 t.ha^{-1} for stimulating the growth of tomato plants in soil with known attributes. Generally, plants respond to increasing concentration of allelochemicals in plant-based products through DRG patterns (Mashela et al. 2017), comprising stimulation phase, neutral phase and inhibition phase (Liu et al. 2003). The MCSP depicts the half-distance stimulation phase, where plant growth is consistently stimulated by allelochemicals (Mashela et al. 2017).

Nutrient Elements Nemafric-manure was shown to increase Ca, Mg and K in leaf tissues of tomato in the current case study, without significant effects on Fe and Zn. Others (Higa and Parr 1994; Hussain et al. 1999) have shown that fermenting plant-based manure inherently improves soil and crop quality. In various cropping systems, it was shown that Nemafric-BL phytonematicide increased Ca, Ma, K, Fe and Zn in leaf tissues of tomato plants (Mashela and Pofu 2017). Consequently, for Fe and Zn which were not affected by Nemafric-manure, it could be that under manure the leaf tissues were already saturated by these micronutrient elements as is usually the case in the neutral phase of the DRG patterns (Liu et al. 2003).

2.4.1.2 Effects on Plant-Parasitic Nematodes

During 2016, Nemafric-manure decreased population densities of various plant-parasitic nematodes (*Meloidogyne* spp., *Helicotylenchus* spp., *Criconemoides* spp., *Pratylenchus* spp.) (Mashela 2017: Unpublished Report). The product also increased population densities of free-living nematodes such as bacterivores, fungivores, herbivores, omnivores and predators. During the 2017 season, similar trends were observed on free-living nematodes and plant-parasitic nematodes except that for reasons that were beyond our comprehension, population densities of *Helicotylenchus* spp. and *Pratylenchus* spp. were rather high (Mashela 2017: Unpublished Report).

2.4.1.3 Effects on *Fusarium* Species

During both seasons, tomato plants on untreated plots continued to have symptoms of *Fusarium* infection, which include wilting from the bottom leaves that spread to the top with speed (Mashela 2017: Unpublished Report). However, on plots with manure the treatments had no significant effects as the manure concentration increased.

2.4.1.4 Effects on Soil Variables

Soil pH Nemafric-manure did not have any significant effect on soil pH, which confirmed observations in Nemafric-BL phytonematicide–soil relations under various conditions (Mashela et al. 2017). Generally, when fermented organic materials with low pH were incorporated into an acid soil, there was an increase in soil pH (Duggan and Wiles 1976), whereas in neutral or basic soils, pH decreased (Gallardo-Lara and Nogales 1987). We do not have an explanation for this observation from literature, except that it could be soil- and climate-specific.

Soil Bulk Density Application of Nemafric-manure reduced (27%) bulk density from 1.38 to 1.01 $g.cm^{-3}$ (Mashela 2017: Unpublished Report) which confirmed observations in other manure trials (Duggan and Wiles 1976). Generally, the application of organic manures is related to the dilution of denser mineral fraction of soil, which is responsible for increasing bulk density (Duggan and Wiles 1976). The reduction in bulk density is related to improved soil health attributes (Brady and Weil 1999), which include an increase in the microbial biomass pool associated with the production of extracellular polysaccharides that link soil aggregates together, thereby improving aggregate stability (Haynes and Naidu 1998). Fermented organic manures are renowned for inducing slow and steady increase in aggregate stability when compared with unfermented organic manures (Monnier 1965).

Soil Nutrients In the current case study, the application of Nemafric-manure did not have significant effects on soil nutrient elements at 64 kg/2 m^2 (Mashela 2017: Unpublished Report), where nutrients were consistently higher. However, it is difficult to recommend the exact amount of organic manures needed for successful crop production as this varies with organic manure type, soil type, crop requirements and prevailing environmental conditions. Practical observation at ZZ2, a semi-organic farming enterprise, demonstrated that a combination of organic manures with inorganic sources of minerals yield higher beneficial effects on production and development compared to their sole application (Taurayi 2011). The latter might be due to the fact that organic manures do not contain all the required plant nutrients in adequate amounts for plant growth. Inorganic fertilisers are leached easily and combining them with organic manures would combine their soluble minerals with the organic fraction. Mineral nutrients would then be released slowly over time, in the form of sub-products of microbial pools and organic acids (Allison 1973).

2.5 Conclusion

Cucurbitacin-containing manure is a potent bioactive allelochemical against various soil-borne pathogens, with the attributes of stimulating plant growth. Along with EM, the product could serve as a suppressive manure that contains nematicidal, insecticidal (chitinase) and fungicidal (chitinase) properties. Since actinomycetes bacteria, which produce chitinase are aerobic and could be easily cultured, the bacteria could be supplemented during packaging of Nemafric-manure in order to enhance the chitinase properties of the product. Also, due to the thermophilic nature of cucurbitacin B, the manure would be suitable for use in soil with increasing temperatures, thereby promoting the adaptation and mitigation activities as viewed in climate-smart agriculture.

References

Allison FE (1973) Soil organic matter and its role in crop production. In: Developments in Soil Science, vol 3. Elsevier, Amsterdam

Arshad J (2006) Foliar application of effective microorganisms on pea as an alternative fertilizer. Agron Sust Dev 26:257–262

Arshad M, Frankenberger WT (1991) Microbial production of plant hormones. In: Keister DL, Cregan PB (eds) The rhizosphere and plant growth. Kluwer Academic, Amsterdam

Brady NC, Weil RR (1999) The nature and properties of soil. Prentice Hall, Upper Saddle River, NJ

Causton DR (1977) A biologist's mathematics: contemporary biology. Edward Arnold Publishers, London

Chen JC, Chiu MH, Nie RL, Cordell GA, Qiu XS (2005) The cucurbitacins and cucurbitane glycosides: structures and biological activities. Nat Prod Rep 22:386–399

Chen C, Kuo TC, Yangs M, Chien T, Chu M, Chen C, Lo H, Jeng S, Chen LO (2014) Identification of cucurbitacins and assembly of a draft genome for *Aquilaria agallocha*. BMC Genomics 15:578–589

Chitwood DJ (2003) Research on plant-parasitic nematode biology conducted by the United States Department of Agricultural Research Services. Pest Manag Sci 59:748–753

Dropkin VH (1969) Necrotic reaction of tomatoes and other hosts resistant to *Meloidogyne*: reversal by temperature. Phytopathology 59:1632–1637

Dudareva N, Klempien A, Muhlemann JK, Kaplan I (2013) Biosynthesis, function and metabolic engineering of plant volatile organic compounds. New Phytol 98:16–32

Duggan JC, Wiles CC (1976) Effects of municipal compost and nitrogen fertilizers on selected soils and plants. Comp Sci Utiliz 17:24–31

FAO (Food and Agricultural Organisation) (2018) Climate-smart agriculture: case studies in 2018, successful approaches from different regions. Food and Agricultural Organisation, Rome

Fujita M, Kato S, Yamada K, Xu HL, Katase K, Umemura H (1997) Applications of effective microorganisms in nature farming. VHI growth and yield of sweet corn as affected by applications of organic fertilizer with effective microorganisms. In: Annual meeting of Japanese Society of Soil Science and Plant Nutrition

Fuller VL, Lilley CJ, Urwin PE (2008) Nematode resistance. New Phytol 180:27–44

Gallardo-Lara F, Nogales R (1987) Effect of the application of town refuse compost on the soil-plant system: a review. Biol Waste 19:35–62

Guo J, Wu G, Bao J, Hao W, Lu J, Chen X (2014) Cucurbitacin B induced ATM-mediated DNA damage causes G2/M cell cycle arrest in a ROS-dependent manner. PLoS One 9:e88140

Haynes RJ, Naidu R (1998) Influence of lime, fertilizer and manure applications on soil organic matter content and soil physical conditions: a review. Nutr Cycl Agroecosys 51:123–137

Higa T, Parr JF (1994) Beneficial and effective microorganisms for a sustainable agriculture and environment. International Nature Farming Research Centre, Atami

Hussain T, Javaid T, Parr JF, Jilanoi G, Haq MA (1999) Rice and wheat production in Pakistan with effective microorganisms. Am J Alternat Agr 14:30–36

Kato S, Yamada K, Fujita M, Xu HL, Katase K, Umemuri H (1997) Applications of effective microorganisms in nature farming. IX. Soil fertility and plant nutrient uptake of sweet corn as affected by applications of organic fertilizer with effective microorganisms added. In: Annual meeting of Japanese Society of Soil Science and Plant Nutrition, 24 Apr 1997, Sizuoka, Proc 43:164

Kaushik U, Aeri V, Mir SR (2015) Cucurbitacins: an insight into medicinal leads from nature. Pharmacogn Rev 9:12–18

Krieger R (2001) Handbook of pesticides toxicology. Academic, San Diego

Kristkova E, Lebeda A, Vinter V, Blahousek O (2003) Genetic resources of genus *Cucumis* and their morphological description. Hort Sci 30:14–42

Läuchli A, Grattan SR (2012) Soil pH extremes. In: Shabala S (ed) Plant stress physiology. CABI, New York

Liu DL, An M (2005) Implementation of card: curve-fitting allelochemical response data. Nonlinear Biol Toxicol Med 3:235–244

Liu DI, Johnson IR, Lovett JV (2003) Mathematical modelling of allelopathy. III: A model for curve-fitting allelochemical dose responses. Nonlinear Biol Toxicol Med 1:37–50

Liu B, Ren J, Zhang Y, An J, Chen M, Chen H, Xu C, Ren H (2015) A new grafted rootstock against root-knot nematode for cucumber, melon and watermelon. Agronomy for sustainable development. Springer/EDP Sciences/INRA 35:251–259

Madaure JT, Mashela PW, De Waele D (2017) Response of entomopathogenic nematode, *Steinernema feltiae* to cucurbitacin-containing Nemafric-BL phytonematicide. Res Crops 18:739–744

Madigan MT, Martino JM (2006) Cell proliferation at 122°C and isotopically heavy CH4 production by a hyperthermophilic methanogen under high-pressure cultivation. PNAS 105:10949–10951

Mafeo TP (2014) Pre-sowing temperature treatment effect on emergence of *Cucumis myriocarpus* seedlings. Afr J Agric Res 9:2028–2030

Maila MY, Mashela PW, Nzanza B (2016) Ex vitro elimination of seed dormancy and in vitro seedling performance of *Cucumis africanus* and *Cucumis myriocarpus*. Acta Agr Scand B-S P 21:340–345

Malherbe ID (1964) Soil fertility. Oxford University Press, London

Marais M, Swart A, Fourie H, Berry SD, Knoetze R, Malan AP (2017) Alternative nematode management strategies. In: Fourie H, Spaull VS, Jones RK, Daneel MS, De Waele D (eds) Nematology in South Africa: a view from the 21st century. Springer Nature, Cham

Mashela PW (2002) Ground wild cucumber fruits suppress numbers of *Meloidogyne incognita* on tomato in microplots. Nematropica 32:13–19

Mashela PW, Pofu KM (2017) Influence of cucurbitacin-containing phytonematicides on selected nutrient elements in leaf tissues of green bean under greenhouse conditions. Acta Agr Scand B-S P 67:743–747

Mashela PW, De Waele D, Pofu KM (2011) Use of indigenous *Cucumis* technology as alternative to synthetic nematicides in management of root-knot nematodes in low-input agricultural farming system: a review. Sci Res Ess 6:6762–6768

Mashela PW, Ndhlala AR, Pofu KM, Dube ZP (2016) Phytochemicals of nematode-resistant transgenic plants. In: Jha S (ed) Transgenesis and secondary metabolism, Reference series in phytochemistry. Springer, Cham

Mashela PW, De Waele D, Dube Z, Khosa MC, Pofu KM, Tefu G, Daneel MS, Fourie H (2017) Alternative nematode management strategies. In: Fourie H, Spaull VS, Jones RK, Daneel MS,

De Waele D (eds) Nematology in South Africa: a view from the 21st century. Springer Nature, Cham

Mkile Z (2001) The use and agronomic effectiveness of kraal manures in the Transkei region of the eastern cape, South Africa. MSc Dissertation, University of Fort Hare, Alice, South Africa

Monnier G (1965) Action des matieres organiques sur la stabilité structurale des sols. Cited in: R.J. Haynes & R. Naidu (eds) Influence of lime, fertilizer and manure applications on soil organic matter content and soil physical conditions: a review. Nutr Cycl Agroecosyst 51:123–137

Mphahlele RR, Mashela PW, Pofu KM (2012) Post-harvest fruit decay inducing pathogen in medicinally important *Cucumis* species indigenous to South Africa. Afr J Agric Res 6:3786–3791

Müller J, Heindl A (2006) Drying of medicinal plants. J Sci Food Agr 86:2543–2550

OECD (2017) (Organisation for economic co-operation and development). Guidance document on botanical active substances used in plant protection products. Series on pesticides no. 90. OECD, Paris

Pofu KM (2012) Potential uses of indigenous *Cucumis africanus* and *Cucumis myriocarpus* as root-knot nematode-resistant rootstocks in watermelon (*Citrullus lanatus*) husbandry. PhD thesis, University of Limpopo, Sovenga

Pofu KM, Mashela PW, Mokgalong NM, Mphosi MS (2011) Alternatives to methyl bromide in suppression of root-knot nematodes in watermelon (*Citrullus lanatus*) production. Acta Hortic (917):227–230

Pofu KM, Mashela PW, Venter S (2019) Dry bean cultivars with the potential for use in potato-dry bean crop rotation systems for managing root-knot nematodes in South Africa. S Afr J Plant Soil 36:315

Prescott LM, Harley JP, Klein DA (2002) Microbiology. The McGraw-Hill, New York

Shadung KG (2016) Quality protocols for Nemarioc0-AL and Nemafric-BL phytonematicides and potential chemical residues in tomato fruits. Thesis, University of Limpopo, Sovenga

Shadung KG, Mashela PW (2016) Suitable organ(s) in wild *Cucumis africanus* for the production of Nemafric-BL phytonematicide. Acta Agric Scand B-S P 66:381–383

Shadung KG, Mashela PW (2017) Influence of storage period on concentration of cucurbitacin B from dried *Cucumis africanus* fruit. Res Crops 18:327–331

Shadung K, Mashela P, Mulaudzi V, Mphosi M, Ncube I (2015) Optimum harvest time of *Cucumis africanus* fruit using concentration of cucurbitacin B as a maturity standard. J Agric Sci 7:181–186

Shadung KG, Mashela PW, Mphosi MS (2016a) Suitable drying temperature for preserving cucurbitacins in fruit of wild cucumber and wild watermelon. HortTechnology 26:816–819

Shadung KG, Mashela PW, Mphosi MS (2016b) Response of cucurbitacin B concentration in Nemafric-BL phytonematicide to increasing storage period. J Stored Prod Postharv Res 7:32–36

Shadung KG, Mashela PW, Mphosi MS, Mulaudzi VL (2017) Study of chemical residues from Nemafric-AL and Nemafric-BL phytonematicides in tomato fruit. Afr J Agric Res 68:1164–1168

Shaik BS, Burrows GE, Urwin NAR, Gopurenko D, Lepschi BJ, Weston LA (2016) The biology and management of prickly paddy melon (*Cucumis myriocarpus* L.), an important summer annual weed in Australia. Crop Prot 92:29–40

Sikora R, Coyne D, Hallmann J, Timper P (2018) Plant-parasitic nematodes in subtropical and tropical agriculture. CABI, New York

Singh RP (ed) (2016) Organic fertilisers: types, production and environmental impact. Nova, New York

Sinha RK, Valani D, Chauhan K, Soni BK, Agarwal S (2016) Earthworms vermicompost: a nutritive biofertiliser and powerful biopesticide for promoting organic farming while protecting farm soils and mitigating global warming. In: Singh RP (ed) Organic fertilisers: types, production and environmental impact. Nova, New York

Stetter K (2006) History of discovery of the first hyperthermophiles. Extremophiles 10:357–362

Taurayi S (2011) An investigation of natuurboerdery (natural farming) approach: a ZZ2 case study. Master thesis, University of Stellenbosch, Stellenbosch, South Africa

Tester CF (1990) Organic amendment effects on physical and chemical properties of a sandy soil. Soil Sci Soc Am J 54:827–831

Thies JA, Levi A (2006) Resistance of *Citrullus lanatus* var. citroides germplasm to root-knot nematodes. J Nematol 38:258–303

Tseke PE, Mashela PW (2017) Efficacy of Nemafric-BL phytonematicide from fresh *Cucumis africanus* fruit on suppression of root-knot nematodes and growth of tomato plants. Res Crops 18:289–293

Van Wyk BE, Wink M (2004) Medicinal plants of the world. Timber Press, Portland, OR

Von Wolferen M, Ajon M, Driessen AJ, Albers SV (2013) How hyperthermophiles adapt to change their lives: DNA exchange in extreme conditions. Extremophiles 17:545–563

Xu HL (2000) Effects of microbial inoculant and organic fertilizers on the growth, photosynthesis and yield of sweet corn. Crop Prod 3:183–214

Yamada K, Kato S, Fujita M, Xu HL, Katase K, Umemura H (1996) An organic fertilizer inoculated with EM used in nature farming practice. Asia Pacific Nature Agriculture Network, Oct 8–12, 1996, Bangkok

Yoganathan S, Sotana MM, Van Averbeke W, Mandiringana OT, Materechera S, Harris PJC, Mnkeni PNS (1998) Kraal manure as a fertiliser in small scale crop production in central Eastern Cape. South Africa ARSRE Int Symp 15:361–368

Chapter 3
Nematode Succession During Composting Process

Mouna Jeridi, Amel Ayari-Akkari, Sazada Siddiqui, and K. K. Chaudhary

Abstract Composting is primarily an aerobic process where the complex organic substrate gets converted into stabilized matter containing soluble minerals, CO_2, and water. This process is mediated by a wide spectrum of microorganisms like bacteria, fungi, arthropods, nematodes, etc. It is well established that nematodes have the capability to influence various soil processes such as decomposition of complex organic matter and nutrient mineralization. Therefore, they have also been considered as important biological indicators of the composting maturity and stability. Understanding the nematode succession during the composting process assumes significance as it can help us effectively assess the compost quality. The present chapter provides an elaborate overview of the succession of nematofauna during the process of composting. Subsequent to providing brief details of soil and compost process, we discuss the structure and community dynamics of nematofauna during the composting process. Various factors governing the nematode succession and their implications on agriculture are also highlighted.

Keywords Soil · Compost · Nematodes community · Nematodes succession · Agriculture

---

Mouna Jeridi and Amel Ayari-Akkari contributed equally to conducting the research and writing the manuscript.

M. Jeridi (✉) · A. Ayari-Akkari
Biology Department, College of Science, King Khalid University, Abha, Saudi Arabia

S. Siddiqui
Department of Biology, College of Science, King Khalid University, Abha, Saudi Arabia
e-mail: sasdeky@kku.edu.sa

K. K. Chaudhary
Department of Plant Protection, Hamelmalo Agricultural College, Hamelmalo, Eritrea

M. K. Meghvansi, A. Varma (eds.), *Biology of Composts*, Soil Biology 58,
https://doi.org/10.1007/978-3-030-39173-7_3

3.1 Introduction

An increasing number of ecologists have started focusing on soil communities, largely because of awareness that not only do soil organisms regulate major ecosystem processes, such as organic matter turnover and nutrient cycling, but also act as important drivers of vegetation change. Nematodes are the most copious metazoans in soil and are surpassed in species assorted variety just by the arthropods. Assessments of nematode diversity in natural and agroecosystems have been founded on both species-level scientific classification and trophic-level guilds, since trophic groups do not act similarly regarding ecological events.

Nematodes of a biotope have frequently been considered as simple collections, yet there is expanding proof that specific plant affiliations have trademark gatherings of species. This notion has been utilized by Bongers (1990) to build up maturity index relating nematode families and site stability and to perceive the group that colonizes diverse sorts of soils. Crucial handicaps on detailed ecological studies of soil nematofauna are an incomplete understanding of trophic groups and their subsets, the need for repeated sampling of occasionally factor populations, and an extreme lack of scientific classification by skilled persons, particularly for the microbial feeder.

Although nematodes seem, by all accounts, to be numerically essential in the composting process, and their diversity and density in mature compost are viewed as crucial to building a vigorous soil food web (Ingham and Slaughter 2004). Research concerning the nematodes function in decomposition processes will in general spotlight on explicit natural processes in the soil (Ruess 2003; Ruess and Ferris 2004; Georgieva et al. 2005; Postma-Blaauw et al. 2005; Wang et al. 2006). Moreover, nematodes demonstrate a few attributes that make them perfect bioindicators of the biological system quality. The nematode community examinations are valuable in evaluating biological system status and roles since nematodes are omnipresent and simple to test (Bongers and Ferris 1999; Ritz and Trudgill 1999; Yeates and Bongers 1999; Neher 2001; Yeates 2003). Furthermore, the species composition of the examples mirrors a wide range of abiotic parameters, for example, substrate texture, climate, biogeography, organic inputs, and both natural and anthropogenic disturbances (Neher 2001; Yeates 2003). The utilization of nematodes as practical pointers depends on the allotment of nematodes to feeding groups (Yeates et al. 1993). Therefore, nematodes are omnipresent in manures and are dynamic in all phases of the fertilizing the soil procedure.

The principal aim of this chapter is to contribute for a better understanding of the nematode succession during the composting process while describing the different types of soil and compost highlighting some of the major groups of nematode fauna that live in soil and discussing their succession during different steps of composting process. In addition, the potential of nematodes for use as a compost maturity bioindicator as well as biological soil indicator and their impact in agriculture are discussed.

3.2 Soil Structure

Soil is a blend of minerals, water, air, and incalculable living beings that are the decaying remains of formerly living things. It is equipped for supporting vegetation and is imperative to life on earth. Soil, as formally determined in the Soil Science Society of America Glossary of Soil Science Terms, is:

(a) The unconsolidated mineral or organic material on the brief of the earth that fills in as natural place for the development of land plants.
(b) The unconsolidated mineral or organic issue on the outside of the earth that has been exposed to and indicates impacts of hereditary and ecological elements of atmosphere, and large scale and microorganisms, following up on parent material over a period.

Soil performs numerous basic activities practically in any ecosystem. Soil can assume seven general occupations:

(a) Soil makes a contribution as media for the development of a wide range of plants.
(b) Soil adjusts the atmosphere by producing and engrossing gases and residues.
(c) Soil gives habitat to creatures that live in the dirt to life forms (for example, bacteria and fungi) that represent the majority of the living things on Earth.
(d) Soil absorbs, holds, discharges, modifies, and decontaminates the greater part of the water in earthly frameworks.
(e) Soil processes recycled supplements, including carbon so that living things can utilize them repeatedly.
(f) Soil fills in as designing media for the development of establishments, roadbeds, dams and structures, and safeguard or decimate antiques of human undertakings.
(g) Soil goes about as a channel to scour water before it moves into an aquifer.

Soil can be sorted into various kinds each with its very own arrangement of attributes. It is made of horizons (O, A, E, B, C, R) and each one presents a particular profile. Most soils have three noteworthy horizons (A, B, C) and some have an organic horizon (O). These distinctive horizons are:

(a) **O-(humus or organic)**: This is the most important horizon primarily containing organic matter from decomposed leaves. The O horizon is thin in some soils, thick in others, and totally absent in others.
(b) **A-(topsoil)**: Generally, minerals from parent material with organic matter consolidated. A good material for plants and different living beings to live.
(c) **E-(eluviated)**: Leached of clay, minerals, and organic matter, leaving a concentration of sand and residue particles of quartz or other resistant materials—missing in some soils, however, frequently found in older soils and forest soils.
(d) **B-(subsoil)**: Wealthy in minerals that filtered down from the A or E horizons and deposited here.
(e) **C-(parent material)**: The deposit at Earth's surface from which the soil established.

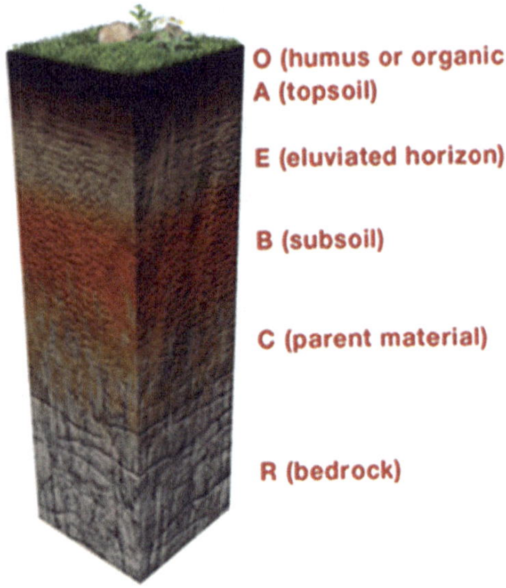

Fig. 3.1 Different types of horizons of soil. Source: Soil Science Society of America Glossary of Soil Science Terms: https://www.soils.org/publications/soils-glossary

(f) **R-(bedrock)**: A pile of rock like granite, basalt, quartzite, limestone, or sandstone that forms the origin material for certain soils—if the bedrock is close enough to the surface to weather. This is not soil and is situated under the C-horizon (Fig. 3.1).

3.3 Composting Process and Its Role in Agriculture

Compost is an organic material that has been disintegrated in a procedure called composting. This procedure recycles different organic matters generally viewed as waste items and produces a soil conditioner called the compost. The compost is wealthy in accessible soil supplements and frequently utilized in finishing, cultivation, and rural field crops, to build and improve soil fertility. Physicochemical qualities of soils are ameliorated by the utilization of compost, which at last increases crop yield (Martens 2000). The utilization of compost in soil has plural advantages. The balanced out organic material in compost improves soil structure. Subsequently, soil air circulation, soil porosity, water holding capacity, and drainage increase (Kuo et al. 2004; Cogger 2005; Baldwin and Greenfield 2006). These advantages show themselves in decreased editing dangers, higher yields, and lower costs on inorganic fertilizers for agriculturalists. At the least difficult level, the composting procedure requires making a mass of green waste, for example, leaves, grass, and nourishment scraps, and trusting that the materials will transform into humus following some months. Nevertheless, composting additionally can happen as a multi-step, intently observed procedure with estimated contributions of water, air, carbon, and nitrogen-rich materials. The decomposition procedure is helped by destroying the plant matter, including water and guaranteeing legitimate air circulation by routinely turning the blend when open heaps or "windrows" are used. Earthworms and fungi further degrade the material. Bacteria expecting oxygen to work (aerobic bacteria) and fungi deal with the procedure by changing over the contributions to carbon dioxide, and ammonium. Likewise, the utilization of vermicomposts in the formation of commercial greenhouse and substitution of little sums into soilless development media has turned out to be a significant practice to expand plant growth, support plant yields, and ameliorate the physical properties of soils (Atiyeh et al. 1999, 2000a, b, 2001, 2002).

Composting can be diverted into two classes as indicated by the idea of the decomposition procedure. In anaerobic composting, deterioration happens where oxygen is not present or in constrained supply. Under this strategy, anaerobic microorganisms create mixes including methane, natural acids, hydrogen sulfide, and different added substances. Without oxygen, these mixes gather and are not processed further. A large number of these mixes have strong scents and some present phytotoxicity. As anaerobic composting is a low-temperature procedure, leaving the weed seeds and pathogens flawless. Additionally, the procedure generally requires more time than aerobic composting. These disadvantages regularly counterbalance the benefits of this procedure, viz., little work included and less supplements lost amid the procedure. Aerobic composting takes place in the occurrence of abundant oxygen. In this procedure, aerobic microorganisms decompose organic matter and produce carbon dioxide, water, ammonia, warmth, and humus, the moderately steady organic finished product. While aerobic composting may deliver intermediate of mixes, for example, organic acids; aerobic microorganisms break down them further. The subsequent compost, with its generally unsteady type of organic matter, has little danger of phytotoxicity. The warmth produced to speed up the decomposition of fats, proteins, and complex carbohydrates, for example, cellulose. Subsequently, the handling time is shorter. Also, this procedure obliterates numerous miniaturized scales living beings that are human or plant pathogens, just as weed seeds, if it experiences adequately high temperature. Albeit more nutrients are lost from the materials by aerobic composting, it is viewed as more proficient and helpful than anaerobic composting for agricultural production (FAO 2003).

As an alternative to avoid landfilling and promote recycling, composting is one of the best known and well-established processes. It is obvious that the particularity of the composting process is considered as a vital factor in the suppressiveness of plant diseases provided by the nematode community occurring during the process. In addition, composting permits the stabilization and sanitation of organic waste via improved aerobic decomposition under managed conditions. Several researches suggest that the usage of compost on land may additionally improve several plants and soil parameters (Fig. 3.2), which could make compost a thrilling alternative for soil restoration functions, in addition to take gain of its fertilizer residences. Compost addition will increase soil natural remember content, which complements aggregation and balance, thereby ameliorating soil shape (Diacono and Montemurro 2010). Soil aggregation, which is basically responsible for soil structure, soil functioning, and agricultural productivity, is mainly depends on soil microbial activities. In

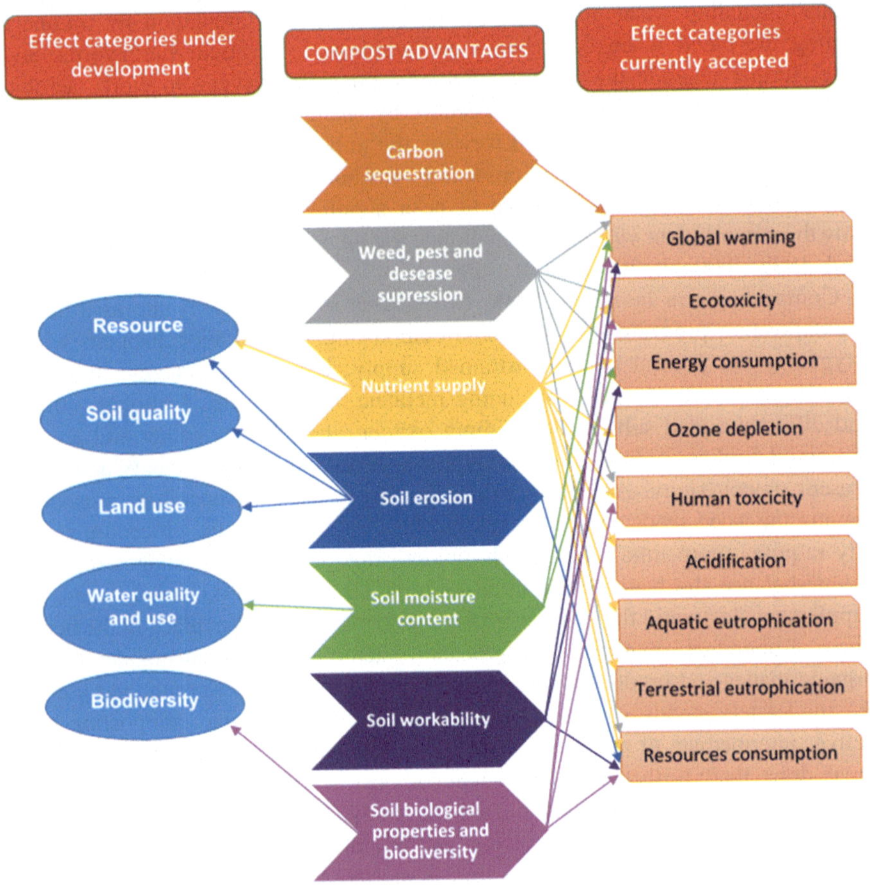

Fig. 3.2 Categories concerned in the evaluation of potential advantages of biowaste compost use-on-land

general, soil aggregation is influenced by various factors such as soil microorganisms and macrofauna, soil mineralogy, soil texture, and environmental conditions. Changes in abundance and structure of soil microbial community is a viable indicator of soil quality changes (Bai et al. 2019).

3.3.1 Composting Process and Soil Suppressiveness

In combination with compost, the effects of fungivorous nematodes on disease suppression were investigated and several studies have demonstrated the capability of fungivorous nematodes to diminish plant diseases and subsequently improve plant health (Grabau and Chen 2016). It additionally exerts a suppressive interest on plant soil-borne pathogens and pests, which includes plant parasitic nematodes (Bailey and Lazarovits 2003). The first systematic study was performed by Rhoades and

Linford (1959), who found that *Aphelenchus avenae* Bastian was efficient in controlling *Pythium* root rot in maize (*Zea mays* L.) under greenhouse conditions. In addition, as example, the impacts of various type of composts were evaluated on the development and yield of okra, planted in plant parasitic nematodes infested soil and the results obtained have showed that the application of *Trichoderma harzianum* incorporated into neem, cassava, sawdust, and *Tithonia* composts have significant effect on the growth and yield of okra; then, it also significantly reduced the erratic nematode population changes (Olabiyi and Oladeji 2014).

3.3.2 Relationship Between Phytopathogenic Nematodes and Tillage of Soil

Nematodes are generally more abundant in the absence of tillage. However, the effects varied with the species of plant parasitic nematodes present and are not always significant (McSorley 1996). Tillage (plowing, weeding, etc.) induces changes in water and heat conditions in the superficial layer of the soil; where nematodes are mainly present, likely to influence their development.

In general, the addition of organic amendments induces a reduction in the number of plant parasitic nematodes due to two main phenomena:

(a) When introducing organic matter, the nematode antagonistic organisms (predators and parasites) can become more abundant.
(b) Some organic acids resulting from decomposition of organic matter are toxic to nematodes (formic, acetic, propionic, or butyric acid).

Moreover, compost contribution, as well as tillage, induces a modification of the physicochemical characteristics of soil as well as a modification of the specific structure of the phytoparasite nematode population. In fact, the soil preparation and the compost contribution make it possible to improve the organic status of the soil compared with other modes of driving, which consequently favors a better development of the cultivated plant. In addition, these treatments induce changes in the specific structure of plant parasitic nematode populations by increasing the abundance of a low pathogenic species that could limit the damage caused by nematodes (Villenave et al. 1998), which demonstrate the importance of soil tillage in the suppression of some diseases.

3.4 Soil Nematodes

3.4.1 Nematode Community Structure

Nematoda, or roundworms, formed a branch of unsegmented worms, are microscopic eelworms. Classified among the ecdysozoaires, they are covered with a thick cuticle. They lead a "free" life (in soils, water, sediments, deadwood, and other forms of necromass where they are detritivores or micro-predators) (Li et al. 2016). Many species have a parasitic life, phytoparasites within fungal organisms, plants or animals. They are formed of nearly a hundred thousand species, capable of causing genuine defects in a wide scope of harvests (Olabiyi and Oladeji 2014). Nematodes are dioecious and have between two and four juvenile stages. They are considered as an evolutionarily successful group of organisms. They are omnipresent in all living spaces that give accessible organic carbon sources making them the planet's most plentiful metazoa. In soil, a few nematodes feed on higher plants, some on parasites (fungi) or microbes (bacteria); others are carnivores or omnivores. They range in reproductive potential from explosive opportunists to conservative survivalists (Bongers and Ferris 1999). They differ in sensitiveness to pollutants and environmental disturbance. Acknowledgment of the roughly 20,000 portrayed species depends basically on morphological and anatomical features supplemented by ecological function (Siddiqi 1986), although molecular markers are increasingly being used for identification and classification (Thomas et al. 1997; Blaxter et al. 1998). Nematode structure and function, particularly concerning feeding behavior, are firmly connected and therefore the feeding behavior of undetermined species can be inferred from their morphology (Yeates et al. 1993; Yeates 1998).

3.4.2 Colonizer–Persister (Cp) Groupings of Nematode Taxa

Following the selection criteria, nematodes are classified by trophic habit (Yeates et al. 1993) and by colonizer–persister (cp) groups (Bongers 1990). The "cp" scale divides nematode families into five gatherings, from microbial feeders with short life cycles (cp1 and 2), these groups present the capacity to enter a nonfeeding, inactive "dauerlarva" survival stage under stressed conditions, to predators and omnivores with long life cycles and significant sensitivity to environmental disruption (cp groups 4 and 5). The efficiency characteristics of general opportunists (cp-2) are less extreme than those of cp-1. Feeding on bacteria seems progressively conscious and the bacterial feeding cp-2 taxa do not have a dauerlarva stage (Sánchez-Moreno et al. 2006). Species in various cp classes contrast in ecological amplitude; lower "cp" groups are represented in terrestrial, freshwater, and marine environmental conditions, while nematode taxa in higher "cp" groups, for the most part, have a smaller environmental abundancy.

3.4.3 Nematode as Biological Soil Indicators

Soil particles vary in size and aggregation throughout the climatic and spontaneous and natural or induced soil changes (Bongers and Ferris 1999). During the 1980s, as concerns extended about the functioning and powerlessness of the soil ecosystem, scientists swung to the appraisal of in situ nematode faunae in ecological investigations (Zullini and Peretti 1986; Kozlowska 1989). The faunal nematode composition proved to be a useful monitor of environmental conditions and ecosystem function in the soil; they play a role of a key indicator in the evaluation of soil food web.

Structure and plenitude of the nematofauna have been utilized as soil health markers in a wide range of environments (Neher 2001; Sithole et al. 2016). Nematodes are functionally diverse and ubiquitous and react promptly to environmental changes. The tight connections between soil characteristics and presence of nematode in various functional guilds (Fiscus and Neher 2002) have been utilized to create soil assessment criteria. In agricultural fields, nematode abundance and diversity are to deduce soil process rates (Ettema 1998; Porazinska et al. 1999), soil functions (Ekschmitt et al. 2003; Yeates 2003), and effects of disturbance on soil fauna (Wardle 1995). In other cases, natural areas nematode assemblages are used to assess the impact of pollution (Korthals et al. 1996a; Gyedu-Ababio et al. 1999) as indicators of enhancement and unsettling influence (Bongers 1990; Ferris et al. 2001; Berkelmans et al. 2003) and to ponder food webs dynamics (Mikola and Setälä 1998; Wardle et al. 2005). Nematode species diversity studies are useful in the detection of indicator species. Certain plant parasites proved good indicators of soil and forest types (Bernard 1992).

In a review on the role of nematodes in soil health, Neher (2001) has made the following observations:

- Nematodes are some of the unpretentious metazoa. They occur in any environment that gives a source of organic carbon, in each soil type, under all climatic conditions, and in habitats that vary from pristine to extremely polluted.
- In soil, nematodes stay in capillary water; their permeable cuticle presents direct touch with their microenvironment. They do not quickly migrate from disturbing situations and lots of species survive dehydration, freezing, or oxygen pressure (despite the fact that others are greater sensitive). The community form is indicative of conditions in the soil horizon that it inhabits.
- Nematodes occupy key positions in soil food webs. They feed on most soil organisms and are food for many others. They also impact plant succession.
- Because nematodes are obvious, their diagnostic internal capabilities may be visible without dissection. They can therefore be identified without biochemical methods.
- There is a visible connection between structure and function: the feeding behavior is effortlessly deduced from the shape of the mouth cavity and pharynx.

Nematodes respond unexpectedly to disturbance and enrichment: increasing microbial activity leads to changes in the percentage of bacterial feeders in the community (Bongers and Ferris 1999). Soil nematode community composition is known to relate to various feature of soil status and has been advocated as an indicator of soil disturbance, nutrient enrichment, and pollution (Bongers 1990; Coleman et al. 1998; Korthals et al. 1996b; Wasilewska 1997; Zhong et al. 2017). Therefore, nematode worm genus richness provided the only indicator to correlate with the complete set of parameters concerning the nematofauna and the microflora. Similarly, the Maturity Index (MI), which is based on the classification of nematodes along a colonizer–persister gradient (Bongers 1990), proved the most important indicator for soil nitrogen rank. The authors concluded that among the alternatives tested, nematode community of interest structure would be seen as the best indicator of an active nematofauna and microflora, and hence as a possible indicator of the decomposition function in the soil.

3.5 Nematodes Succession During Composting Process

The composting process needs the interference of several organisms. The taxa that are pivotal to this process are bacteria, algae, fungi, isopoda, acari, nematoda, and protozoa (Ingham et al. 1985; Steel et al. 2010). These wide ranges of organisms constitute a higher complex community that changes during the composting process. Likewise, nematodes exist in a large numbers in soil, they interfere in the composting procedure, and their assorted variety and thickness in mature manure are viewed as inevitable to building a vigorous soil food web (Ingham and Slaughter 2004). Then, a significant advantage of using nematodes to evaluate compost maturity is their settled status as environmental indicators (Bongers and Ferris 1999; Neher 2001; Yeates 2003) of ecosystem processes such as organic fertilization (Ferris and Bongers 2006). In addition, nematode assemblages are often highly diverse, depending to ecosystems, which is a problematic because species diversity measurement depends on sampling intensity, which varies considerably among a survey done on the relation between soil nematode species diversity and the kind of soil cores related to various ecosystems (Bernard 1992; Bloemers et al. 1997; Freckman and Virginia 1997).

Since nematodes as present in soil from the beginning of the composting process, they can survive during the first phase of heat peaks like dauer stages or eggs. It is important to note that nematode community structure changes continuously during this process. According to Steel et al. (2010), the ordered exploration of the nematode groups uncovered three clear successional stages after the heat peak (abiotic factors). Beginning with the bacterial-feeding enrichment opportunists (cp-1) (Rhabditidae, Panagrolaimidae, Diplogastridae) come behind the bacterial-feeding general opportunists (cp-2) (Cephalobidae) and the fungal-feeding general opportunists (Aphelenchoididae), noticing that after the intervention of opportunistic bacterial-feeding predator nematodes (*Mononchoides* sp.) which become the dominant species. At the end of the process, at the most mature stage, the fungal-feeding Anguinidae (e.g., *D. filimus*) was the most plentiful group. This investigation was based on the effect of the abiotic factors and conditions in the succession of nematodes activity in the soil during the composting process. In fact, many authors confirmed these observations. Bongers (1990) showed that during the first phase (on Day 15) the dominance of the bacterial-feeding and enrichment opportunistic nematodes which was also affirmed by many others (Ettema and Bongers 1993; Bardgett and Cook 1998; Wasilewska 1997). However, in these last investigations, the bacterial-feeding/predators were more conspicuously present in the early cooling phase (second phase of composting process). Moreover, Steel et al. (2013) confirmed that nematode populates the compost and explained the origin of the compost colonizers. Similarities among studies (Steel et al. 2010, 2013) feature a solid sign that the portrayed patterns of nematode community succession are universal for compost, evenly they are relatively independent of the environmental conditions of the composting process, because shielding them from contact with insects or soil has almost no impact. A positive correlation was reported between free-living nematode population and composting process while a negative correlation was reported between plant parasitic nematode and free-living nematode population. Therefore, omnivorous and free-living nematodes were beneficial in decomposition, nitrogen mineralization, and microbial community development (Pokharel et al. 2015; Treonis et al. 2018). Similarly, Moura and Franzener (2017) suggested a positive role of bacteriophage nematodes in organic matter decomposition and raising carbon mineralization. Presence of the bacterivore's nematode in compositing soil reflects the fertility status of compositing soil while excess soil stresses are represented by the presence of predators–omnivores nematodes (Steel et al. 2018).

3.6 Conclusions and Future Perspectives

Nematodes are among the most essential groups of soil microbial communities having direct linkages with various soil processes. The organic matter decomposition in the soil is also significantly influenced by nematode structure and dynamics. Therefore, nematode activity monitoring is one of the key aspects keeping in view its implications in agriculture. Taking cues from this, various researchers have attempted to identify and discuss the pattern of nematode succession during organic matter decomposition. Similar studies have also been made in the context of composting process. As evident from the discussion, nematode succession in composting is a very important phenomenon. Generating data on the pattern of nematode succession in composting would certainly increase our understanding about the role and effect of nematodes on the overall composting process. Furthermore, this would help in devising strategies for assessment of compost maturity and stability. In fact some researchers have attempted to propose a nematode-based index of compost maturity for the purpose of determining the biological maturity of the finished product. However, owing to intricacies associated with the complex interactions of nematode community with other microorganisms present in decomposing organic matter, there are still many unknowns with regard to universal applicability of such indices under varying abiotic and biotic factors. It is a very challenging task that would require a thorough analysis of functional attributes at various composting phases using modern molecular tools and techniques.

Acknowledgements The authors extend their appreciation to the Deanship of Scientific Research (DSR), King Khalid University, Abha, Saudi Arabia for their help in this chapter.

References

Atiyeh RM, Subler S, Edwards CA, Metzger JD (1999) Growth of tomato plants in horticultural media amended with vermicompost. Pedobiologia 43:724–728

Atiyeh RM, Dominguez J, Subler S, Edwards CA (2000a) Changes in biochemical properties of cow manure during processing by earthworms (*Eisenia andrei*, Bouché) and the effects on seedling growth. Pedobiologia 44:709–724

Atiyeh RM, Edwards CA, Subler S, Metzger JD (2000b) Earthworm processed organic wastes as components of horticultural potting media for growing marigolds and vegetable seedlings. Compost Sci Util 8:215–233

Atiyeh RM, Edwards CA, Subler S, Metzger JD (2001) Pig manure vermicompost as a component of a horticultural bedding plant medium: effects on physicochemical properties and plant growth. Bioresour Technol 78:11–20

Atiyeh RM, Arancon NQ, Edwards CA, Metzger JD (2002) The influence of earthworm-processed pig manure on the growth and productivity of marigolds. Bioresour Technol 81:103–108

Bai N, Zhang H, Li S, Zheng X, Zhang J, Zhang H, Zhou S, Sun H, Lv W (2019) Long-term effects of straw and straw-derived biochar on soil aggregation and fungal community in a rice–wheat rotation system. Peer J 6:e6171. https://doi.org/10.7717/peerj.6171

Bailey KL, Lazarovits G (2003) Suppressing soil-borne diseases with residue management and organic amendments. Soil Tillage Res 72:169–180

Baldwin KR, Greenfield JK)2006) Composting on organic farms. Publication no. AG-659W-01, North Carolina Cooperative Extension Service, NC State University

Bardgett RD, Cook R (1998) Functional aspects of soil animal diversity in agricultural grasslands. Appl Soil Ecol 10:263–276

Berkelmans R, Ferris H, Tenuta M, van Bruggen AHC (2003) Effects of long-term crop management on nematode trophic levels other than plant feeders disappear after 1 year of disruptive soil management. Appl Soil Ecol 23:223–235

Bernard EC (1992) Soil nematode biodiversity. Biol Fertil Soils 14:99–103

Blaxter ML, De Ley P, Garey JR, Liu LX, Scheldeman P, Vierstraete A, Vanfleteren JR, Mackey LY, Dorris M, Frisse LM, Vida JT, Thomas WK (1998) A molecular evolutionary framework for the phylum Nematoda. Nature 392:71–75

Bloemers GF, Hodda M, Lambshead PJD, Lawton JH, Wanless FR (1997) The effects of forest disturbance on diversity of tropical soil nematodes. Oecologia 111:575–582

Bongers T (1990) The maturity index: an ecological measure of environmental disturbance based on nematode species composition. Oecologia 83:14–19

Bongers T, Ferris H (1999) Nematode community structure as a bioindicator in environmental monitoring. Trends Ecol Evol 14:224–228

Cogger CG (2005) Potential compost benefits for restoration of soils disturbed by urban development. Compost Sci Util 13:243–251

Coleman DC, Hendrix PE, Odum (1998) EP Ecosystem health: an overview. Soil Chemistry and Ecosystem Health, SSSA Special Publ No 52: 1–20

Diacono M, Montemurro F (2010) Long-term effects of organic amendments on soil fertility. A review. Agron Sustain Dev 30:401–422

Ekschmitt K, Stierhof T, Dauber J, Kreimes K, Wolters V (2003) On the quality of soil biodiversity indicators: abiotic and biotic parameters as predictors of soil faunal richness at different spatial scales. Agric Ecosyst Environ 98:273–283

Ettema CH (1998) Soil nematode diversity: species coexistence and ecosystem function. J Nematol 30:159–169

Ettema CH, Bongers T (1993) Characterization of nematode colonization and succession in disturbed soil using the maturity index. Biol Fertil Soils 16:79–85

Ferris H, Bongers T (2006) Nematode indicators of organic enrichment. J Nematol 38:3–12

Ferris H, Bongers T, de Goede RGM (2001) A framework for soil food web diagnostics: extension of the nematode faunal analysis concept. Appl Soil Ecol 18:13–29

Fiscus DA, Neher DA (2002) Distinguishing sensitivity of free-living soil nematode genera to physical and chemical disturbances. Ecol Appl 12:565–575
Food and Agriculture Organization (2003) On farm composting methods. Food and Agriculture Organization of the United Nation, Rome
Freckman DW, Virginia RA (1997) Low diversity Antarctic soil nematode communities: distribution and response to disturbance. Ecology 78:363–369
Georgieva S, Christensen S, Stevnbak K (2005) Nematode succession and microfauna–microorganism interactions during root residue decomposition. Soil Biol Biochem 37:1763–1774
Grabau ZJ, Chen S (2016) Influence of long-term corn–soybean crop sequences on soil ecology as indicated by the nematode community. Appl Soil Ecol 100:172–185
Gyedu-Ababio TK, Furtenberg JP, Baird D, Vanreusel A (1999) Nematodes as indicators of pollution: a case study from the Swartkops River system. S Afr Hydrobiologia 397:155–169
Ingham RE, Slaughter MD (2004) The soil foodweb – soil and composts as living ecosystems. In: International conference on soil and compost eco-biology, 15–17 Sept 2004, León, Spain
Ingham RE, Trofymow JA, Ingham ER, Coleman DC (1985) Interactions of bacteria, fungi, and their nematode grazers: effects on nutrient cycling and plant growth. Ecol Monogr 55:119–140
Korthals GW, de RGM G, Kammenga JE, Bongers T (1996a) The Maturity Index as an instrument for assessment of soil pollution. In: van Straalen NM, Krivolutsky DA (eds) Bioindicator systems for soil pollution. Kluwer Academic, Dordrecht, pp 85–93
Korthals GW, van de Ende A, van Megen H, Lexmond TM, Kammenga JE, Bongers T (1996b) Short-term effects of cadmium, copper, nickel and zinc on soil nematodes from different feeding and life-history strategy groups. Appl Soil Ecol 4:107–117
Kozlowska J (1989) The effect of sewage sediments on communities of soil nematodes. Pol Ecol Stud 15:27–39
Kuo S, Ortiz-Escobar ME, Hue NV, Hummel RL (2004) Composting and compost utilization for agronomic and container crops. Rec Res Dev Environ Biol 1:451–513
Li N, Pan FJ, Han XZ, Zhang B (2016) Development of soil food web of microbes and nematodes under different agricultural practices during the early stage of pedogenesis of a Mollisol. Soil Biol Biochem 98:208–216
Martens DA (2000) Management and crop residue influence soil aggregate stability. J Environ Qual 29:723–727
McSorley R (1996) Impact of crop management practices on soil nematode populations. Annu Proc Soil Crop Sci Soc Fla 55:63–66
Mikola J, Setälä H (1998) Relating species diversity to ecosystem functioning: mechanistic backgrounds and experimental approach with a decomposer food web. Oikos 83:180–194
Moura GS, Franzener G (2017) Biodiversity of nematodes biological indicators of soil quality in the agroecosystems. Arq Inst Biol 84:1–8
Neher D (2001) Role of nematodes in soil health and their use as indicators. J Nematol 33:161–168
Olabiyi TI, Oladeji OO (2014) Assessment of four compost types on the nematode population dynamics in the soil sown with okra. Int J Org Agric Res Dev 9:146–155
Pokharel R, Marahatta SP, Handoo ZA, Chitwood DJ (2015) Nematode community structures in different deciduous tree fruits and grape in Colorado, USA and impact of organic peach and apple production practices. Eur J Soil Biol 67:59–68
Porazinska DL, Duncan LW, Mcsorley R, Graham JH (1999) Nematode communities as indicators of status and processes of a soil ecosystem influenced by agricultural management practices. Appl Soil Ecol 13:69–86
Postma-Blaauw MB, de Vries FT, de Goede RGM, Bloem J, Faber JH, Brussaard L (2005) Within-trophic group interactions of bacterivorous nematode species and their effects on the bacterial community and nitrogen mineralization. Oecologia 142:428–439
Rhoades HL, Linford MB (1959) Control of Pythium root rot by the nematode *Aphelenchus avenae*. Plant Dis Rep 43:323–328
Ritz K, Trudgill DL (1999) Utility of nematode community analysis as an integrated measure of the functional state of soils: perspectives and challenges. Plant Soil 212:1–11

Ruess L (2003) Nematode soil faunal analysis of decomposition pathways in different ecosystems. Nematology 5:179–181
Ruess L, Ferris H (2004) Decomposition pathways and successional changes. Nematol Monogr Perspect 2:547–556
Sánchez-Moreno S, Minoshima H, Ferris H, Jackson LE (2006) Linking soil properties and nematode community composition: effects of soil management on soil food webs. Nematology 8:703–715
Siddiqi MR (1986) Tylenchida parasites of plants and insects. Commonwealth Agricultural Bureau
Sithole NJ, Magwaza LS, Mafongoya PL (2016) Conservation agriculture and its impact on soil quality and maize yield: a south African perspective. Soil Tillage Res 162:55–67
Steel H, de la Pena E, Fonderie P, Willekens K, Borgonie G, Bert W (2010) Nematode succession during composting and the potential of the nematode community as an indicator of compost maturity. Pedobiologia 53:181–190
Steel H, Verdoodt F, Cerevkov A, Couvreur M, Fonderie P, Moens T, Bert W (2013) Survival and colonization of nematodes in a composting process. Invertebr Biol 132:108–119
Steel H, Moens T, Vandecasteele B, Hendrickx F, De Neve S, Neher DA, Bert W (2018) Factors influencing the nematode community during composting and nematode-based criteria for compost maturity. Ecol Indic 85:409–421
The Soil Science Society of America Glossary. https://www.soils.org/publications/soils-glossary. Accessed 20 Apr 2019
Thomas WK, Vida JT, Frisse LM, Mundo M, Baldwin JG (1997) DNA sequences from formalin-fixed nematodes: integrating molecular and morphological approaches to taxonomy. J Nematol 29:250–254
Treonis AM, Unangst SK, Kepler RM, Buyer JS, Cavigelli MA, Mirsky SB, Maul JE (2018) Characterization of soil nematode communities in three cropping systems through morphological and DNA metabarcoding approaches. Sci Rep 8:2004
Villenave C, Fernandes P, Niane-Badiane A, Sène M, Perez P, Ganry F, Oliver R (1998) Influence of tillage and compost on communities of phytoparasitic nematodes, XVIe Congrès Mondial des Sciences du Sol, Montpellier (France), 20–26 août 1998. AISS; AFES. Montpellier: CIRAD, 2 p. Congrès Mondial de Science du Sol. 16, Montpellier, France, 20–26 Aug 1998. http://www.vertcarbone.fr/wp-content/uploads/2018/03/Compost-et-nemathodes.pdf
Wang KH, McSorley R, Kokalis-Burelle N (2006) Effects of cover cropping, solarization, and soil fumigation on nematode communities. Plant Soil 286:229–243
Wardle DA (1995) Impacts of disturbance on detritus food webs in agro-ecosystems of contrasting tillage and weed management practices. In: Begon M, Fitter AH (eds) Advances in ecological research, vol 26. Academic, New York, NY, pp 105–185
Wardle DA, Williamson WM, Yeates GW, Bonner KI (2005) Trickle-down effects of aboveground trophic cascades on the soil food web. Oikos 111:348–358
Wasilewska L (1997) Soil invertebrates as bioindicators, with special reference to soil-inhabiting nematodes. Russ J Nematol 5:113–126
Yeates GW (1998) Feeding in free-living soil nematodes: a functional approach. In: Perry NR, Wright DJ (eds) The physiology and biochemistry of free-living and plant parasitic nematodes. CAB International, Wallingford, UK, pp 245–269
Yeates GW (2003) Nematodes as soil indicators: functional and biodiversity aspects. Biol Fertil Soils 37:199–210
Yeates GW, Bongers T (1999) Nematode diversity in agroecosystems. Agric Ecosyst Environ 74:113–135
Yeates GW, Bongers T, de Goede RGM, Freckman DW, Georgieva SS (1993) Feeding habits in soil nematode families and genera—an outline for soil ecologists. J Nematol 25:315–331
Zhong S, Zeng H, Jin Z (2017) Influences of different tillage and residue management systems on soil nematode community composition and diversity in the tropics. Soil Biol Biochem 107:234–243
Zullini A, Peretti E (1986) Lead pollution and moss-inhabiting nematodes of an industrial area. Water Air Soil Pollut 27:403–410

Chapter 4
Review on Physiological Effects of Vermicomposts on Plants

Gederts Ievinsh

Abstract Vermicompost is a type of organic fertilizer produced by earthworms and their symbiotic microorganisms. In addition to nutrient-rich organic fraction, vermicomposts also contain plant-available soluble minerals as well as substances with plant growth-regulating activity. In order to promote understanding on beneficial effects of vermicompost in different farming systems, the aim of the present review is to analyze possible mechanisms of vermicompost action on plants. After overview of plant growth-affecting activity of vermicomposts at the level of seed germination and vegetative growth, detailed analysis is dedicated to physiological effects associated with mineral nutrition: changes in both soil mineral nutrient availability and their uptake. Further, additional plant growth-regulating activity is considered, with emphasis on plant hormonelike compounds and humic substances. Information on activation of plant metabolism by vermicompost application is given at the level of photosynthesis-related parameters and defense responses. Approaches of evaluation of vermicompost quality based on physiological criteria are described, followed by summary on use of vermicompost in different farming systems.

Keywords Humic substances · Mineral nutrition · Organic fertilizer · Physiological effects · Plant growth promoting-activity

4.1 Introduction

Vermicompost represents a type of organic fertilizer produced by a concerted action of earthworms and their symbiotic microorganisms. During a complex interaction, organic matter from any organic waste material is fragmented, decomposed and stabilized, leading to formation of finely structured peat-like material with high porosity and significant water-holding capacity (Huang et al. 2014). Most

G. Ievinsh (✉)
Department of Plant Physiology, Faculty of Biology, University of Latvia, Riga, Latvia
e-mail: gederts.ievins@lu.lv

M. K. Meghvansi, A. Varma (eds.), *Biology of Composts*, Soil Biology 58,
https://doi.org/10.1007/978-3-030-39173-7_4

importantly, vermicomposts contain a significant concentration of plant-available soluble mineral nutrients, high nutrient fraction in a form of organic matter as well as presence of various microorganisms ensuring high rate of mineralization, leading to gradual release of additional amount of plant-available nutrients in soil (Joshi et al. 2015). Besides providing of mineral nutrients, vermicomposts have additional beneficial effects on plant growth.

The majority of studies so far have analyzed the effect of vermicompost on some yield-related parameters of cultivated plants irrespective of particular physiological target mechanism (as reviewed in Joshi et al. 2015; Hussain and Abbasi 2018). In addition, the effect of vermicompost on seed germination has been particularly considered. Benefits of vermicompost use are based on both direct and indirect effects of its application on plants, as well as on the additional improvement of soil properties with potential long-term benefits and soil sustainability. Lack of understanding of specific mechanisms of the beneficial influence of vermicompost application has led to inability to predict the outcome of vermicompost application in a scientifically sound manner.

Therefore, the aim of the present chapter was to analyze different possible physiological mechanisms of vermicompost action of plants, in order to promote understanding on the beneficial effects of vermicompost application in different farming systems.

4.2 An Overview of Plant Growth-Affecting Activity of Vermicomposts

4.2.1 Seed Germination

Different aspects of seed germination and seedling establishment are usually positively affected by low doses of vermicompost application. These include total seed germination, germination index, speed of seedling emergence, and seedling vigor. A number of studies have indicated only the positive effect of vermicomposts on seed germination. As an example, in *Pinus pinaster* there was a 16% increase in germination percentage (Lazcano et al. 2010).

However, several studies have shown variable effect of vermicomposts on seed germination. In a comparative study, effect of different feedstock combinations (paper + kitchen waste, kitchen + yard waste, cattle manure + yard waste) vermicomposted for different duration (45, 68, 90 days) was analyzed on germination of three cultivated species (radish, marigold, upland cress) (Warman and AngLopez 2010). In general, seed germination was significantly inhibited either by treatment with vermicompost water extract (12.5 and 25%) or by direct sowing in vermicompost/soil mixes (5 and 10% vermicompost supplement), but both genotype-, duration-, and feedstock-specific effects were evident. Even samples vermicomposted for 90 days could cause a decrease of marigold seed germination from 64 to 15% at supplement rate as low as 5%. The authors of the study hypothesized that phytotoxicity of vermicompost was caused by organic substances rather than soluble salts because electrical conductivity did not correlate with germination inhibition percentage. Seed germination of winter rye was significantly inhibited by substrate (quartz sand) substitution (20–50%) with sewage sludge vermicompost irrespective on mineral nutrient availability (Karlsons et al. 2016).

In a comparative study on the effect of cow manure vermicompost on seed germination of 16 crop species, it was shown that substrate substitution with increasing doses of vermicompost results in a near-linear decrease of germination percentage (Ievinsh 2011). However, the effect on seed germination of vermicompost water extract used for seed imbibition was neutral or slightly stimulative at low concentrations, and inhibiting at moderate and high concentrations, showing a linear decrease of germination percentage with increasing extract concentration. In addition, it appeared that seeds from species or cultivars with lower germination capacity tended to have higher stimulation of germination by vermicompost extract. As an example, seeds from two cultivars of celery with a germination percentage of 0 and 3% in control conditions, were stimulated by 20% vermicompost extract imbibition up to 7 and 13%.

It is possible that one of the reasons for the high genetic diversity of germination responses to vermicompost treatment is related to differences in seed endosperm reserves: in seeds with a high potential to sustain prolonged seedling growth biologically active substances in vermicompost can more likely result in inhibitory effects.

Mineral nutrients usually have a negative effect on seed germination and during early seedling growth. Therefore, any beneficial effects of vermicompost on stimulation of seed germination could be due to the presence of hormonelike substances in vermicompost preparations.

4.2.2 *Plant Vegetative Growth*

When vermicomposts are used as basic fertilizers in conditions of mineral nutrient shortage, any positive effect on plant growth, biomass accumulation, and yield can be related to the direct effect of an increase in plant-available mineral elements. Therefore, this aspect will be analyzed further separately in Sect. 4.3. For recent reviews of general results on the effect of vermicomposts on growth, yield, and quality of cultivated plants, readers are asked to focus on Joshi et al. (2015) and Hussain and Abbasi (2018) as both papers provide a comprehensive list of representative publications on these important aspects.

Besides field production, due to a necessity to find an alternative for use of peat moss as a substrate in potting media in horticulture, several studies have investigated potential of vermicompost. It was shown that the addition of vermicompost

Fig. 4.1 Effects of increasing rates of mineral-fertilized peat moss substitution with vermicompost on the growth of Swedish turnip seedlings. Courtesy G. Ievinsh

stimulated emergence and growth of tomato seedlings, and positively altered tomato fruit quality even after transplantation into equally fertilized soil (Zaller 2007). Unfortunately, significant differences in mineral composition in growth substrate resulting from different degrees of substitution with higher nutrient level-containing vermicompost were not taken into the account.

Negative effects of vermicompost on plant growth have long been ignored in the scientific literature. One of the reasons appears to be different sensitivity of cultivated plants to vermicompost overdose, diversity in general mineral nutrient availability in the substrate used as a base for cultivation, as well as extreme variability of plant growth-affecting characteristics of different vermicompost preparations. Growth retardation usually has been documented at >60% vermicompost substitution rate, as in a study with *Linum usitatissimum* (Makkar et al. 2017). However, more dramatic negative effects have been encountered, as in the case of Swedish turnip (*Brassica napus* var. *napobrassica*) seedlings cultivated in a mineral-enriched neutralized peat moss substituted with different doses of cow manure vermicompost (Fig. 4.1). A significant negative effect on seed germination and seedling growth was evident as low as at 20% vermicompost substitution. In general, seed germination, as well as early stages, of seedling growth is more sensitive to the negative effects of vermicomposts on plant growth. As an example, vermicompost substitution of mineral-enriched neutralized peat moss resulted in near-linear growth inhibition of basil (*Ocimum basilicum*) seedlings by increasing vermicompost substitution rates, with significant effect already at 5% substitution (Fig. 4.2a). The negative effect clearly diminished with time, due to gradual vermicompost-induced growth acceleration at higher rates of substitution (Fig. 4.2b and c). Fifty-seven days after the start of the treatment, plants at 5% substitution had higher dry matter in comparison to control, but optimum growth was achieved at 10% substitution, while plants in 20 and 30% substitution also had higher biomass than that in control.

Several mechanisms are suggested as responsible for the negative effects of vermicomposts on plant growth at high substrate substitution rates. One of them is the accumulation of toxic ammonium concentration negatively affecting nutrient uptake (Atiyeh et al. 2000). This is further supported by the fact that vermicomposts produced from animal wastes usually show more toxic effects at relatively lower rates of application (Domínguez 2004). Given the high total concentration of soluble salts in vermicomposts, it is also possible that negative effects are related to osmotic mechanisms.

Fig. 4.2 Growth and development of basil plants as affected by different doses of vermicompost. (**a**) 7 days; (**b**) 20 days; and (**c**) 50 days after the start of the treatment. Courtesy G. Ievinsh

Significant differences in physiological effects of vermicompost due to the use of various cultivated plant species and even cultivars have been reported. Genotype-dependent effects of vermicompost have been first described for tomato seedlings (Zaller 2007), tomato, marigold, pepper and cornflower plants (Bachman and Metzger 2008), and in progenies of *Pinus pinaster* (Lazcano et al. 2010). Since then, it has been widely established that even within some species of cultivated plants extreme genotype-dependence of vermicompost effects can be seen. For example, when six tomato cultivars were compared with respect to the effect of vermicompost water extract on seed germination, germination of the most sensitive cultivar was already inhibited by almost 60% at 10% treatment, but germination of most positively affected cultivar showed 10% stimulation at the same concentration (Ievinsh 2011). Also, optimum dose of soil substitution with vermicompost significantly differed between two cultivars of *Linum usitatissimum* (Makkar et al. 2017).

An important aspect of studies with vermicomposts is its comparison with other types of organic fertilizers. It is difficult to compare the effects of composts and vermicomposts on plant growth and yield because usually these products have different chemical composition (Fornes et al. 2013) or their mineral composition is artificially balanced (Doan et al. 2013). As an extreme example, in a study with tomato transplants, composted cow manure and pig manure vermicompost were compared for their suitability to replace a peat-based substrate (Lazcano et al. 2009). Doses of compost above 50% caused plant mortality, while plants even in 100% vermicompost showed increased biomass as compared to peat substrate. The choice of commercial compost used seems to be inappropriate for this type of experiment, as this product contained 2.4 g kg^{-1} Na and 7.2 g kg^{-1} Cl, obviously causing osmotic stress and ion toxicity on the background of 8.3 g kg^{-1} K and 8.4 g kg^{-1} Mg. In the particular vermicompost product, all these concentrations were significantly lower, for instance, ten times for Cl and five times for Na.

When identical starting organic material was used for the preparation of compost and vermicompost, it was evident that these techniques had a significantly different effects on both physical and chemical properties of the final products (Frederickson et al. 2007). In particular, vermicompost had higher total concentration of several mineral nutrients (N, P, K, Cu), but composts had higher concentration of water-soluble K and electrical conductivity. Effect of compost and vermicompost in organic greenhouse production of herbs using organically certified soil has been compared recently (Ievinsh et al. 2020). Both products were made from the same feedstock, cow manure and grass biomass, and had highly similar concentration of all plant-available mineral nutrients and total soluble salt concentration. The exception was significantly higher Ca, Zn, and Cu concentrations in vermicompost and higher K and Fe concentrations in compost. At identical amendment rate, both growth and biomass production of *Dracocephalum moldavica*, *Melissa officinalis*, *Nepeta cataria*, and *Thymus vulgaris* were more stimulated by vermicompost treatment. For *Nepeta cataria*, the highest biomass within 60 days of cultivation was achieved at 30% amendment rate and was 511 ± 24 and 763 ± 22 g per plant, for compost and vermicompost treatment, respectively (Fig. 4.3).

Fig. 4.3 Effects of different amendment rates with compost and vermicompost on growth and development of *Nepeta cataria* plants cultivated in organically certified soil. *C* compost, *VC* vermicompost. Courtesy G. Ievinsh

In spite of identical concentration of nitrogen, vermicompost superiority over compost can be related to differences in the balance between chemical forms of nitrogen, ammonium, and nitrate, as vermicomposts contain mostly nitrate, but composts contain mostly ammonium (Lazcano et al. 2008). These aspects are analyzed in detail in Sect. 4.3.

Larger plant growth-affecting capacity of vermicomposts versus composts can be due also to differences in microbial composition (Lazcano et al. 2008). Vermicomposts generally contain both higher diversity and particular numbers of microorganisms, possibly due to the significant contribution of earthworm symbionts (Grantina-Ievina et al. 2013). On the other hand, many rhizosphere microorganisms produce plant hormones (Arshad and Frankenberger 1991; Baca and Elmerich 2003). Already more than 30 years ago, it was shown that earthworm activity is related to the production of substances with auxin- and cytokinin-like activity in soil (Krishnamoorthy and Vajranabhaiah 1986). Effects of vermicompost related to hormonelike activity are analyzed in Sect. 4.5.1.

4.3 Physiological Effects Associated with Mineral Nutrition: Changes in Soil Mineral Nutrient Availability

It can be expected that plant mineral nutrition will improve following soil amendment with vermicompost mainly due to the supply of plant-available forms of mineral nutrients. While usually containing several mineral nutrients in high concentration in plant-available forms, vermicomposts usually are not fully balanced with respect to plant mineral requirements. Therefore, attempts have been made to improve physiological quality of produced vermicomposts by adding mineral nutrients. More reliable procedure can be balancing mineral nutrients at the level of feedstock preparation before vermicomposting, as rock phosphate for use of

Table 4.1 Characteristics of vermicompost, bat guano, and soil samples used in comparison to the optimum level of minerals for cultivated plants (Osvalde 2011)

	Vermicompost		Bat guano			
	Total[b]	Plant-available[a]	Total[b]	Plant-available[a]	Soil plant available[a]	Optimum for cultivated plants[a]
N	2.08 (%)	910	1.06 (%)	882	76	120
P	0.49 (%)	4578	5.93 (%)	39,240	447	60
K	1.24 (%)	12,400	0.30 (%)	1250	120	150
Ca	1.64 (%)	13,100	13.32 (%)	73,500	720	800
Mg	0.55 (%)	3650	0.37 (%)	2120	70	50
S	0.3 (%)	250	3.25 (%)	2400	11	50
Fe	2940	775	6800	800	805	30
Mn	250	236	1840	1250	130	1.5
Zn	74	64	1080	650	4.95	1
Cu	11	4.5	280	10.5	2.0	0.5
Mo	1.9	0.04	–	0.98	0.07	0.02
B	14	7.4	–	6.2	0.2	0.2
Na	0.032 (%)	290	0.330 (%)	2400	7.5	–
Cl	–	800	–	16,750	0	–
Organic matter (%)	–	46.5	–	12.7	–	–
pH (units)	–	7.67	–	6.60	4.86	–
EC ($mS\ cm^{-1}$)	–	24.8	–	53.1	–	–

[a]Mineral concentration expressed as $mg\ l^{-1}$
[b]Mineral concentration expressed as % or $mg\ kg^{-1}$

vermicompost in organic cultivation (Mupondi et al. 2018) or clinoptilolite zeolite (Zarrabi et al. 2018), but this approach could be cumbersome in conditions of practical production. Optimization of vermicomposting by inorganic additives was recently reviewed by Mupambwa and Mnkeni (2018). Another opportunity may be the production of more balanced vermicompost products by using additives of other types of organic fertilizers. A study was performed with winter rye (*Secale cereale*) and potato (*Solanum tuberosum*) plants cultivated in organically certified soil in controlled conditions, with various rates of organic fertilizer prepared using vermicompost from starchless potato pulp and grass with addition of different concentration (10, 20, and 30%) of bat guano from Madagascar (Grantina-Ievina and Ievinsh 2015). The soil was acidic, had relatively high P and was deficient in N, K, Ca, Mg, S, Cu, and B (Table 4.1). Vermicompost had a higher plant-available concentration of K and Mg, but guano had relatively high degree of mineralization and was rich in P, Ca, S, Mn, Zn, Cu, and Mo. Addition of bat guano to vermicompost-containing organic fertilizer significantly enhanced the stimulating effect of vermicompost on growth and development of both winter rye and potato plants (Fig. 4.4). Additional effect of guano was concentration dependent and was

Fig. 4.4 Effects of increasing dose of bat guano (GU) in organic fertilizer together with vermicompost (VC) on the growth of potato plants in organically certified soil. Courtesy G. Ievinsh

more pronounced at higher rates of application of organic fertilizer, indicating positive role of increased amount of plant-available nutrients in conditions of nutrient limitation. Plant growth-promoting effect was clearly associated with high microbial activity found in bat guano leading to additional beneficial effects of the resulting organic fertilizer.

In many vermicomposts, inorganic nitrogen is mostly present in the form of nitrate. Thus, rabbit manure vermicompost contained 1303 mg kg^{-1} N-NO_3 and only 15 mg kg^{-1} N-NH_4^+ (Lazcano et al. 2010), and that from cow manure had 220 mg kg^{-1} N-NO_3 and 53 mg kg^{-1} N-NH_4^+ (Zhao et al. 2017). In contrast, pig manure vermicompost contained 297 mg kg^{-1} N-NH_4^+ and 104 mg kg^{-1} N-NO_3 (Lazcano et al. 2009). It is evident that maturation of composts and, especially, vermicomposts lead to increased conversion of ammonium into nitrate in a process of nitrification (Cáceres et al. 2018). Thus, raw cattle manure contained 610 mg kg^{-1} N-NH_4^+ and 19 mg kg^{-1} N-NO_3, which increased up to 1235 mg kg^{-1} N-NH_4^+ and 721 mg kg^{-1} N-NO_3 in a composted manure, but vermicompost made from respective compost had 191 mg kg^{-1} N-NH_4^+ and 829 mg kg^{-1} N-NO_3 (Lazcano et al. 2008). Aspects of ammonium toxicity in plants are analyzed in detail elsewhere (i.e., Britto and Kronzuker 2002).

4.4 Physiological Effects Associated with Mineral Nutrition: Changes in Mineral Nutrient Uptake

It has been long suggested that application of vermicomposts results in additional effects on mineral nutrition besides supply of mineral elements both in plant-available and organic forms. These effects can be both soil and physiology related.

Fine particulate structure of vermicomposts evidently forms the rather large immobilizing surface, supporting their slow-release potential and protecting nutrients from leaching. It is well known that several ions are easily immobilized in certain structures of organic matter, making them less available for plants. Of course, remobilization of captured elements is also possible in certain situations, as due to changes in soil redox potential, pH or release of root exudates. The last mechanism has been mostly associated with presence of humic substances. Therefore, vermicompost application to soil with low content of organic matter can significantly change availability of certain minerals for plants, but the beneficial nature of particular changes can be highly variable. From a physiological point of view, stimulation of uptake of particular mineral nutrients by vermicompost treatment could be related to activation of transport systems in a plasmatic membrane of root cells, but the importance of this mechanism has not been proven so far.

Improved nutrient uptake per unit root is suggested as one of the mechanisms involved in vermicompost-associated plant growth stimulation. Presence of microorganism-produced organic acids and substances with siderofore-like activity in vermicompost can likely participate in mineral acquisition by roots (Pii et al. 2015).

The main problems of testing vermicompost effects on plant mineral nutrition can be summarized as follows (Karlsons et al. 2016): (1) when substrate with relatively high level of plant-available mineral nutrients (as with chemical fertilizer at optimum level) are substituted with increasing doses of vermicompost with relatively lower level of mineral nutrients, negative effects on plant growth can be caused by gradual decrease of overall mineral supply; (2) when substrate with relatively low level of plant-available mineral nutrients (as common soil below optimum level) are substituted with increasing doses of vermicompost, positive effects on plant growth are caused by greater mineral availability. Thus, without appropriate controls (plants with identical mineral nutrient availability), it is difficult to distinguish if vermicompost treatment stimulates uptake of particular mineral elements.

When inert (quartz sand) substrate with optimum level of all mineral nutrients was substituted with 10% sewage sludge vermicompost, this resulted in 98% increase of shoot dry mass of winter rye plants, but 20% substitution ensured additional 42% increase (Karlsons et al. 2016). Comparable results were obtained in a similar study with beetroot plants (Fig. 4.5). Data on mineral element concentrations in plant tissues supported the view that growth stimulation of plants by vermicompost treatment occurs without stimulation of mineral nutrient uptake.

It is commonly suggested that humic substances in organic fertilizers, including vermicomposts, are responsible for stimulation of mineral uptake. Several studies in controlled conditions indeed have supported stimulation of accumulation of several mineral nutrients in plants treated with humic substances (Noroozisharaf and Kaviani 2018). However, these effects are usually difficult to reproduce in field conditions, and very often negative effects prevail (Osvalde et al. 2012, 2016).

Fig. 4.5 Effect of vermicompost treatment on the growth of beetroot plants. Plants were cultivated either in quartz sand (S) of quartz sand substituted with an optimum dose of mineral nutrients (SM) and amended with different doses of vermicompost (VC). Courtesy G. Ievinsh

4.5 Vermicompost Substances with Plant Growth-Regulating Activity

4.5.1 Plant Hormones

Hormonelike substances (indole acetic acid, gibberellins, cytokinins) with significant effect on plant growth have been isolated from vermicomposts already during the early studies with vermicomposts (Tomati et al. 1988; Chen and Aviad 1990). Indirect evidence on association of plant growth-stimulating activity in composts and vermicomposts with microorganism-derived hormonelike substances is provided by data on correlation between a number of certain groups of microorganisms and plant growth promotion. Thus, the total number of cultivable filamentous fungi and fungal diversity had positive effect on plant growth stimulating activity when a large number of different vermicomposts and composts were compared (Grantina-Ievina et al. 2013). In contrast, growth-inhibiting activity positively correlated with a total number of coliforms and a total number of bacteria. When particular fungal genera were considered, *Aspergillus* (on Mycosel agar) and *Verticillium* (on Rose Bengal agar with chloramphenicol) had medium strong or weak positive correlation, respectively, with growth stimulating activity. Several fungal genera [*Pseudallescheria, Mucor, Verticillium* (on Mycosel agar), *Trichoderma, Penicillium*, and *Paecilomyces*; in an order of decreasing correlation strength] showed negative correlation with stimulating activity. These data are further supported by results from experiments where particular microbiological isolates from vermicompost have growth stimulating activity on plants (Gopalakrishnan et al. 2015; Sreevidya et al. 2016).

Several plant hormones were extracted, identified, and quantified by spectrophotometric methods in various samples of composts and vermicomposts (Ravindran et al. 2016). Maximum concentration found for indole-3-acetic acid, cytokinin (indicated as "kinetin"), and gibberellic acid was 7.37, 2.8, and 5.7 mg kg^{-1}, respectively, in vermicompost made from cow manure plus leaf litter with addition of fermented tannery waste. More recently, presence of plant hormones in vermicomposts has been proven by more precise analytical methods. Thus, several natural cytokinins have been identified in a vermicompost tea using liquid chromatography–mass spectrometry in a concentration that could affect plant growth (Zhang et al. 2014). By means of an ultrahigh-performance liquid chromatography–tandem mass spectrometry diversity of natural plant growth regulators was analyzed in different batches of garden waste vermicompost leachate (Aremu et al. 2015). The major hormonal ingredient found was indole-3-acetic acid (0.55–0.77 pmol mL^{-1}), in addition to various amounts of four isoprenoid-type cytokinins (with 60% of N^6-isopentenyladenine), 18 types of gibberellins and 6 types of brassinosteroids in relatively low amounts. It was summarized recently that a wide variety of microorganism-containing biological fertilizers could provide plants with plant hormones and hormonelike substances, with a practical importance for both conventional and organic plant cultivation (Wong et al. 2016).

4.5.2 Humic Substances

At physiological level, scientifically proven significant effect of humic substances isolated from vermicompost has been associated only with stimulation of root elongation and formation of lateral roots (Canellas et al. 2002). However, it is a common practice to associate beneficial effects of vermicomposts with high content of humic acids in these products, because several studies indeed showed significant stimulation of growth in some model systems (Atiyeh et al. 2002; Arancon et al. 2003, 2006). However, effects are dependent on experimental system used, appear to be highly variable and difficult to reproduce. Usually, stimulation of water uptake and accumulation by humic substances are described, with only little effect on accumulation of dry matter (Piccolo et al. 1993). In addition, effect of humic substances usually becomes less predictable with a shift from a simple (seed germination, seedling growth) to more complicated (soil with different mineral nutrient availability) model systems. As already more than 15 years ago it was hypothesized that only direct effect of humic acids might be associated with adsorption of plant hormones in their molecular structure (Atiyeh et al. 2002), explaining high variability of plant growth affecting activity between different humic substances-containing products.

When the experimental design allows distinguishing between effects of mineral nutrients, humic acids, and other substances with growth-promoting activity, it is evident that positive effect of vermicompost is associated with a complex action of different vermicompost components. In a study with industrial hemp (*Cannabis sativa*) seedlings effect of vermicompost-derived water-soluble (fulvic acid) and water-insoluble (humic acid) was compared with that of whole vermicompost water extract and mineral nutrients of the water extract over a wide range of concentrations (Ievinsh et al. 2017). The results were equalized with respect to the relative amount of vermicompost used for the preparation of the particular product used for treatment. All these vermicompost-derived products had a significant concentration-dependent stimulative effect on hemp seed germination and seedling growth, but their relative effect and optimum concentration of action were significantly different. Thus, stimulation of seed germination was mostly due to the summed activity of humic and fulvic acids, as mineral nutrients were inhibitory even at relatively low concentrations. Linear growth of both hypocotyl and radicle at medium concentration was stimulated by unidentified soluble components of vermicompost extract, as in this concentration range humic substances were already inhibitory but mineral nutrients had no significant effect. In contrast, increase of hypocotyl fresh mass was mostly due to the action of mineral nutrients.

It was suggested that it is necessary to distinguish between possible physiological effects of different humic substances, at least these made from "dead" organic sources (coal, leonardite, peat) as opposed to "living" organic material containing microorganisms (biological waste, compost, vermicompost) (Ievinsh et al. 2017), as humic substances from "living" sources could have higher biological activity (Xu et al. 2012). A meta-analysis on effects of humic substances on plants was performed, using results from 81 studies (Rose et al. 2014). One of the main conclusions was that compost-derived humic substances significantly outperform these isolated from lignite and peat in terms of plant growth stimulation. As analysis of data supporting or refusing particular aspects of possible effects of humic substances on plants is outside the scope of the present review, the readers are directed to several available contradictory-oriented discussions in this respect (Canellas et al. 2015; Olaetxea et al. 2018 for the "pros" side and Hartz and Bottoms 2010 for the "contra" side).

From a physiological point of view, any positive effect of growth-promoting chemical substances (plant hormonelike activity) can only occur at adequate supply of mineral nutrients. In this respect, it is important to stress out that stimulation of root growth without any positive effect on shoot growth, usually associated with effect of humic acids, more likely indicates an adaptive response to the mineral nutrient shortage. It has been shown in numerous studies that plants in mineral-deficient conditions allocate more resources for root growth and development as a mechanism to capture more nutrients from soil (Marschner et al. 1996). Consequently, there is no reason to believe that treatment with humic substances will show any positive effect on plant yield in conditions of mineral nutrient shortage. Only in the case of adequate nutrient reserves in a form of organic matter and high level of microbial activity leading to mineralization plants treated with humic substances could benefit from more vigorous root growth. Another possibility is related to some adaptogenic activity of humic substances, where positive effect of their application can be seen only when plant growth is limited by some detrimental environmental factor(s). As an example, both plant growth and physiological status can be recovered by the application of humic substances in a situation of moderate water shortage (Lotfi et al. 2018).

4.6 Activation of Metabolic Processes

4.6.1 Photosynthesis-Related Parameters

As leaf chlorophyll concentration is roughly related to the general nitrogen status of a plant, any positive changes in chlorophyll due to vermicompost treatment can be related to additional N supply for plants, especially in conditions of nutrient shortage. Indeed, there are numerous reports supporting the positive effect of vermicompost application on leaf chlorophyll concentration (i.e., Pirdashti et al. 2010; Yadav and Garg 2015; Hosseinzadeh et al. 2016).

Similar to chlorophyll concentration, increase in photosynthetic gas exchange parameters, photosystem II activity and other photosynthesis-related parameters in plants cultivated in the vermicompost-amended substrate with inadequate nutrient availability most likely reflect physiological activation due to general improvement of mineral nutrient supply (Hosseinzadeh et al. 2016; Srivastava et al. 2018). However, the majority of practically oriented studies do not discriminate between different states of nutrient availability for plants. When vermicompost is used at optimum mineral nutrient supply, increase in photosynthesis-related parameters could be related to some type of metabolic activation. Consequently, increased photosynthesis is not a prerequisite but rather a consequence of vermicompost-dependent stimulation of growth and/or enhancement of mineral nutrition status. In contrast, increase of photosynthesis-related parameters by application of vermicompost or humic substances in water-limited plants can lead to recovery of growth rate comparable to that of optimally watered plants (Lofti et al. 2018).

4.6.2 Defense Responses

Activity of different enzymes associated with antioxidative metabolism has been measured to show that treatment with vermicompost improves enzymatic scavenging capacity of reactive oxygen species in plant tissues. Following vermicompost amendment, superoxide dismutase activity increased in leaves of strawberry plants (Zuo et al. 2018); and superoxide dismutase, ascorbate peroxidase, catalase, and glutathione peroxidase activity in borago (Afkari 2018). Similarly, humic acids isolated from vermicompost stimulated the activity of antioxidative enzymes (ascorbate peroxidase, catalase, superoxide dismutase) both in roots and leaves of rice plants (García et al. 2012).

Decrease of biochemical indicator of oxidative membrane damage, malondialdehyde concentration, has been noted in several studies, further supporting the idea of improved antioxidative capacity in tissues of vermicompost-treated plants (Xu et al. 2016; Zuo et al. 2018). However, in situations when high vermicompost application rates result in significant decrease in plant growth in comparison to control, increase in lipid peroxidation has been noted (Srivastava et al. 2018).

From a practical point of view, induced resistance is a component of compost- and vermicompost-associated soil suppressiveness (Yogev et al. 2010). Induced biochemical defense-related responses due to vermicompost application can possibly lead to resistance against abiotic factors, pathogens and herbivores, showing its adaptogenic potential. It is not clear whether defense responses induced by vermicompost application are regulated by the same signaling pathways as in the case of beneficial soil microorganisms (Van der Ent et al. 2009), but it is highly likely that plant growth-promoting rhizobacteria from vermicomposts are implicated in activation of defense system in vermicompost-treated plants, leading to enhanced activity of antioxidative enzymes and concentration of phenolic substances (Abdelkrim et al. 2018). In several studies, increase of concentration of secondary metabolites (mostly phenolic substances) in plant tissues has been found as a result of vermicompost treatment (Arancon et al. 2007; Edwards et al. 2010; Gholami et al. 2018).

4.7 Evaluation of Vermicompost Quality Based on Physiological Criteria

There are no scientifically based quantifiable criteria for estimation of quality of vermicompost. Usually, simple physical (bulk density, water retention capacity, electrical conductivity) or chemical (pH, macronutrient concentration, organic carbon content) parameters are measured, which have only negligible or minor direct effect on plant growth. As these parameters alone cannot explain beneficial (and sometimes, detrimental) effects of vermicompost on plants, more physiology-related criteria of vermicompost quality are clearly necessary. Besides plant growth-affecting activity of vermicompost preparations, other mostly microbiology-related aspects are of great practical importance when working with vermicomposts (content of plant growth-promoting microorganisms, presence of potential plant, animal, and human pathogens), but analysis of these aspects are outside the scope of the present review (Grantina-Ievina et al. 2013).

Very often different phytotoxicity test systems are used for evaluation of general quality of organic fertilizers, e.g., for measurement of compost maturity (Kopec et al. 2013; Nakhshiniev et al. 2014), but there are no tests available for quantitative estimation of positive growth-related effects on plants. Therefore, we have designed and verified the usefulness of a simple test system based on relative changes of vegetable seedling growth for general estimation of plant growth-affecting activity of vermicomposts and other potential plant growth stimulants (Grantina-Ievina et al. 2013). The system comprises soilless cultivation of vegetable seeds in darkness at room temperature in closed Petri dishes on a filter paper in the presence of a test solution (at least three different concentrations of water extract of the tested product), monitoring growth of both hypocotyl and radicle after 5–6 days and estimation of relative plant growth affecting activity in comparison to control as summed growth activation and summed growth inhibition. Possible additional positive control can include mineral nutrient solution with a composition similar to that in tested solutions. Use of four vegetable species (beetroot, Swedish turnip, carrot, and tomato) allows to partially eliminating genotype specificity of the effects, as these species showed different responses to vermicompost treatment (Ievinsh 2011). For example, by this system, it is possible to estimate and compare physiological growth-related effects of vermicomposts as related to feedstock characteristics, vermicompost maturation time, extraction time, etc. (Fig. 4.6). The test system has been widely used for the estimation of potential plant growth-affecting activity for a number of vermicomposts and composts (Grantina-Ievina et al. 2013), sapropel-containing fertilizer products (Grantina-Ievina et al. 2014), algal stimulators, fly ash-based fertilizers, etc. In addition, attempts have been made to estimate potential adaptogenic activity (induced metabolic activation) of a wide range of biological and organic fertilizers and plant growth stimulators, based on changes of thermotolerance of photosystem II activity as monitored by relative changes of chlorophyll *a* fluorescence parameters after high-temperature treatment (Šenberga et al. 2012).

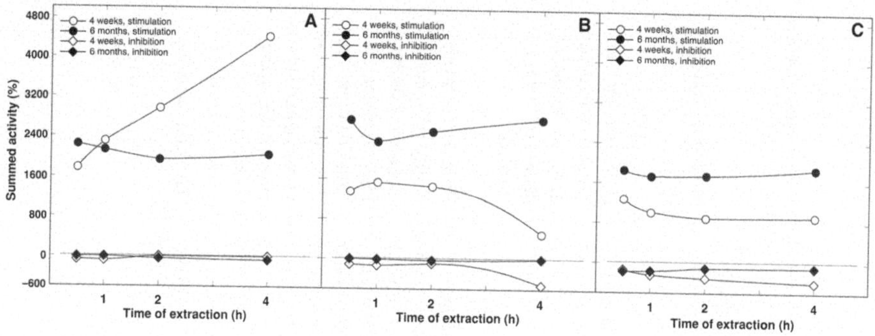

Fig. 4.6 Summed growth stimulating and growth inhibiting activity of vermicompost water extracts as affected by the type of feedstock (**a**, 80% cow manure and 20% organic waste; **b**, 80% garden waste and 20% organic waste; **c**, 40% cow manure and 60% organic waste), maturation time and time of extraction. Growth-affecting activity was measured with beetroot, Swedish turnip, carrot, and tomato seedlings, both for radicle and hypocotyl, at four different concentrations. Courtesy G. Ievinsh

4.8 Practical Implications: Use in Different Farming Systems

Both relatively high concentration of plant-available minerals and non-mineralized concentration of plant nutrients in vermicomposts are of great importance for organic horticultural and agricultural production.

Recently, we briefly summarized the main possible direct and indirect physiological effects of vermicompost on plants (Grantina-Ievina et al. 2015). Beneficial growth- and resistance-related physiological effects of vermicompost could be further analyzed with respect to various constituents and their different mechanisms of action, taking into the account the needs of various types of farming (organic versus conventional). Through the analysis presented in this chapter, it is evident that the major physiological effects of vermicomposts are related to the activity of earthworms and specific microorganisms during vermicomposting and after application of the final product, leading to the presence of plant-available mineral nutrients and biologically active substances, including hormones (Table 4.2). For needs of organic farming, vermicomposts can be used as basal soil fertilizers, providing readily available mineral nutrients as well as organic matter as a source for further plant nutrients through the activity of microorganisms. Due to rather diverse range of concentration for particular mineral elements in different vermicomposts, efforts are necessary to balance the mineral composition of applied products according to the plant needs. When used in conventional farming, vermicomposts can substitute chemical fertilizers. Additional direct benefits of vermicomposts to plants in both farming systems are related to the presence of diverse substances with plant growth- and defense-affecting (adaptogenic) activity.

Table 4.2 The main constituents of vermicomposts, their direct and indirect physiological effects and benefits from application in different farming systems

	Minerals	Organic matter (general)	Humic substances	Plant hormones and hormonelike substances	Other biologically active substances	Microorganisms
Concentration or amount	Low to moderate	High	High	Moderate to high	Moderate to high	High
Main characteristics	Unbalanced with respect to plant needs. Depends on feedstock. High in NO_3-N, low in NH_4^+-N	High water holding capacity. Source of nutrition for microorganisms	Structure and features depend on the source of feedstock	Mainly auxins and cytokinins of microbial origin	Not clear, mostly phenolic substances	High diversity and variability in dependence on feedstock, degree of maturation, etc.
Fertilizer benefits in organic farming	As a basal soil fertilizer provides necessary elements. Need to be balanced by additives or mixes with other types of organic fertilizers	Indirect effect through improving soil properties. Direct long-term effect from mineralization and increase in plant-available elements	Not clear. Stimulate root growth. Possible improvement of mineral uptake	Directly stimulates growth, beneficial only for well-fertilized plants	Induce defense responses, increasing resistance to abiotic factors, pathogens, herbivores/ possible direct antimicrobial activity	Promote the availability of mineral elements through mineralization and solubilization. Produce hormones and hormonelike substances. Compete with harmful microorganisms
Fertilizer benefits in conventional farming	As an additional soil fertilizer, substitutes chemical fertilizers	The same as above	Not clear. Possible improvement of mineral uptake	Directly stimulates growth	The same as above	The same as above

Benefits from use in plant protection in organic farming	When sprayed on foliage can work as a leaf fertilizer	No	Not clear, but can possibly induce defense responses	Not clear	The same as above	No
Possible negative consequences	Some elements can reach toxic levels. Increase of total soil salinity. Can contain heavy metals	Decrease of plant availability of certain elements	Effect will depend on soil properties and mineral availability. Can result in growth inhibition in high concentration	Can lead to decreased vigor at low mineral nutrient availability	Can inhibit growth	Presence of potentially harmful microorganisms

4.9 Conclusion

Analysis of physiological effects of vermicomposts and vermicompost-containing products leads to a better understanding of major possible mechanisms of action and allows estimating benefits from the use of vermicomposts in different farming systems. It is evident that large variability in quality-related characteristics of vermicomposts makes any direct comparison of different products extremely difficult and any quantifiable effects can be predicted only with low probability. Production and use of organic fertilizers with more strongly predictable or even controlled parameters should be a future challenge. More studies focusing on the addition of inorganic constituents or microbial strains during vermicomposting, or combining vermicomposts with other types of organic or biological fertilizers to ensure mineral- and microorganism-related beneficial effects are clearly necessary.

References

Abdelkrim S, Jebara SH, Saadani O, Jebara M (2018) Potential of efficient and resistant plant growth-promoting rhizobacteria in lead uptake and plant defence stimulation in *Lathyrus sativus* under lead stress. Plant Biol 20:857–869

Afkari A (2018) An investigation to the vermicompost efficacy on the activity of antioxidant enzymes and photosynthetic pigments of borage (*Borago officinalis* L.) under salinity stress conditions. Russ Agric Sci 44:310–317

Arancon NQ, Lee S, Edwards CA, Atiyeh R (2003) Effects of humic acids derived from cattle, food and paper-waste vermicomposts on growth of greenhouse plants. Pedobiologia 47:741–744

Arancon NQ, Edwards CA, Lee S, Byrne R (2006) Effects of humic acids from vermicomposts on plant growth. Eur J Soil Biol 42:S65–S69

Arancon NQ, Edwards CA, Yardim EN, Oliver TJ, Byrne RJ, Keeney G (2007) Suppression of two-spotted spider mite (*Tetranychus urticae*), mealy bug (*Pseudococcus* sp) and aphid (*Myzus persicae*) populations and damage by vermicomposts. Crop Protect 26:29–39

Aremu AO, Stirk WA, Kulkarni MG, Tarkowská D, Turečková V, Gruz J, Šubrtová M, Pěnčík A, Novák O, Doležal K, Strnad M, Van Staden J (2015) Evidence of phytohormones and phenolic acids variability in garden-waste-derived vermicompost leachate, a well-known plant growth stimulant. Plant Growth Regul 75:483–492

Arshad M, Frankenberger WT (1991) Microbial production of plant hormones. In: Keister DL, Cregan PB (eds) The rhizosphere and plant growth. Beltsville Symposia in agricultural research, vol 14. Springer, Dordrecht, pp 327–334

Atiyeh RM, Subler S, Edwards CA, Bachman G, Metzger JD, Shuster W (2000) Effects of vermicomposts and compost on plant growth in horticultural container media and soil. Pedobiologia 44:579–590

Atiyeh RM, Edwards CA, Metzger JD, Lee S, Arancon NQ (2002) The influence of humic acids derived from earthworm-processed organic wastes on plant growth. Bioresour Technol 84:7–14

Baca BE, Elmerich C (2003) Microbial production of plant hormones. In: Elmerich C, Newton WE (eds) Associative and endophytic nitrogen-fixing bacteria and cyanobacterial associations, nitrogen fixation: origins, applications, and research progress, vol 5. Springer, Dordrecht, pp 113–143

Bachman GR, Metzger JD (2008) Growth of bedding plants in commercial potting substrate amended with vermicompost. Bioresour Technol 99:3155–3161

Britto DR, Kronzuker HJ (2002) NH_4^+ toxicity in higher plants: a critical review. J Plant Physiol 159:567–584
Cáceres R, Malińska K, Marfà O (2018) Nitrification within composting: a review. Waste Manag 72:119–137
Canellas LP, Olivares FL, Okorokova AL, Façanha RA (2002) Humic acids isolated from earthworm compost enhance root elongation, lateral root emergence, and plasma membrane H^+-ATPase activity in maize roots. J Plant Physiol 130:1951–1957
Canellas LP, Olivares FL, Aguiar NO, Jones DL, Nebbioso A, Mazzei P, Piccolo A (2015) Humic and fulvic acids as biostimulants in horticulture. Sci Hortic 196:15–27
Chen Y, Aviad T (1990) Effects of humic substances on plant growth. In: MacCarthy P, Clapp CE, Malcolm RL, Bloom PR (eds) Humic substances in soil and crop sciences: selected readings. ASA and SSSA, Madison, pp 161–186
Doan TT, Ngo PT, Rumpel C, Nguyen BV, Jouquet P (2013) Interactions between compost, vermicompost and earthworms influence plant growth and yield: a one-year greenhouse experiment. Sci Hortic 160:148–154
Domínguez J (2004) State-of-the art and new perspectives on vermicomposting research. In: edwards CA (ed) Earthworm ecology, 2nd edn. CRC, Boca Raton, FL, pp 401–424
Edwards CA, Arancon NQ, Vasko-Bennett M, Askar A, Keeney G (2010) Efficacy of aqueous extracts from vermicomposts on attacks by cucumber beetles (*Acalymma vittatum*) (Fabr.) on cucumber and tobacco hornworm (*Manduca sexta*) (L.) on tomatoes. Pedobiologia 53:141–148
Fornes F, Mendoza-Hernandez D, Belda RM (2013) Compost versus vermicompost as substrate constituents for rooting shrub cuttings. Spanish J Agric Res 11:518–528
Frederickson J, Howell G, Hobson AM (2007) Effect of pre-composting and vermicomposting on compost characteristics. Eur J Soil Biol 43:S320–S326
García AC, Santos LA, Izquierdo FG, Sperandio MVL, Castro RN, Berbara RLL (2012) Vermicompost humic acids as an ecological pathway to protect rice plant against oxidative stress. Ecol Eng 47:203–208
Gholami H, Saharkhiz MJ, Fard FR, Ghani A, Nadaf F (2018) Humic acid and vermicompost increased bioactive components, antioxidant activity and herb yield of chicory (*Cichorium intybus* L.). Biocatal Agric Biotechnol 14:286–292
Gopalakrishnan S, Srinivas V, Alekhya G, Prakash B, Kudapa H, Varshney RK (2015) Evaluation of Streptomyces sp. obtained from herbal vermicompost for broad spectrum of plant growth-promoting activities in chickpea. Org Agr 5:123–133
Grantina-Ievina L, Ievinsh G (2015) Microbiological characteristics and effect on plants of the organic fertilizer from vermicompost and bat guano. Res Rural Develop 1:95–101
Grantina-Ievina L, Andersone U, Berkolde-Pīre D, Nikolajeva V, Ievinsh G (2013) Critical tests for determination of microbiological quality and biological activity in commercial vermicompost samples of different origins. Appl Microbiol Biotechnol 97:10541–10554
Grantina-Ievina L, Karlsons A, Andersone-Ozola U, Ievinsh G (2014) Effect of freshwater sapropel on plants in respect to its growth-affecting activity and cultivable microorganism content. Zemdirbyste-Agriculture 101:355–366
Grantina-Ievina L, Nikolajeva V, Rostoks N, Skrabule I, Zarina L, Pogulis A, Ievinsh G (2015) Impact of green manure and vermicompost on soil suppressiveness, soil microbial populations and plant growth in conditions of organic agriculture of northern temperate climate. In: Meghvansi MK, Varma A (eds) Organic amendments and soil suppressiveness in plant disease management. Springer, Switzerland, pp 381–399
Hartz TK, Bottoms TG (2010) Humic substances generally ineffective in improving vegetable crop nutrient uptake or productivity. HortScience 45:906–910
Hosseinzadeh SR, Amiri H, Ismaili A (2016) Effect of vermicompost fertilizer on photosynthetic characteristics of chickpea (*Cicer arietinum* L.) under drought stress. Photosynthetica 54:87–92
Huang K, Li F, Wei Y, Fu X, Chen X (2014) Effects of earthworms on physicochemical properties and microbial profiles during vermicomposting of fresh fruit and vegetable wastes. Bioresour Technol 170:45–52

Hussain N, Abbasi SA (2018) Efficacy of the vermicomposts of different organic wastes as "clean" fertilizers: state-of-the-art. Sustainability 10:1205
Ievinsh G (2011) Vermicompost treatment differentially affects seed germination, seedling growth and physiological status of vegetable crop species. Plant Growth Regul 65:169–181
Ievinsh G, Vikmane M, Ķirse A, Karlsons A (2017) Effect of vermicompost extract and vermicompost-derived humic acids on seed germination and growth of industrial hemp. Proc Latv Acad Sci B 71:286–292
Ievinsh G, Andersone-Ozola U, Zeipina S (2020) Comparison of efficiency of compost and vermicompost as soil amendments for organic production of four herb species. Biol Agric Hortic [under review]
Joshi R, Singh J, Vig AP (2015) Vermicompost as an effective organic fertilizer and biocontrol agent: effect on growth, yield and quality of plants. Rev Environ Sci Biotechnol 14:137–159
Karlsons A, Osvalde A, Andersone-Ozola U, Ievinsh G (2016) Vermicompost from municipal sewage sludge affects growth and mineral nutrition of winter rye (*Secale cereale*) plants. J Plant Nutr 39:765–780
Kopec M, Gondek K, Baran A (2013) Assessment of respiration activity and ecotoxicity of composts containing biopolymers. Ecotoxicol Environ Saf 89:137–142
Krishnamoorthy RV, Vajranabhaiah SN (1986) Biological activity of earthworm casts: an assessment of plant growth promoter levels in casts. Proc Indian Acad Sci Anim Sci 95:341–351
Lazcano C, Gómez-Brandón M, Domínguez J (2008) Comparison of the effectiveness of composting and vermicomposting for the biological stabilization of cattle manure. Chemosphere 72:1013–1019
Lazcano C, Arnold J, Tato A, Zaller JG, Domínguez J (2009) Compost and vermicompost as nursery pot components: effects on tomato plant growth and morphology. Spanish J Agric Res 7:944–951
Lazcano C, Sampedro L, Zas R, Domínguez J (2010) Vermicompost enhances germination of the maritime pine (*Pinus pinaster* Ait.). New For 39:387–400
Lotfi R, Kalaji HM, Valizadeh GR, Behrozyar EK, Hemati A, Gharavi-Kochebagh P, Ghassemi A (2018) Effects of humic acid on photosynthetic efficiency of rapeseed plants growing under different watering conditions. Photosynthetica 56:962–970
Makkar C, Singh J, Parkash C (2017) Vermicompost and vermiwash as supplement to improve seedling, plant growth and yield in *Linum usitassimum* L. for organic agriculture. Int J Recycl Org Waste Agricult 6:203–218
Marschner H, Kirkby EA, Cakmak I (1996) Effect of mineral nutritional status on shoot-root partitioning of photoassimilates and cycling of mineral nutrients. J Exp Bot 47:1255–1263
Mupambwa HA, Mnkeni PNS (2018) Optimizing the vermicomposting of organic wastes amended with inorganic materials for production of nutrient-rich organic fertilizers: a review. Environ Sci Pollut Res 25:10577–10595
Mupondi LT, Mnkeni PNS, Muchaonyerwa P, Mupambwa HA (2018) Vermicomposting manure-paper mixture with igneous rock phosphate enhances biodegradation, phosphorus bioavailability and reduces heavy metal concentrations. Heliyon 4:e00749
Nakhshiniev B, Biddinika MK, Gonzales HB, Sumida H, Yoshikawa K (2014) Evaluation of hydrothermal treatment in enhancing rice straw compost stability and maturity. Bioresour Technol 151:306–313
Noroozisharaf A, Kaviani M (2018) Effect of soil application of humic acid on nutrients uptake, essential oil and chemical compositions of garden thyme (*Thymus vulgaris* L.) under greenhouse conditions. Physiol Mol Biol Plants 24:423–431
Olaetxea M, De Hita D, Garcia CA, Fuentes M, Baigorri R, Mora V, Garnica M, Urrutia O, Erro J, Zamarreño AM, Berbara RL, Garcia-Mina JM (2018) Hypothetical framework integrating the main mechanisms involved in the promoting action of rhizospheric humic substances on plant root- and shoot-growth. Appl Soil Ecol 123:521–537
Osvalde A (2011) Optimization of plant mineral nutrition revisited: the roles of plant requirements, nutrient interactions, and soil properties in fertilization management. Environ Exp Biol 9:1–8

Osvalde A, Karlsons A, Čekstere G, Maļecka S (2012) Effect of humic substances on nutrient status and yield of onion (*Allium cepa* L.) in field conditions. Proc Latv Acad Sci B 66:192–199

Osvalde A, Karlsons A, Cekstere G, Vojevode L (2016) The effect of vermicompost-derived humic substances on nutrient status and yield of organic potato in field conditions. Acta Hortic 1142:277–283

Piccolo A, Celano G, Pietramellara G (1993) Effects of fractions of coal-derived humic substances on seed germination and growth of seedlings (*Lactuga sativa* and *Lycopersicon esculentum*). Biol Fertil Soils 16:11–15

Pii Y, Mimmo T, Tomasi N, Terzano R, Sesco S, Crecchio C (2015) Microbial interactions in the rhizosphere: beneficial influences of plant growth-promoting rhizobacteria on nutrient acquisition process. A review. Biol Fertil Soils 51:403–415

Pirdashti H, Motaghian A, Bahmanyar MA (2010) Effects of organic amendments application on grain yield, leaf chlorophyll content and some morphological characteristics in soybean cultivars. J Plant Nutr 33:485–495

Ravindran B, Wong JWC, Selvam A, Sekaran G (2016) Influence of microbial diversity and plant growth hormones in compost and vermicompost from fermented tannery waste. Bioresour Technol 217:200–204

Rose MT, Patti AF, Little KR, Brown AL, Jackson WR, Cavagnaro TR (2014) A meta-analysis and review of plant -growth response to humic substances: practical implications for agriculture. In: Sparks DL (ed) Advances in agronomy, vol 124. Elsevier, Amsterdam, pp 37–89

Šenberga A, Andersone U, Ievinsh G (2012) Increased thermotolerance of photosystem II activity in *Phaseolus vulgaris* by *Trichoderma harzianum*, vermicompost extract and humic acid treatments. Abstract of the 18th FESPB/EPSO Plant Biology Congress 2012, Freiburg

Sreevidya M, Gopalakrishnan S, Kudapa H, Varshney RK (2016) Exploring plant growth-promotion actinomycetes from vermicompost and rhizosphere soil for yield enhancement in chickpea. Braz J Microbiol 47:85–95

Srivastava V, Gupta SK, Singh P, Sharma B, Singh RP (2018) Biochemical, physiological, and yield responses of lady's finger (*Abelmoschus esculentus* L.) grown on varying ratios of municipal solid waste vermicompost. Int J Recyl Org Waste Agric 7:241–250

Tomati U, Grappeli A, Galli E (1988) The hormone-like effect of earthworm casts on plant growth. Biol Fertil Soils 5:288–294

Van der Ent S, Van Wees SCM, Pieterse CMJ (2009) Jasmonate signaling in plant interactions with resistance-inducing beneficial microbes. Phytochemistry 70:1581–1588

Warman PR, AngLopez MJ (2010) Vermicompost derived from different feedstocks as a plant growth medium. Bioresour Technol 101:4479–4483

Wong WS, Tan SN, Ge L, Chen X, Letham DS, Wong JWH (2016) The importance of phytohormones and microbes in biostimulants: mass spectrometric evidence and their positive effects on plant growth. Acta Hortic 1148:49–59

Xu D-B, Wang Q-J, Wu Y-C, Yu G-H, Shen Q-R, Huang Q-W (2012) Humic-like substances from different compost extracts could significantly promote cucumber growth. Pedosphere 22:815–824

Xu L, Yan D, Ren X, Wei Y, Zhou J, Zhao H, Liang M (2016) Vermicompost improves the physiological and biochemical responses of blessed thistle (*Silybum marianum* Gaertn.) and peppermint (*Mentha haplocalyx* Briq) to salinity stress. Ind Crops Prod 94:574–585

Yadav A, Garg VK (2015) Influence of vermi-fortification on chickpea (*Cicer arietinum* L.) growth and photosynthetic pigments. Int J Recycl Org Waste Agriculture 4:299–305

Yogev A, Raviv M, Hadar Y, Cohen R, Wolf S, Gil L, Katan J (2010) Induced resistance as a putative component of compost suppressiveness. Biol Control 54:46–51

Zaller JG (2007) Vermicompost as a substitute for peat in potting media: effects on germination, biomass allocation, yields and fruit quality of three tomato varieties. Sci Hortic 112:191–199

Zarrabi M, Mohammadi AA, Al-Musawi TJ, Saleh HN (2018) Using natural clinoptilolite zeolite as an amendment in vermicomposting of food waste. Environ Sci Pollut Res Int 25:23045–23054

Zhang H, Tan SN, Wong WS, Ng CYL, Teo CH, Ge L, Chen X, Yong JWH (2014) Mass spectrometric evidence for the occurrence of plant growth promoting cytokinins in vermicompost tea. Biol Fertil Soils 50:401–403

Zhao H-T, Li T-P, Zhang Y, Hu J, Bai Y-C, Shan Y-H, Ke F (2017) Effects of vermicompost amendment as a basal fertilizer on soil properties and cucumber yield and quality under continuous cropping conditions in a greenhouse. J Soils Sedminets 17:2718–2730

Zuo Y, Zhang J, Zhao R, Dai H, Zhang Z (2018) Application of vermicompost improves strawberry growth and quality through increased photosynthesis rate, free radical scavenging and soil enzymatic activity. Sci Hortic 233:132–140

Chapter 5
Interaction of Earthworm Activity with Soil Structure and Enzymes

Sazada Siddiqui

Abstract The earthworms have been substantially affected by weather, characteristics of soil, industrial and agronomic actions, and pollution of environment in relation to their biomass and inhabitants' dynamics in soil. Many studies relating to soil earthworms stipulated the evidence that they accelerate the changes in biological, chemical, and physical properties of soil and also impact aeration of soil and drainage due to their dynamic actions of casting, burrowing, and feeding. These actions have also significant roles in alterations of minerals and nutrients available in soil for plant. The fact that synergies in interaction of earthworm activities, soil structure, and enzymatic activities are significant signs of soil productiveness has been given an escalating consideration in soil science. The goal of this review chapter is to elucidate the relations between earthworms and soil enzymes at different levels in soil, types of earthworm, soil enzymes, kind of soil enzymes and their roles in maintaining soil health, and factors affecting soil structure. In addition, as the significant illustrative feature of soil life, the earthworms are profoundly affected by traditional agrarian procedures such as the application of chemical fertilizers and intensive tillage. So, special attention has been given on the queries of how agronomic practices govern the associations between soil earthworm and enzyme activities.

Keywords Earthworms · Soil structure · Soil enzymes

5.1 Introduction

Aristotle called earthworms as "the intestines of the earth." The name Earthworm denotes to an explicit group of metamerically segmented invertebrates of class Oligochaeta of phylum Annelida. They are soft bodied, elongated, and bilaterally

S. Siddiqui (✉)
Department of Biology, College of Science, King Khalid University, Abha, Saudi Arabia
e-mail: sasdeky@kku.edu.sa

M. K. Meghvansi, A. Varma (eds.), *Biology of Composts*, Soil Biology 58,
https://doi.org/10.1007/978-3-030-39173-7_5

symmetrical and found in a diversity of soil types and environments in which there is organic matter and enough moisture accessible. Some earthworm species live in perpetual burrows deep in the soil, some live in compost, and some live in the intermediate ground and make an intricate net of burrows. By making burrows and casts, earthworms have a direct impact on soil structure. Burrowing controls the physical properties of soil whereas casting enriches soil aggregation. The activity of earthworm leads to the decrease in soil bulk density by making pores, and thus they make the soil porous and enhance the water holding capacity of the soil. By increasing water holding capability and accessibility of nut improving soil structure, earthworm (*Eisenia foetida*) compost sturdily affects soil fertility (Landgraf et al. 1999). It has recommended that earthworms amplify the velocity of decomposition of organic residues (Vinceslas-Akpa and Loquet 1997). They also yield various bioactive humic substances (Masciandaro et al. 1999).

Earthworms are considered as soil engineers in temperate and tropical ecosystems since they alter the configuration of soil properties and consequently impact soil microorganisms, a parameter of soil organic matter and development of plant (Lavelle 1997). Earthworms are found near to rhizosphere region (root zone of plants), which is abundant in organic matter. Earthworms are a precise applicant for the biotic decay of organic materials. As per the present existing information, the prospective role of earthworms in monitoring the multiplicity of micro vegetation in soil organization and in regulating soil structure and soil enzymes and need advance research. The scope of the current study is to appraise the leading role of earthworms in monitoring soil structure and soil enzyme. However, few studies are available on the significance of micro soil vegetation in agroecology procedures. The study of the mechanism of earthworm activity and how they impact soil structure and soil enzymes is required. Here we have to soothe to argue various mechanisms, with a functional approach rather than a descriptive one, by which earthworm activity can render the environment less disease favorable and the host plant less susceptible. The present chapter discusses the relations between earthworms and soil enzymes at different levels in soil, types of earthworm, soil enzymes, kind of soil enzymes and their role in maintaining soil health, and factors affecting soil structure. In addition, the influence of agronomic practices on the associations between soil earthworm and enzyme activities has been elucidated through literature metaanalysis.

5.2 Earthworm Ecotypes

Earthworms are alienated into *four types*, entitled as *ecotypes*. Each ecotype designates a diverse ecological alliance established on its behavior.

5.2.1 Compost Earthworms

Compost earthworms are found in regions rich in decaying flora or compost, and they choose warm and humid environments. They can swiftly consume the decaying flora or compost and can replicate very speedily. They are stripy and bright red in color. Since compost earthworms can eradicate impurities from the soil, they are used to help dispose of unwanted things and thus purify the soil. *Eisenia fetida* and *Dendrobaena veneta* are the two kinds of compost earthworm species.

5.2.2 Epigeic Earthworms

Epigeic earthworms are 1–2.5 mm in diameter, live and feed on the surface of the soil in forest litter. Epigeic earthworms do not make burrows, but they live and feed on the forest litter. Epigeic earthworms are not stripped, but they are reddish brown or bright red in color. *Dendrodrilus rubidus, Dendrobaena attemsi, Dendrobaena octaedra, Lumbricus festivus, Heliodrilus oculatus, Lumbricus castaneus, Lumbricus rubellus, Lumbricus friendi, Eiseniella tetraedra, and Satchellius mammalis are* epigeic earthworm species.

5.2.3 Endogeic Earthworms

Endogeic earthworms are 2–4.5 mm in diameter, live in 10–15 cm soil surface, and they feed on the soil. They make the widespread system of horizontal burrows through the soil to search for food, and they instantly replenish with their casts. Endogeic earthworms are frequently, green, pink, blue, gray, or pale colors. *Murchieona muldali, Octolasion lacteum Octolasion cyaneum, Apporectodea icterica, Ap. caliginosa, Ap. rosea, and Allolobophora chlorotica* are endogeic earthworm species.

5.2.4 Anecic Earthworms

Anecic earthworms are 4–8 mm in diameter, feed on the leaves on the soil surface and rotten organic residues. They make perpetual more or less vertical burrows in the soil. They also cast on the soil surface. In the UK, anecic earthworms are the largest earthworm species. In anecic species, the head is darkly colored, and they have paler tails. Anecic earthworm species include *Lumbricus terrestris* and *Ap. longa.*

5.3 Soil Enzymes

Soil enzymes are a collection of enzymes whose normal occupants are the soil and are continually playing a vital role in sustaining soil health, physical and chemical properties, fertility, and soil ecology. These soil enzymes have some significant biochemical utilities in the general procedure of decaying biological matter in the soil structure (Sinsabaugh et al. 1991). These enzymes play a critical role in agronomy by the decay of biological wastes, biological material creation, nutrient cycling, and catalyzing various dynamic responses essential for the life procedures of microorganisms in soils and the balance of soil system (Dick 1997). All types of soils hold a set of enzymes that regulate soil metabolic procedures (McLaren 1975) that depend on its chemical, physical, biochemical, and microbiological properties. Different types of soils have different quantity of arrangement, biological material content, and strength of organic procedures and movement of its alive organisms. These enzymes are released from animals (Kanfer et al. 1974), organic compounds and microorganisms (Richmond 1991), plants (Miwa et al. 1937), and soils (Ganeshamurthy et al. 1995). These enzymes include phosphatase, amylase, urease, protease, cellulases, chitinase, dehydrogenase, β-glucosidase, and arylsulfatases.

Categories of soil enzymes:

(a) Inducible: Inducible enzymes are present in a very less amount in the soil, when its substrate is present they rapidly rise in concentration for example (Amidase).
(b) Constitutive: Constitutive enzymes are present almost in a fixed amount in a cell, and they are not affected by adding any specific substrate—genes constantly expressed for example (Pyrophosphatase).

5.3.1 Types of Soil Enzymes and Their Role in Maintaining Soil Health

Here, various important enzymes playing role in maintaining soil health are discussed.

5.3.1.1 Dehydrogenase

Soil dehydrogenases are vital agents of oxidoreductase enzyme class (Gu et al. 2009). In soils, dehydrogenase enzyme is most important and is generally used as an indicator of microorganisms activity (Salazar et al. 2011), since, in all living microbial cells, dehydrogenases are found intracellular (Yuan and Yue 2012). Also, dehydrogenases are strongly connected with microbial oxidoreduction procedures (Moeskops et al. 2010). In the biotic oxidation of soil biological matter by relocating hydrogen from organic substrates to inorganic acceptors, dehydrogenases play an important role (Zhang et al. 2000). Several precise dehydrogenases relocate hydrogen to either nicotinamide adenine dinucleotide phosphate or nicotinamide adenine dinucleotide (Subhani et al. 2001). Active dehydrogenases can use both other compounds and O_2 as terminal electron acceptors, but maximum dehydrogenases are produced by anaerobic microbes (Brzezińska et al. 2001).

5.3.1.2 β-Glucosidase

The prime and familiar enzyme in soils is glucosidase (Tabatabai 1994a, b). In the literature, it is mostly reported that BG (β-glucosidase-EC3.2.1.21; old name: cellobiase) is an immobilized enzyme and also considered as an indicator of soil management effects (Bandick and Dick 1999). For the decay of plant residues and soil biological matter, a significant role is played by β-glucosidase. Glucosidase is named as per the kind of bond that it hydrolyzes. In soil, glucosidase plays a vital role in catalyzing the hydrolysis and biodegradation of some β-glucosidase found in vegetal remains decaying in the biome (Martinez and Tabatabai 1997). Glucose is the final product, which is a vital source of carbon for microbes in the soil (Esen 1993). β-Glucosidase is an indicator of soil quality and also gives a reflection of past biological activity, can be used to distinguish managing effect on soils and the capacity of the soil to stabilize the soil organic matter (Ndiaye et al. 2000). With growing soil microbial biomass, an enhanced β-glucosidase activity has been reported, which reveals the soil's capability to break plant remains and increase the accessibility of nutrients for succeeding crops.

5.3.1.3 Cellulase

Fungi, bacteria, and protozoans mostly yield cellulase enzymes, which catalyze cellulolysis. By decaying insoluble cellulose to soluble sugars, cellulases play a vital role in the global carbon cycle, and thus they are essential enzymes. Cellulases divided into three kinds such as endoglucanases, exocellulases, and processive endoglucanases based on the diverse structures and unique mode of action. In the biosphere, cellulose is the plentiful rich biological compound encompassing nearly 50% of the biofuel manufactured by photosynthetic fixation of carbon dioxide (Eriksson et al. 1990). In most cultivated soils, development, and subsistence of microbes rely on the carbon source confined in the cellulose found in the soils (Deng and Tabatabai 1994). For microbial use, cellulose in plant remains has to be decayed into cellobiose, oligosaccharides, and glucose by cellulase enzymes for carbon to be free as a source of energy (White 1982). A collection of enzymes that catalyze cellulose decay are called cellulases; polysaccharides are made up of b-1,4-linked glucose units (Deng and Tabatabai 1994). Plant debris is the main source of celluloses in the soil, and some quantity might be derived from bacteria and fungi in soils (Richmond 1991). Activities of cellulases in agronomic soils are exaggerated by a number of causes such as pH of soil, temperature, abiotic conditions, chemical structure of biological matter and its position in the soil profile perspective (Alf and Nannipieri 1995), trace elements from fungicides (Arinze and Yubedee 2000), quality of plant debris or biological matter, and mineral elements of soil (Deng and Tabatabai 1994). Increased stimulatory effect of cellulases in black soil than red soil stated by Srinivasulu and Rangaswamy (2006). In the presence of cellulose, chitin prompts the production of chitinase and further cell wall lytic enzymes that stimulate the discharge of intramural β-glucosidase in the medium. Cellulases activity can be used to determine the initial suggestion of a few chemical and physical properties of soil, and hence facilitate agrarian soil management policies.

5.3.1.4 Urease

Plants are the main source for originating soil urease enzymes (Polacco 1977) and microbes are responsible for originating both extra and cellular enzymes (Mobley and Hausinger 1989). The urease that is originated from the plants and microbes is quickly degraded in soil by proteolytic enzymes (Zantua and Bremner 1977). Due to this role, urease activities in soils have received a lot of attention since it was first reported by Rotini (1935), a process considered vital in the regulation of N supply to plants after urea fertilization. Through NH_3 volatilization, it outcomes in a prompt N loss in the air (Simpson and Freney 1988). The enzyme that is liable for the hydrolysis of urea fertilizers that are applied on the soil into ammonia and carbon dioxide with a simultaneous rise in soil pH is urease enzyme (Byrnes and Amberger 1989). In soil, several causes affect urease activity like soil depth, cropping history, soil amendments, biological matter content of the soil, heavy metals, and ecological dynamics like temperatures (Yang et al. 2006).

5.3.1.5 Proteases

Proteases, in the soil, are mostly related to organic and inorganic colloids (Nannipieri et al. 1996). In the soil, proteases play an essential role in the mineralization of N (Nitrogen) (Ladd and Jackson 1982) and control the quantity of plant available N and growth of the plant. The number of proteases, which is an extracellular enzyme activity, might reveal not only the biotic capability of soil for the enzymatic change of the substrate that is not dependent on the magnitude of microbial action. Proteases also have an essential role in the ecosystem of microbes in the biome (Burns 1982).

5.3.1.6 Phosphatases

Phosphatases are a broad class of phosphohydrolases that catalyzes the hydrolysis of anhydrides of phosphoric acid and esters (Makoi and Ndakidemi 2008). Phosphatases are good pointers of soil richness, and they play an essential role in phosphorous cycles, in soil environments (Speir and Ross 1978) as proof shows that they are associated with P stress and plant growth. Stability, the rate of synthesis and release of phosphatases enzyme is influenced by soil pH. When phosphorus deficiency occurs in soil, the release of acid phosphatase from plant roots is enhanced to enrich the remobilization and solubilization of p-(phosphate), therefore affecting the capacity of the plant to manage P-stressed conditions (Versaw and Harrison 2002). Considering the changing aspects of enzyme actions in these systems is essential for envisaging their relation as their actions control plant growth and nutrient uptake.

5.3.1.7 Arylsulfatases

Arylsulfatases are found extensively in soils (Ganeshamurthy et al. 1995) and in nature (Dodgson et al. 1982). In the soil, arylsulfatases are liable for the hydrolysis of sulfate esters (Kertesz and Mirleau 2004). Arylsulfatases are released by bacteria as a reply to sulfur constraint into the outside atmosphere (McGill and Colle 1981). Arylsulfatases play an essential role in the hydrolysis of aromatic sulfate esters (R–O–SO3) to sulfate and phenols (R–OH) and or sulfate sulfur (SO42 or SO4–S) (Tabatabai 1994a, b). Several environmental factors affect soil like pH changes in the soil solution (Acosta-Martınez and Tabatabai 2000), heavy metal pollution (Tyler 1981), organic matter content and its type (Sarathchandra and Perrott 1981), as the activity perseverance of extracellular arylsulfatases and concentration to particle surfaces in soils. Since sulfur is essential in the nutrition of plants, a more in-depth study of the role of arylsulfatases in the mobilization of sulfur in agrarian soils is important.

5.3.1.8 Amylase

An enzyme that catalyzes the hydrolysis of starch into sugars is called an amylase. Amylase is hydrolyzing starch enzymes (Ross 1976). Amylase is made up of a- and b-amylase (Thoma et al. 1971). A-amylase is produced by animals, plants, and microorganisms while b-amylase is produced mostly by plants. Amylase is usually found in soils and plants. Conversion of starch like substrates into oligosaccharides or glucose is done by amylase while the conversion of starch into maltose is done by b-amylase (Thoma et al. 1971). Different factors like the type of vegetation, soil types, environmental and cultural practices affect the activities, and roles of a- and b-amylase (Ross 1975). Plants affect amylase enzyme activities of soil by providing enzymes from their residues or by furnishing substrates for the synthetic actions of microbes. The more in-depth study is essential to the importance of these enzymes in the soil. Amylase has an extensive variety of properties and activities and is extensively scattered in soils (Ladd and Butler 1972). Inside most plant tissues, starch is a major carbon compound. Starch increases during active photosynthesis and drops as it is enzymatically changed into sugar. In soil, hydrolytic depolymerization of polysaccharides is catalyzed by amylase (Tu and Miles 1976). Starch-hydrolyzing enzymes action relies on the kind of substrate, and they are generally inducible and extracellular (Alexander 1977). The critical breakdown of composite polysaccharides, as well as starch to an accessible kind of glucose, is induced by soil amylase (Singaram and Kumari 2000). Throughout litter decay, the creation of these extracellular enzymes from microorganisms might be affected by moisture, pH, and temperature, the involvement of substrate (Sinsabaugh and Linkins 1987). During litter decay, variations in amylase activity were ascribed to fluctuations in microbial populations (Ross and Roberts 1973). On treatment of soil with pesticides, herbicides, and insecticides, enhanced amylase activity was reported (Tu 1982). Effluents discharge from Cotton ginning mills (Narasimha 1997), from pressmud plus paper mills (Chinnaiah et al. 2002) and from paper and pulp mills (Kannan and Oblisami 1990) also leads to enhanced amylase activity. On the other hand, the activity of amylase decreases when soil was treated with chlorothalonil (Singh et al. 2002), imidacloprid (Tu 1995), and dimethoate (Mandic et al. 1997).

5.3.1.9 Chitinase

Chitin is the key structural component of shells of crustaceans, fungal cell walls, exoskeletons of other arthropods and insects. Chitinase enzyme may be used against insect pests as an insecticide and phytopathogenic fungi as an abiotic fungicide. For the decay and hydrolysis of chitin, chitinase enzymes are liable. Chitinase enzyme is useful for agronomy and is produced by numerous organisms like microbes and plants (Deshpande 1986). It is useful in controlling soil-borne diseases like *Rhizoctonia solani* and *Sclerotium rolfsii* in cotton and beans, respectively (Shlomo et al. 1995). The cause of the decay of pathogenic fungi cell walls is by lytic enzyme chitinase (Singh et al. 1999). It plays an important role in the biotic regulation of pests.

5.3.2 Importance of Soil Enzymes for Maintaining Soil Structure and Health

Soil enzymes are very important for maintaining soil structure and health. They are subtle pointers of environmental change. When biological matter decays, it discharges nutrients in the soil. Soil enzymes are involved in recognition of microbial action. Soil enzymes are used for assessing soil fertility, quality, biological cycling of several components in soil (S, C, N), and amount of pollution (SO_4, heavy metals). They are also studied for determining various successional phases of an environment. Besides, they prompt decay of pesticide residues thereby contributing to soil health.

5.3.2.1 Factors Affecting Soil Structure and Health

Soil structure denotes the shape, size, and arrangement of spaces and solids, continuousness of holes and voids and their capability to hold and conduct fluids and inorganic as well as organic substances, and capacity to sustain dynamic root growth and enlargement (Lal 1991). Satisfactory soil structure and high cumulative constancy are essential to augmenting agronomic yield, increasing porosity, improving soil fertility, and declining credibility. Since Darwin's time (1881), common statements about the significance of earthworms for the fertility of soil have been stated. Distribution of earthworms is affected by various soil-related farm management applications that underline the collaborative nature of the association among earthworms and soil fertility. Lumbricidae is a family of earthworms, which includes most of the earthworm species. Inside the Lumbricidae family, there is an excessive deviation in the environmental necessities and effects of various species, few of them live and feed on the superficial layer of soil whereas others burrow intensively into the lower soil levels. In the subtropics and tropics, numerous species belong to the Megascolecidae, but there is less information about their effects on the soils and their ecosystem (Lal 1978). Brady (1974) defines "soil fertility" to be the intrinsic ability of soil to source nutrients to plants in sufficient quantities and appropriate amounts. Earthworms have a significant effect on biological and physical attributes, which in turn affect the nutrient sources to plants. These interactions are explained in Fig. 5.1.

Soil structure is affected by different factors such as plant growth, weather and external factors, environmental impacts, and soil organic carbon pool.

Plant Growth

Plant growth is affected by soil structure by prompting root distribution and the capacity to take up nutrients and water (Pardo et al. 2000). Soil structure increases water storage, and it expedites water and oxygen permeation. Fertilizer withholding in the soil medium and its efficacy in plants is decreased by improved water transmission through the soil (Franzluebbers 2002). Decreased air and water availability to roots, swift recycling of nutrients, and crusting is caused by disruption of soil structure by compaction.

Weather and External Factors

Soil structure is impelled by weather and landscape position by features such as height, rainfall, temperature, and slope gradient. Impact of the meteorological conditions is regulated by soil properties, for example, minerals, soil organic carbon, texture, and organisms. Due to variations in temperature and humidity systems and wet-dry cycles, that can reorient elements (Singer et al. 1992) probably causing amplified segregation of SOC in aggregates and also enhanced aggregation; weather

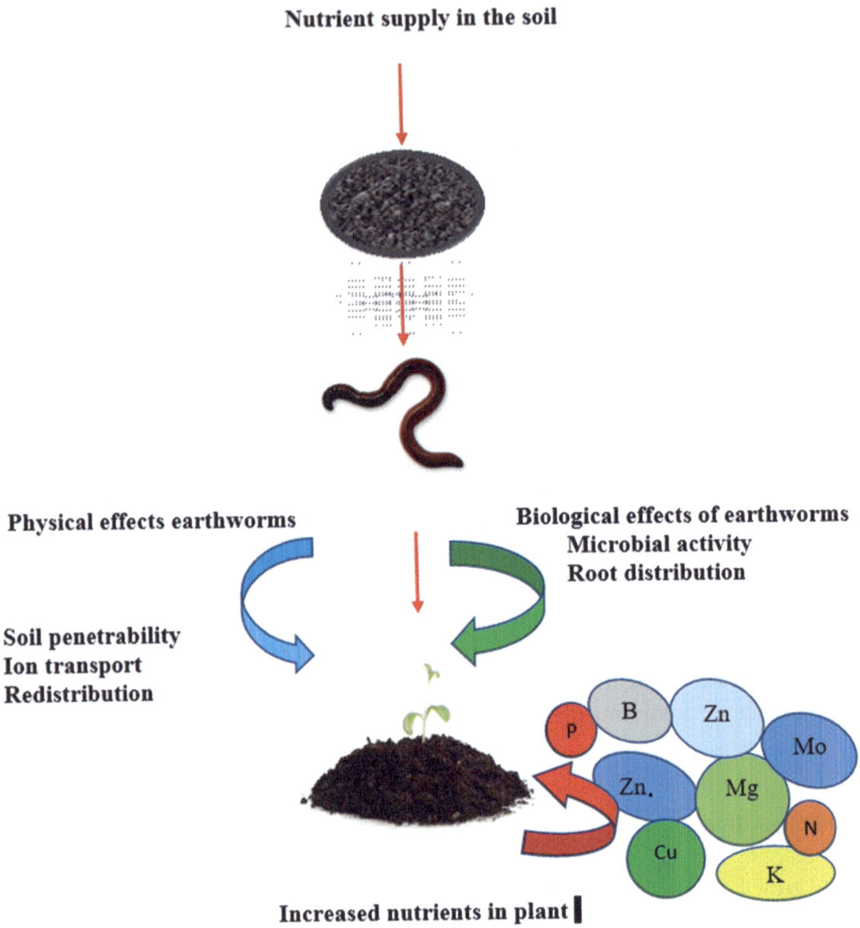

Fig. 5.1 Interaction of earthworms with abiotic and biotic factors in soil

also affects soil aggregation. Disintegration rates get transformed because of changes in humidity and temperature levels that affect biotic and microbes' action. Because of the impact of multiple causes, the correlation between temperature and putrefaction is very flexible. Higher rate of breathing and organic activity is caused by higher temperatures, and higher standing stock of soil organic carbon is caused by lower temperatures. Dry and warm soils have less inaccessible soil organic carbon than frosty and damp soils (Franzluebbers et al. 2001). Aggregation is affected by dry conditions. In a dry atmosphere, various factors, for example, are crusting, earthworms and carbonates may enhance aggregate strength (Boix-Fayos et al. 2001). Freeze–thaw phases affect aggregation in humid, temperate areas (Dalal and Bridge 1996). Water penetration is diminished by crusting. Crusting also decreases erosion and detachment that has the following effect on aggregation (Amezketa 1999). Reduced aggregation and structural development and enhanced erosion are caused by a diminution in soil moisture. Management practices, for example, mulching, irrigation, and cover cropping can modify moisture and temperature systems. Exposure to wind, sun, and air is enhanced by conventional tillage. Altering soils with humic matters decreases dispersion produced by wet–dry successions (Piccolo et al. 1997).

Environmental Impacts

The exact effect on the soil structure of elevated atmospheric carbon dioxide is not well defined. The surge in photosynthesis and consequent rise in roots, photosynthate, and microbial communities is caused by increased environmental carbon dioxide. The rise in environmental CO_2 may cause a surge in aggregation and SOM, in atmospheres with ample nutrient provisions whereas the equilibrium amid plant requirements, carbon turnover, and accessibility of nutrients might constraint the effect in nutrient-restrained atmospheres. The critical procedures of variations in carbon intake and decay are fluctuations in dominant species and carbon distribution. Deviations in plant species affect microscopic compounds comprised in soil structural growth and microbial inhabitant allocation (Dukes and Hungate 2002). Proliferation in soil organic matter and microbes in the rhizosphere is caused by the distribution of photosynthate to root and its exudates (Swift 2001). Enhanced glomalin, arbuscular mycorrhizal fungi hyphal lengths, and WSA with *Sorghum* is caused by augmented levels of CO_2 (Rillig and Steinberg 2002). Lesser hyphal density is caused by enhanced environmental carbon dioxide because of reduced oxygen diffusion in soils (Schack-Kirchner et al. 2000).

Augmenting the Soil Organic Carbon Pool

Maximization of soil organic carbon (SOC) return to the soil, surge in SOC pool and enhanced productivity is caused by soil management techniques that reduce disturbance. Proliferation in SOC pool and aggregation is caused by the suitable use of soil modifications, for example, compost, manure, lime, and fertilizer. The upsurge in SOC pool is caused by viable agricultural techniques, for example, crop rotations, cover crops, and mixed cropping. The escalation in SOC storage is caused by management techniques, which decrease putrefaction rates and CO_2 discharges.

Effect of Casting and Burrowing by Earthworm Species

Root penetration is modified by the digging of earthworms (Whalley and Dexter 1994). Movement of water and solute to deeper soil horizons is also improved by digging (Kung et al. 2000). Aeration in the soil, the storage capacity of water in soil and water infiltration is enhanced by burrowing of earthworms (Devliegher and Verstraete 1997). Correspondingly when residues were amalgamated into the soil, the percentage of steady microaggregates in macroaggregates was more celebrated in soils having earthworms. This recommends that small microaggregates be broken by earthworms and integrated into macroaggregates. The latest finding reveals that on the stable aggregate formation, there was a substantial impact of activity of earthworms and residue application (Bossuyt et al. 2006). When residues are assimilated in soils, higher than 2 mm, aggregates were formed as compared to when remains are kept on the soil surface have earthworms. For higher aggregate formation, earthworms were also essential. The density of cast is generally greater than the adjacent bulk density of soil, and they are rich in cations and organic matter. During the passage through the gut of earthworm, microaggregates are created. As a cast, egestion of inorganic and organic matrices takes place, and the strengthening of bonds within inorganic and organic elements occurs due to the reduction of water in the cast (Shipitalo and Protz 1989). Increase in structure, cumulative tensile strength, and accumulation of soil are done by the creation of casts by earthworms (McKenzie and Dexter 1987). Mostly, the stability of cast is greater than bulk soil, but in some cases, the stability of cast could be lower than adjacent soil (Shipitalo and Protz 1988).

Casts made by earthworms are either made on the soil surface or inside the soil. There are various structures, sizes, shapes, and arrangements of casts made on the surface of the soil and hence the roles of casts on the structure of soil could be diverse. Flattened units make spherical casts whereas granular casts are made of a buildup of minor textured pellets. Structure, shape and size of casts depend on the earthworm species. Usually, smaller casts are formed by small species whereas larger species form bigger casts (Lee 1985). Relying on the aging of casts, effects of earthworms on soil erosion, and detachment may be constructive or adverse. Casts having higher age are generally new stable than the neighboring soil whereas new casts are extremely vulnerable to dispersion (Blanchart et al. 1999).

Effect of Burrows Made by Earthworms on the Structure of the Soil

During movement and feeding, earthworms horizontally and vertically burrow soil. Continuity and burrowing size depend on earthworm species and properties of soil. Burrowing size could be greater than 0.1 cm in diameter (Lee 1985). In comparison with endogeic species, the anecic species, for example, *L. terrestris L.* inclines to construct vertical and huge burrows in soil (Bastardie et al. 2005). Endogeic species, for example, *Octolasion tyrtaem* Savigny make horizontal burrows having depths of 2000 mm as they consume soil. Epigeic species such as *L. rubellus* Hoffmeister can make burrows having a depth of 2000 mm on the soil surface in case there are unfavorable conditions on the soil surface. Soil drying is enhanced by earthworms having higher burrowing action. As compared to epigeic species, both anecic and endogeic earthworm species may increase the drying process of soil. Soil–water transport in the soil is influenced by burrows (Ernst et al. 2009).

5.4 Influences of Agricultural Practices on Earthworm Interactions of Enzymes

Fundamental agrarian techniques, for example, pesticide, tillage, and fertilization have a significant influence on soil inhabitants, soil enzymes, and earthworm's activities.

5.4.1 *Tillage*

By making their galleries and holes in subterranean soil layers, soil tillage influences earthworm populations. When tillage is done once a year, the influence of tillage on earthworms is less damaging as compared to birds, which feed on earthworms. Also, populations of earthworm were reduced rigorously when severe, and numerous tillage practices were done whereas an increase in their population was reported in case of no-till management practices (Johnson-Maynard et al. 2007). Subsequently, an accustomed increase in enzyme activities in soil and decline in earthworm inhabitations are the two significant consequences of soil tillage whereas exhaustive tillage practices have a deleterious influence on both soil enzyme activities and earthworm populations.

5.4.2 *Pesticides*

Use of pesticides is the prime concern in developing agronomic nations. On earthworm inhabitations, the outcome of pesticides is supplementary to the kind of pesticides used (Bauer and Roembke 1997). No adversarial outcome on earthworms is displayed by lesser concentrations of herbicides. Less fatal outcome on earthworm subsistence is displayed by triazine herbicides. By changing their nurturing behaviors and mineralization of biological resources in the soil, the effects of herbicides on earthworms emerge indirectly (Edwards and Thompson 1973). Most of the fungicides including carbamates are lethal to the earthworm's inhabitations. Also avermectins, insecticides having carbamates for example carbofuran, methiocarb and carbaryl with organophosphate such as phorate are poisonous to earthworm populations and it results in substantial reduction in earthworm inhabitations as they enter the soil (Edwards 1984). Furthermore, earthworms are proficient in bioaccumulating pesticide deposits and hence decrease the pesticide concentrations in soil (Tarrant et al. 1997). Proliferation in produce and plant biomass is caused by the usage of pesticides for the protection of plants. In the rhizosphere, it outcomes in progressing enzyme synthesis. Soil microorganisms reduce the synthesis of enzymes as they are undesirably affected by pesticides and their remains. On the contrary, in the soil environment, since pesticides get dispersed and are used as a source of energy and carbon, microorganism inhabitations and activities of their enzymes may get enhanced (Niemi et al. 2009).

5.4.3 Fertilizers

Fertilizers are of two kinds, organic and inorganic. They are required to deal with the precise requirements of nutrition and hence increases produce achieved per unit area. Earthworm inhabitations are found to increase many times due to the usage of solid constituents acquired from animal and plant instigated remains (Leroy et al. 2007). Liquid constituents of cattle wastes might affect in decline in earthworm inhabitations because of ammonia and salt constituents when used devoid of composting. Earthworms are influenced by chemical fertilizers indirectly. Plant yield is usually increased by chemical fertilizers and proliferation in the permeation of biological material of plant remains that persisted in the soil after produce and hence it results in an escalation in earthworm inhabitations. However, earthworm inhabitations in cultivated soils are harmfully affected by exhaustive usage of ammonia-based fertilizers (Edwards and Lofty 1982). Enzyme activities in the soil start to proliferate by the use of organic fertilizers instigated from animal and plant remains (Ros et al. 2006). There is no adverse consequence on enzyme activities in soil by fluid livestock fertilizers with greater ammonia and salt components. Correspondingly, enzyme activities in the soil are enhanced by fertilizers based on ammonia. Enzyme activities in the soil and earthworm inhabitations are also affected by sewage sludges, which are used as an organic nutrient supplier (Le Bayon and Binet 2006). *Eisenia fetida* species was unable to stay alive due to higher concentrations of ammonia and soluble salts when their nurturing environs have 50% or higher sewage sludge (Kizilkaya et al. 2009). Several researchers found that after application of sewage to soil increased activities of soil enzyme were reported (Banerjee et al. 1997), although several others articulated the contradictory interpretation, stating that applications of sewage sludge introverted activity of soil enzymes (Knight et al. 1997). These conflicting outcomes in the collected works are possible may be the differences in application rates and chemical characteristics of sewage sludge such as heavy metal content and C/N ratio (Tam and Wong 1990). Kizilkaya and Hepsen (2004) reported that enzyme activities of earthworm excrement were found to be higher than those of control soil (with no sewage sludge amendment). Heavy metals adversely affect soil biotic systems (Giller et al. 1998). A substantial reduction in soil biomass and their enzymatic action is caused by higher metal concentrations (Karaca et al. 2010). Earthworm inhabitations and enzymatic actions in their excretion are enhanced by the natural soil organic matter and by the addition of organic constituents, for example, municipal deposits to agrarian soils (Kizilkaya and Hepsen 2007). Several different heavy metals are found in human-made organic discarded products and consequently may prevent enzymatic actions in earthworm excretions based on the concentration of heavy metals (Kizilkaya et al. 2004).

5.5 Conclusion

For sustaining healthy and fertile soil in biomes, it is essential to know the probable roles of enzymes in soil. Soil enzymes have a substantial effect on soil biology, nutrient uptake, and growth in plants and management of the ecosystem. Exercises that can positively influence the enhancement in the growth of plants and the execution of the friendly biotic ecosystems for sustaining further living creatures are the researches concentrating on the findings of new enzymes from microbes multiplicity in the soil. For biotic catabolism of organic soil and mineral elements, enzymes are the direct intermediaries. Activities of soil enzymes are frequently associated with the activity of microbes, organic matter of soil, and physical properties of soil, including secure processes, and fluctuates more readily than other constraints. However, for the evaluation of the productivity of soil, the activity of microbes and constraining effects of contaminants on enzymatic activities of soil can be utilized.

Acknowledgments The author would like to express her gratitude to King Khalid University, Abha, Saudi Arabia for providing administrative and technical support. We are also thankful to Deanship of Scientific Research, King Khalid University. We are also grateful to Head of the Department of Biology, College of Science, King Khalid University, Abha, Saudi Arabia for providing facilities to carry out the current work.

References

Acosta-Martıinez V, Tabatabai MA (2000) Enzyme activities in a limed agricultural soil. Biol Fertil Soils 31:85–91

Alexander M (1977) Introduction to soil microbiology, 2nd edn. Wiley, New York

Alf K, Nannipieri P (1995) Cellulase activity, methods in applied soil microbiology and biochemistry. Academic, London

Amezketa E (1999) Soil aggregate stability: a review. J Sustain Agr 14:83–151

Arinze AE, Yubedee AG (2000) Effect of fungicides on Fusarium grain rot and enzyme production in maize (*Zea mays* L.). Global J Appl Sci 6:629–634

Bandick AK, Dick RP (1999) Field management effects on soil enzyme activities. Soil Biol Biochem 31:1471–1479

Banerjee MR, Burton DL, Depoe S (1997) Impact of sewage sludge application on soil biological characteristics. Agric Ecosyst Environ 66:241–249

Bastardie F, Capowiez Y, Renault P, Cluzeau D (2005) A radio-labelled study of earthworm behaviour in artificial soil cores in term of ecological types. Biol Fertil Soils 41:320–327

Bauer C, Roembke J (1997) Factors influencing the toxicity of two pesticides on three lumbricid species in laboratory tests. Soil Biol Biochem 29:705–708

Blanchart E, Albrecht A, Alegre J, Duboisset A, Gilot C, Pashanasi B, Lavelle P, Brussaard L (1999) Effects of earthworms on soil structure and physical properties. In: Lavelle P, Brussaard L, Hendrix P (eds) Earthworm management in tropical agroecosystems. CABI, Wallingford, pp 149–172

Boix-Fayos C, Calvo-Cases A, Imeson AC (2001) Influence of soil properties on the aggregation of some Mediterranean soils and the use of aggregate size and stability as land degradation indicators. Catena 44:47–67

Bossuyt H, Six J, Hendrix PF (2006) Interactive effects of functionally different earthworm species on aggregation and incorporation and decomposition of newly added residue carbon. Geoderma 130:14–25

Brady NC (1974) The nature and properties of soils. Macmillan, New York, p 639

Brzezińska M, Stępniewski W, Stępniewska Z, Przywara G (2001) Effect of oxygen deficiency on soil dehydrogenase activity in a pot experiment with triticale cv. Jago vegetation. Int Agrophys 15:145–149

Burns RG (1982) Enzyme activity in soil: location and possible role in microbial ecology. Soil Biol Biochem 14:423–427

Byrnes BH, Amberger A (1989) Fate of broadcast urea in a flooded soil when treated with N-(nbutyl) thiophospheric triamide, a urease inhibitor. Fertil Res 18:221–231

Chinnaiah U, Palaniappan M, Augustine S (2002) Rehabilitation of paper mill effluent polluted soil habitat: an Indian experience. Poster presentation, paper no. 770, symposium no. 24. 17th WCSS

Dalal RC, Bridge BJ (1996) Aggregation and organic matter storage in sub-humid and semi-arid soils. In: Carter MR, Stewart BA (eds) Structure and organic matter storage in agricultural soils. CRC, Boca Raton, FL, pp 263–307

Deng SP, Tabatabai MA (1994) Cellulase activity of soils. Soil Biol Biochem 26:1347–1354

Deshpande MV (1986) Enzymatic degradation of chitin and its biological applications. J Sci Ind Res 45:273–281

Devliegher W, Verstraete W (1997) Microorganisms and soil physico-chemical conditions in the drilosphere of *Lumbricus terrestris*. Soil Biol Biochem 29:1721–1729

Dick RP (1997) Soil enzyme activities as integrative indicators of soil health. In: Pankhurst CE, Doube BM, Gupta VVSR (eds) Biological indicators of soil health. CABI, Wellingford, pp 121–156

Dodgson KS, White G, Fitzgerald JW (1982) Sulphatase enzyme of microbial origin. Afr J Biotechnol 1:156–159

Dukes JS, Hungate BA (2002) Elevated carbon dioxide and litter decomposition in California annual grasslands: which mechanisms matter? Ecosystems 5:171–183

Edwards CA (1984) Changes in agricultural practice and their impact upon soil organisms. In: Jenkins D (ed) Proceedings of symposium No. 13, The impact of agriculture on wildlife, agriculture and the environment, UK, pp 46–65

Edwards CA, Lofty JR (1982) Nitrogenous fertilizers and earthworm populations in agricultural soils. Soil Biol Biochem 14:515–521

Edwards CA, Thompson AR (1973) Pesticides and the soil fauna. Residue Rev 45:1–79

Eriksson KEL, Blancbette RA, Ander P (1990) Biodegradation of cellulose. In: Eriksson KEL, Blanchette RA, Ander P (eds) Microbial and enzymatic degradation of wood and wood components. Springer, New York, pp 89–180

Ernst G, Felten D, Vohland M, Emmerling C (2009) Impact of ecologically different earthworm species on soil water characteristics. Eur J Soil Biol 45:207–213

Esen A (1993) b-glucosidases: overview. In: Esen A (ed) b-glucosidases and molecular biology. American Chemical Society, Washington, DC, pp 9–17

Franzluebbers AJ (2002) Water infiltration and soil structure related to organic matter and its stratification with depth. Soil Till Res 66:197–205

Franzluebbers AJ, Haney RL, Honeycutt CW, Arshad MA, Schomberg HH, Hons FM (2001) Climatic influences on active fractions of soil organic matter. Soil Biol Biochem 33:1103–1111

Ganeshamurthy AM, Singh G, Singh NT (1995) Sulphur status and response of rice to sulphur on some soils of Andaman and Nicobar Islands. J Indian Soc Soil Sci 43:637–641

Giller KE, Witter E, McGrath SP (1998) Toxicity of heavy metals to microorganisms and microbial processes in agricultural soils: a review. Soil Biol Biochem 30:1389–1414

Gu Y, Wag P, Kong C (2009) Urease, invertase, dehydrogenase and polyphenoloxidase activities in paddy soils influenced by allelopathic rice variety. Eur J Soil Biol 45:436–441

Johnson-Maynard JL, Umiker KJ, Guy SO (2007) Earthworm dynamics and soil physical proper ties in the first three years of no-till management. Soil Till Res 94:338–345

Kanfer JN, Mumford RA, Raghavan SS, Byrd J (1974) Purification of b-glucosidase activities from bovine spleen affinity chromatography. Anal Biochem 60:200–205

Kannan K, Oblisami G (1990) Influence of pulp and paper mill effluents on soil enzyme activities. Soil Biol Biochem 22:923–927

Karaca A, Cetin SC, Turgay OC, Kizilkaya R (2010) Effects of heavy metals on soil enzyme activities. In: Sherameti I, Varma A (eds) Soil heavy metals, Soil biology, vol 19. Springer, Berlin, pp 237–262

Kertesz MA, Mirleau P (2004) The role of soil microbes in plant sulphur nutrition. J Exp Bot 55:1939–1945

Kizilkaya R, Hepsen S (2004) Effect of biosolid amendment on enzyme activities in earthworm (*Lumbricus terrestris*) casts. J Plant Nutr Soil Sci 167:202–208

Kizilkaya R, Hepsen S (2007) Microbiological properties in earthworm *Lumbricus terrestris* L. cast and surrounding soil amended with various organic wastes. Commun Soil Sci Plan 38:2861–2876

Kizilkaya R, Askin T, Bayrakli B, Saglam M (2004) Microbiological characteristics of soils contaminated with heavy metals. Eur J Soil Biol 40:95–102

Kizilkaya R, Hepsen S, Akca I˙, Bayrakli B, Askin T, Turkmen C (2009) Determination of total and mobile Pb fractions during vermicomposting in sewage sludge. International symposium on environment. 20–23 May 2009. Kyrgyzstan – Turkey Manas University, Faculty of Engineering, Bishkek, Kyrgyz Republic

Knight BP, McGrath MJ, Doran JW, Cline RG, Harris RF, Schuman GE (1997) Biomass carbon measurements and substrate utilization patterns of microbial populations from soils amended with cadmium, copper, or zinc. Appl Environ Microbiol 63:39–43

Kung K-JS, Steenhuis TS, Kladivko EJ, Gish TJ, Bubenzer G, Helling CS (2000) Impact of preferential flow on the transport of adsorbing and non-adsorbing tracers. Soil Sci Soc Am J 64:1290–1296

Ladd JN, Butler JHA (1972) Short-term assays of soil proteolytic enzyme activities using proteins and dipeptide derivatives as substrates. Soil Biol Biochem 4:19–30

Ladd JN, Jackson RB (1982) In: Stevenson FJ (ed) Nitrogen in agricultural soils. American Society of Agronomy, WI, pp 173–228

Lal R (1978) Influence of within- and between-row mulching on soil temperature, soil moisture, root development and yield of maize (*Zea mays* L.) in a tropical soil. Field Crops Res 1:127–139

Lal R (1991) Soil structure and sustainability. J Sustain Agric 1:67–92

Landgraf MD, Alves MR, Da Silva SC, Rezende MOD (1999) Characterization of humic acids from vermicompost of cattle manure composting for 3 and 6 months. Quim Nova 22:483–486

Lavelle P (1997) Faunal activities and soil processes: adaptive strategies that determine ecosystem function. Adv Ecol Res 27:93–132

Le Bayon RC, Binet F (2006) Earthworms change the distribution and availability of phosphorous in organic substrates. Soil Biol Biochem 38:235–246

Lee KE (1985) Earthworms, their ecology and relationships with soils and land use. Academic, Orlando, p 356

Leroy BLM, Van den Bossche A, Neve SD, Reheul D, Moens M (2007) The quality of exogenous organic matter: short-term influence on earthworm abundance. Eur J Soil Biol 43:S196–S200

Makoi JHJR, Ndakidemi PA (2008) Selected soil enzymes: examples of their potential roles in the ecosystem. Afr J Biotechnol 7:181–191

Mandic L, Dukic D, Govedarica M, Stamenkovic S (1997) The effect of some insecticides on the number of amylolytic microorganisms and azotobacter in apple nursery soil. Czeckoslevensko Vocarstvo 31:177–184

Martinez CE, Tabatabai MA (1997) Decomposition of biotechnology by-products in soils. J Environ Qual 26:625–632

Masciandaro G, Ceccanti B, Garcia C (1999) Soil agro-ecological management: fert irrigation and vermicompost treatments. Bioresour Technol 59:199–206

McGill WB, Colle CV (1981) Comparative aspects of cycling of organic C, N, S and P through soil organic matter. Geoderma 26:267–286

McKenzie BM, Dexter AR (1987) Physical properties of casts of the earthworms *Aporrectodea rosea*. Biol Fertil Soils 5:152–157
McLaren AD (1975) Soil as a system of humus and clay immobilised enzymes. Chem Scr 8:97–99
Miwa T, Ceng CT, Fujisaki M, Toishi A (1937) Zur Frage der Spezifitat der Glykosidasen. I. Verhalted vonb-d-glucosidases verschiedener Herkunft gegenuberdenb-d-Glucosiden mit verschiedenen Aglykonen. Acta Phytochim 10:155–170
Mobley HLT, Hausinger RP (1989) Microbial urease: significance, regulation and molecular characterization. Microbiol Rev 53:85–108
Moeskops B, Buchan D, Sleutel S, Herawaty L, Husen E, Saraswati R, Setyorini D, De Neve S (2010) Soil microbial communities and activities under intensive organic and conventional vegetable farming in West Java, Indonesia. Appl Soil Ecol 45:112–120
Nannipieri P, Sequi P, Fusi P (1996) Humus and enzyme activity. In: Piccolo A (ed) Humic substances in terrestrial ecosystems. Elsevier, New York, pp 293–328
Narasimha G (1997) Effect of effluent of cotton ginning industry soil microbial activities. M. Phil. Dissertation, Sri Krishnadevaraya University, Anantapur, India
Ndiaye EL, Sandeno JM, McGrath D, Dick RP (2000) Integrative biological indicators for detecting change in soil quality. Am J Alternative Agr 15:26–36
Niemi RM, Heiskanen I, Ahtiainen JH, Rahkonen A, Mantykoski K, Welling L, Laitinen P, Ruuttunen P (2009) Microbial toxicity and impacts on soil enzyme activities of pesticides used in potato cultivation. Appl Soil Ecol 41:293–304
Pardo A, Amato M, Chiaranda FQ (2000) Relationships between soil structure, root distribution and water uptake of chickpea (*Cicer arietinum* L). Plant growth and water distribution. Eur J Agron 13:39–45
Piccolo S, Agius E, Lu B, Goodman S, Dale L, De Robertis EM (1997) Cleavage of Chordin by the Xolloid metalloprotease suggests a role for proteolytic processing in the regulation of Spemann organizer activity. Cell 91:407–416
Polacco JC (1977) Is nickel a universal component of plant ureases? Plant Sci Lett 10:249–255
Richmond PA (1991) Occurrence and functions of native cellulose. In: Haigler CH, Weimer PJ (eds) Biosynthesis and biodegradation of cellulose. Marcel Dekker, New York, pp 5–23
Rillig MC, Steinberg PD (2002) Glomalin production by an arbuscular mycorrhizal fungus: a mechanism of habitat modification? Soil Biol Biochem 34:1371–1374
Ros M, Pascual JA, Garcia C, Hernandez MT, Insam H (2006) Hydrolase activities, microbial biomass and bacterial community in a soil after long-term amendment with different composts. Soil Biol Biochem 38:3443–3452
Ross DJ (1975) Studies on a climosequence of soils in tussock grasslands-5. Invertase and amylase activities of topsoils and their relationships with other properties. New Zealand J Sci 18:511–518
Ross DJ (1976) Invertase and amylase activities in ryegrass and white clover plants and their relationships with activities in soils under pasture. Soil Biol Biochem 8:351–356
Ross DJ, Roberts HS (1973) Biochemical activities in soil profile under hard beech forest. Invertase, and amylase activities and relationships, with other properties. New Zealand J Sci 16:209–224
Rotini OT (1935) La trasformazione enzimatica dell'urea nel terreno. Ann Labor Ric Ferm Spallanrani 3:143–154
Salazar S, Sanchez L, Alvarez J, Valverde A, Galindo P, Igual J, Peix A, Santa Regina I (2011) Correlation among soil enzyme activities under different forest system management practices. Ecol Eng 37:1123–1131
Sarathchandra SU, Perrott KW (1981) Determination of phosphatase and arylsulphatase activity in soils. Soil Biol Biochem 13:543–545
Schack-Kirchner H, Wilpert KV, Hildebrand EE (2000) The spatial distribution of soil hyphae in structured spruce-forest soils. Plant Soil 224:195–205
Shipitalo MJ, Protz R (1988) Factors influencing the dispersibility of clay in worm casts. Soil Sci Soc Am J 52:764–769
Shipitalo MJ, Protz R (1989) Chemistry and micromorphology of aggregation in earthworm casts. Geoderma 45:357–374

Shlomo P, Fanya I, Ilan C (1995) Control of Rhizoctonia solani and Sclerotium rolfsii in the greenhouse using endophytic Bacillus spp. Eur J Plant Pathol 101:665–672
Simpson JR, Freney JR (1988) Interacting processes in gaseous nitrogen loss from urea applied to flooded rice fields. In: Pushparajah E, Husin A, Bachik AT (eds) Proceedings of international symposium on urea technology and utilization. Malaysian Society of Soil Science, Kuala Lumpur, pp 281–290
Singaram P, Kamalakumari K (2000) Effect of continuous application of different levels of fertilizers and farm yard manure on enzyme dynamics of soil. Mad Agric J 87:364–365
Singer MJ, Southard RJ, Warrington DJ, Janitzky P (1992) Stability of synthetic sand clay aggregates after wetting and drying cycles. Soil Sci Soc Am J 56:1843–1848
Singh PP, Shin YC, Park CS, Chung YR (1999) Biological control of Fusarium wilt of cucumber by chitinolytic bacteria. Phytopathology 89:92–99
Singh BK, Allan W, Denis JW (2002) Degradation of chlorpyrifos, fenamiphos, and chlorothalonil alone and in combination and their effects on soil microbial activity. Environ Toxicol Chem 21:2600–2605
Sinsabaugh RL, Linkins AE (1987) Inhibition of the *Trichoderma viridae* cellulase complex by leaf litter extracts. Soil Biol Biochem 19:719–725
Sinsabaugh RL, Antibus RK, Linkins AE (1991) An enzymic approach to the analysis of microbial activity during plant litter decomposition. Agric Ecosyst Environ 34:43–54
Speir TW, Ross DJ (1978) Soil phosphatase and sulphatase. In: Burns RG (ed) Soil enzymes. Academic, London, pp 197–250
Srinivasulu M, Rangaswamy V (2006) Activities of invertase and cellulase as influenced by the application of tridemorph and captan to groundnut (*Arachis hypogaea*) soil. Afr J Biotechnol 5:175–180
Subhani A, Changyong H, Zhengmiao Y, Min L, El-ghamry A (2001) Impact of soil environment and agronomic practices on microbial/dehydrogenase enzyme activity in soil: a review. Pak J Biol Sci 4:333–338
Swift RS (2001) Sequestration of carbon by soil. Soil Sci 166:858–871
Tabatabai MA (1994a) Soil enzymes. In: Weaver RW, Angle JS, Bottomley PS (eds) Methods of soil analysis, part 2. Microbiological and biochemical properties, SSSA Book Series No. 5. Soil Science Society of America, Madison, WI, pp 775–833
Tabatabai MA (1994b) Soil enzymes. In: Mickelson SH (ed) Methods of soil analysis, Part 2. Microbiological and biochemical properties. Soil Science Society of America, Madison, WI, pp 775–833
Tam NFY, Wong YS (1990) Respiration studies on the decomposition of organic waste-amended colliery spoil. Agric Ecosyst Environ 32:25–38
Tarrant KA, Field SA, Langton SD, Hart ADM (1997) Effects on earthworm populations of reducing pesticide use in arable crop rotations. Soil Biol Biochem 29:657–661
Thoma JA, Spradlin JE, Dygert S (1971) Plant and animal amylases. In: Boyer PD (ed) The enzymes, 5th edn. Academic, New York, pp 115–189
Tu CM (1982) Influence of pesticides on activities of amylase, invertase and level of adenosine triphosphate inorganic soil. Chemosphere 2:909–914
Tu CM (1995) Effect of five insecticides on microbial and enzyme activities in sandy soil. J Environ Sci Health 30:289–306
Tu CM, Miles JRW (1976) Interactions between insecticides and soil microbes. Res Rev 64:17–65
Tyler G (1981) Heavy metals in soil biology and biochemistry. In: Paul EA, Ladd JN (eds) Soil biochemistry, vol 5. Marcel Dekker, New York, pp 371–414
Versaw WK, Harrison MJ (2002) A chloroplast phosphate transporter, PHT2; 1, influences allocation of phosphate within the plant and phosphate-starvation responses. Plant Cell 14:1751–1766
Vinceslas-Akpa M, Loquet M (1997) Organic matter transformations in lignocellulosic waste products composted or vermicomposted (*Eisenia fetida* Andrei): chemical analysis and 13C CPMAS NMR spectroscopy. Soil Biol Biochem 29:751–758
Whalley WR, Dexter AR (1994) Root development and earthworm movement in relation to soil strength and structure. Arch Agron Soil Sci 38:1–40

White AR (1982) Visualization of cellulases and cellulose degradation. In: Brown RM (ed) Cellulose and other natural polymer systems: biogenesis, structure, and degradation. Plenum, New York, pp 489–509

Yang Z, Liu S, Zheng D, Feng S (2006) Effects of cadmium, zinc and lead on soil enzyme activities. J Environ Sci 18:1135–1141

Yuan B, Yue D (2012) Soil microbial and enzymatic activities across a chronosequence of Chinese pine plantation development on the loess plateau of China. Pedosphere 22:1–12

Zantua MI, Bremner JM (1977) Stability of urease in soils. Soil Biol Biochem 9:135–140

Zhang BG, Li GT, Shen TS, Wang JK, Sun Z (2000) Changes in microbial biomass C, N, and P and enzyme activities in soil incubated with the earthworms *Metaphire guillelmi* or *Eisenia fetida*. Soil Biol Biochem 32:2055–2062

Chapter 6
Survival of Pathogenic and Antibiotic-Resistant Bacteria in Vermicompost, Sewage Sludge, and Other Types of Composts in Temperate Climate Conditions

Lelde Grantina-Ievina and Ieva Rodze

Abstract A great variety of organic waste materials are common in Europe and northern temperate climate conditions, depending from waste sources, agricultural practices, characteristics of national industries, waste treatment practices, and economic circumstances. Waste can contain various harmful and undesirable substances, such as pathogenic organisms and antibiotic residues; the application of waste to the soil as fertilizer can spread antibiotic resistance genes into the environment. Pathogenic organisms that are dangerous to human health include verotoxin-producing *Escherichia coli*, *Salmonella enterica*, *Enterococcus*, *Clostridium perfringens*, *Listeria monocytogenes*, *Campylobacter coli*, *C. jejuni*, and *Neisseria meningitidis*. It has proven to be difficult to reduce the numbers of spore-forming bacteria (*Clostridium* spp. and *Bacillus* spp.) and antibiotic resistance genes, especially in the anaerobic digestion of manure. This review aims to provide an overview of the most common waste sources and waste treatment methodologies suitable for Northern temperate climate conditions in the presence of pathogenic bacteria and antibiotic resistance genes. Vermicomposting with black soldier fly *Hermetia illucens*, thermal hydrolysis, thermophilic anaerobic digestion, wet oxidation, pyrolysis and roto-autoclaving of the waste, and drying and pelleting of the digestate have been identified as novel approaches in waste treatment.

Keywords Municipal waste · Sludge · Wood waste · Potato pulp · Human and animal faeces · Animal waste · Pathogenic bacteria

L. Grantina-Ievina (✉) · I. Rodze
Animal Disease Diagnostic Laboratory, Institute of Food Safety, Animal Health and Environment "BIOR", Riga, Latvia
e-mail: lelde.grantina-ievina@bior.lv; ieva.rodze@bior.lv

M. K. Meghvansi, A. Varma (eds.), *Biology of Composts*, Soil Biology 58,
https://doi.org/10.1007/978-3-030-39173-7_6

6.1 Introduction

Waste types can be specific to particular climatic conditions due to the waste sources, agricultural practices, characteristics of national industries, waste treatment practices, and economic circumstances. In temperate climate conditions, particular waste types, such as animal manure, have to be collected and stored during winter until land application is possible. In Nordic countries and other European countries, the anaerobic digestion of animal manure for biogas production is very common and has great potential to further increase in the future.

Waste contains various harmful and undesirable substances, such as pathogenic organisms and antibiotic residues, and can spread antibiotic resistance genes in the environment. Pathogenic organisms that are dangerous to human health include verotoxin-producing *Escherichia coli*, *Salmonella enterica*, *Enterococcus*, *Clostridium perfringens*, *Listeria monocytogenes*, *Campylobacter coli*, *C. jejuni*, and *Neisseria meningitidis*. A literature search found comprehensive information on the allowable limits of chemical pollutants in various waste types; however, microbiological quality characteristics still have to be standardized among European Union countries.

Composting, vermicomposting by earthworms, anaerobic digestion, or a combination of these methods have been shown to reduce (fully or at least partially) the levels of undesired organisms or substances in the waste. It has proven to be difficult to reduce the numbers of spore-forming bacteria (*Clostridium* spp. and *Bacillus* spp.) and antibiotic resistance genes, especially in the anaerobic digestion of manure. Drying of the digestate and pelleting may be promising solutions to the problem of spore-forming bacteria, but composting has shown good results in reducing the presence of antibiotic resistance genes. In several investigations in Asia, vermicomposting by black soldier fly *Hermetia illucens* can significantly reduce the abundance of antibiotic resistance genes and pathogenic bacteria; however, this composting method still has to be investigated in temperate climate conditions in order to evaluate its suitability for this purpose. Thermal hydrolysis, thermophilic anaerobic digestion, wet oxidation, pyrolysis, and roto-autoclaving have been considered as other novel approaches in waste treatment.

This review aims to provide an overview of the most common waste sources and waste treatment methodologies that are suitable for northern temperate climate conditions in the presence of pathogenic bacteria and antibiotic resistance genes.

6.2 Main Types of Organic Waste and Compost in Northern Temperate Climate Conditions

A great variety of organic waste materials are common in Europe and northern temperate climate conditions: organic fractions of municipal solid waste or organic waste from households, sludge from wastewater treatment plants, sawdust, tree bark, wood chips, wood ash, tree leaves, pulp and paper mill sludge, agricultural waste, hay, grass (lawn cuttings), garden waste, potato pulp from starch production, human and animal feces, slaughterhouse waste, brewery waste, molasses, substrates from mushroom cultivation, aquaculture sludge, other industrial organic waste products, shredded newspapers, and other waste paper, among others (Ryckeboer et al. 2003; Grantina-Ievina et al. 2013; Insam et al. 2014; Fernández-Delgado Juárez et al. 2015; Martinez-Sanchez et al. 2015; Nielfa et al. 2015; Faubert et al. 2016; Nigussie et al. 2016; Andreev et al. 2017; Bloem et al. 2017; Franke-Whittle et al. 2018; Kouba et al. 2018). The climate of Nordic countries can be characterized by fluctuating temperatures among seasons and seasonal variations in waste composition (Sundberg et al. 2011). Particular waste types, such as animal manure, has to be collected and stored during winter because application to arable land is limited for part of the year (Ruuskanen et al. 2016). In a large study about locally available composts in Austria, Germany, Italy, and Switzerland, it was shown that microbial populations in the composts were more affected by the geographical origin than by the waste type (Franke-Whittle et al. 2018).

Organic waste can be treated by composting, vermicomposting, anaerobic digestion, or a combination of these methods. For example, sewage sludge can be used for vermicomposting in its fresh form or after conventional composting (Villar et al. 2016). An emerging technology is the use of black soldier fly *Hermetia illucens* for waste treatment (Lohri et al. 2017; Salomone et al. 2017; Lalander et al. 2019). Organic solid waste can be anaerobically digested for biogas production (Ariunbaatar et al. 2014; Cesaro and Belgiorno 2014; Insam et al. 2014; Stoknes et al. 2016) or used for vermicomposting. In the biogas production process, a digester residue is formed as a side product that can be directly used as a soil fertilizer (Stoknes et al. 2016), but its quality aspects are important (Coelho et al. 2018). Biogas production is common, for example, in Latvia, where currently there are 60 biogas production plants that use wastewater sludge (5%), municipal waste from landfills (10%), and manure and agricultural sludge (85%) (Kalnina et al. 2018). The theoretical potential of a small European Union (EU) country, such as Latvia, is to have as many as 122 biogas plants. For other Northern European countries, the estimated numbers of biogas plants could be as follows: 66 in Estonia, 160 in Finland, 212 in Lithuania, 128 in Norway, and 345 in Sweden (Scarlat et al. 2018). However, direct land application or land spreading of animal manure or food waste is still common in many countries as well as in the EU (Lohri et al. 2017).

Composting is the biodegradation of solid phase organic materials in a controlled self-heating, aerobic process consisting of mesophilic and thermophilic phases (Ryckeboer et al. 2003). The traditional composting process can be divided into four phases:

1. An initial phase or first mesophilic phase (10–42 °C), which may last for only a few hours or a couple of days.
2. A thermophilic phase (45–70 °C), lasting a few days up to several weeks (particularly for food waste) or even months (particularly for wood waste). The optimal temperature for most thermophilic microorganisms is 60 °C, which is useful for the elimination of pathogenic organisms; regular aeration or frequent turning of the material can prolong this phase.

3. A second mesophilic phase during which mesophilic microorganisms, which are often dissimilar from those of the first mesophilic phase, recolonize the substrates.
4. A maturation and stabilization phase (curing), which can last for several weeks to several months. In this phase, lignin-humus complexes are formed (Tuomela et al. 2000; Mondini and Insam 2003; Ryckeboer et al. 2003).

Other authors distinguish only two phases: the thermophilic and maturation phases (Lazcano et al. 2008).

Vermicomposting is the microbial decomposition of organic matter fractions within the gut of the earthworm (Flack and Hartenstein 1984). The earthworm cast is biologically stabile, homogenous material with lower levels of contaminants and smaller volume than the starting material (Mitchell 1997; Ndegwa et al. 2000). The duration of vermicomposting was reported to be as short as 1 month in northern climates with incubation temperatures of 22–26 °C to convert the initial material (sewage sludge, pine bark) to earthworm cast (Haimi and Huhta 1986), and up to 4.5 months using aquaculture sludge (Kouba et al. 2018). Vermicomposting consists of two stages: an active stage and maturation. In the active phase, earthworms process the substrate; however, in the last step, earthworms migrate away from the processed material towards unprocessed layers. Vermicomposting always occurs in mesophilic temperatures (Lazcano et al. 2008). Various materials can be added to the basic substrate for vermicomposting. For example, biochar from willow woodchips or sewage sludge-derived biochar can be added to the composting material (sewage sludge mixed with straw) in order to increase the production rates of vermicomposting (Malińska et al. 2016 2017).

New vermicomposting systems are becoming common in certain European countries, such as France, where urine-diverting vermicomposting toilets are used (Hill and Baldwin 2012; Lalander et al. 2013). Vermicompost produced in this system can be used as a soil amendment; however, in the case of crops that are consumed without thermal treatment, an additional sanitation step is needed (Lalander et al. 2013).

Anaerobic digestion is "a biological process that converts complex substrates into biogas and digestate by microbial action in the absence of oxygen through four main steps, namely hydrolysis, acidogenesis, acetogenesis and methanogenesis" (Ariunbaatar et al. 2014). According to the EU Regulation EC1772/2002, municipal solid waste, food waste, and slaughterhouse waste need to be pasteurized or sterilized before and/or after anaerobic digestion. Instead of pasteurization and sterilization, various pretreatment methods can be used, including mechanical, thermal, chemical, and biological, or a combination of several pretreatment methods (Ariunbaatar et al. 2014; Jain et al. 2015).

Mechanical substrate treatment is the disintegration and grinding of material in order to reduce the particle size and increase the surface between the contact of bacteria and substrate (Ariunbaatar et al. 2014). Mechanical treatment can be performed using various methods, including sonication by vibrating probes, lysis centrifuge, liquid shear, collision, high-pressure homogenizer, maceration, liquefaction, electroporation, and ultrasonic pretreatment with low-frequency sound waves (Jain et al. 2015).

It has been proven that mesophilic temperatures (35 °C) cannot eliminate pathogenic indicator bacteria (*E. coli* and *Salmonella* spp.). Significantly better results can be achieved with thermophilic digestion at temperatures of 55 °C and higher (Smith et al. 2005). Temperatures for thermal treatment range from 50 to 250 °C; this treatment can eliminate pathogens and reduce the water content of solid wastes. At temperatures higher than 150 °C or 170 °C, various compounds can be formed that cannot be used by microorganisms, including products of Maillard reactions, which occur between carbohydrates and amino acids and form difficult-to-degrade complex substrates, or components that are inhibitory to the microorganisms, such as phenolic compounds or melanoidins. Pretreatment at 70 °C for 60 min is sufficient for pathogen elimination and is also required according to EC1772/2002 (Ariunbaatar et al. 2014). Low-temperature thermal pre-treatment (<100 °C) has proven to be effective for pathogen inactivation in the anaerobic digestion of poultry sludge in small experimental reactors (Ruiz-Espinoza et al. 2012). Thermal treatments can be performed by steam, electric, or microwave heating (Jain et al. 2015; Gil et al. 2018).

In the chemical pretreatment, the destruction of organic compounds is achieved by the use of strong acids, alkalis, and oxidants. One of the chemical pretreatment methods is ozonation, which makes the substrate more accessible to microorganisms and also reduces the number of pathogens (Ariunbaatar et al. 2014). An earlier study reported that ozonation (applying 20 mg O_3 g^{-1} for 2 h) was not as effective for pathogen inactivation as was alkaline treatment (by adding lime up to pH 12 for 24 h) (Carballa et al. 2009).

Biological pretreatment can be anaerobic or aerobic, and some enzymes can also be used, such as peptidase, carbohydrolase, and lipase. Conventional biological pretreatment is composting or microaeration. Conversely to conventional biological pretreatment, a two-stage anaerobic digestion system consists of a hydrolytic-acidogenic stage followed by a methanogenic stage, which is suitable for pathogen removal. A variation of this method, temperature-phased anaerobic digestion, consists of primary digestion at thermophilic (55 °C) or hyper-thermophilic (80 °C) temperatures followed by mesophilic secondary digestion. The product of this method is a pathogen-free digestate that can be used as a soil amendment (Ariunbaatar et al. 2014). Several types of waste in northern temperate climate conditions contain recalcitrant substances such as lignocellulose, cellulose, hemicellulose, and lignin. For the pretreatment of these substances, active microorganisms (bacteria, fungi, and various consortia) can be used (Wagner et al. 2018). Methods such as thermo-chemical pretreatment, thermo-mechanical pretreatment, physico-chemical pretreatment, or various other pretreatments can be combined for a two-stage anaerobic digestion (Tyagi and Lo 2011; Ariunbaatar et al. 2014).

Composts can be used not only as fertilizers to improve soil quality and the growth of agricultural crops (Gutiérrez-Miceli et al. 2007), but also, as in the case of vermicompost, as animal feed (Spencer and Garcia 1995; Spencer et al. 1998). In addition, earthworm biomass can be used as animal feed, such as in aquaculture or as feed for other livestock (Atlason et al. 2017; Kouba et al. 2018). Therefore, their quality aspects are important.

6.3 Prevalence of Pathogenic Bacteria in Sewage Sludge and Biowaste of Biogas Plants Used as Raw Materials for Further Composting or Directly as Soil Amendments

Various pathogenic bacteria are present in the waste that is used for composting. In a research study of mainly incoming food wastes at composting plants in Sweden, Finland, and Norway, the presence of pathogenic bacteria was demonstrated by clone library sequencing and COMPOCHIP microarray analysis. The following pathogenic bacteria were found to be common in waste samples: *Escherichia coli*, *Salmonella enterica*, *Enterococcus*, *Clostridium*, *Listeria*, and *Neisseria meningitidis* (Sundberg et al. 2011). In a large study in Finland, sewage sludge from municipal wastewater plants was composted by various methods; indicator bacteria such as *Clostridium perfringens*, *E. coli*, and enterococci were detected during composting, as well as *C. perfringens* in the end-product, when quicklime or peat addition was used for sanitation purposes or after mesophilic anaerobic digestion (Rimhanen-Finne et al. 2004). When composts are amended to arable land, pathogenic bacteria can survive in the soil for a period of days to several months (Bloem et al. 2017).

Verotoxin-producing *Escherichia coli* (VTEC) can cause diarrhea, hemorrhagic colitis, and hemolytic uremic syndrome in humans. Ruminants are considered to be natural reservoirs for these bacteria, and ruminants and other farm animals can excrete large amounts of VTEC in their feces (Vernozy-Rozand et al. 2002). Dairy cattle feces, slurry, and manure often contain these bacteria: the prevalence was shown to range from 8 to 11%, or up to 17% in pig feces, in France. In the various steps of 20 municipal wastewater treatment plants in France and 10 waste storage lagoons, VTEC genes were detected by polymerase chain reaction even more often, in up to 80% of samples (Vernozy-Rozand et al. 2002). Genes specific to pathogenic *E. coli* have been detected in various samples of top-soil improvers in Italy: municipal sewage sludges, pig manure, green waste, and household waste (Tozzoli et al. 2017). To reduce the numbers of pathogenic bacteria, sewage sludge can be treated in various ways; for example, it can undergo thermophilic anaerobic digestion, mesophilic anaerobic digestion, composting, or sedimentation. In Swedish wastewater treatment plants, all of these measures reduced the numbers of colony-forming units of various pathogenic bacteria, such as *Salmonella* spp., *Listeria monocytogenes*, *Campylobacter coli*, and *C. jejuni* and VTEC serotype O157, but they still had high prevalence rates at the end of treatment. For example, *Salmonella* spp. were found in 55% of the treated sludge samples; however, the prevalence of other pathogenic bacteria was reduced by several fold (Sahlström et al. 2004).

Animal manure and sewage sludge are often used as fertilizer on fields. In an artificial inoculation study in the United Kingdom, it was determined that *E. coli* O157, *Salmonella* (serotypes Typhimurium and Enteritidis), and *Campylobacter jejuni* survived in stored dairy cattle slurries for up to 3 months at ambient temperatures. *Listeria monocytogenes* had survival times up to half a year in these conditions. The shorter survival times of the pathogenic bacteria were observed in mixed and undisturbed solid manure heaps where temperatures greater than 55 °C were reached. When later this manure was applied on arable lands, it was detected that *L. monocytogenes* survived for at least 1 month after application, but the survival times of other pathogenic bacteria were much shorter (Nicholson et al. 2005). Similar results have been obtained in other field studies in the United Kingdom with organic wastes artificially inoculated with *E. coli* O157:H7 originating from slaughterhouses, wastewater treatment plants (raw and treated sewage), creameries, and farms (bovine slurry) intended for land application (Avery et al. 2005).

Biogas production is becoming increasingly common in Europe. For biogas production, animal manure can be used in mixtures with municipal waste. As required by European Economic Community regulation 1774/2002, pre-pasteurization at 70 °C for 60 min is performed in the biogas production process. The numbers of indicator bacteria (*Escherichia coli, Enterococcus* spp., and coliforms) and bacterial pathogens (*Salmonella, Listeria, Campylobacter,* and VTEC O157) were shown to be successfully reduced during pasteurization, but the numbers of spore-forming bacteria (*Clostridium* spp. and *Bacillus* spp.) did not decrease. Unfortunately, regrowth of indicator and pathogenic bacteria was observed in later steps of biogas production after pasteurization (Bagge et al. 2005). *C. perfringens* and *E. faecalis* are more heat resistant than other bacteria; therefore, they can be used as indicator bacteria in waste treatment by anaerobic digestion (Watcharasukarn et al. 2009).

As possible solutions for waste treatment to reduce the presence of pathogenic bacteria, the following methods were suggested in an earlier study: composting, anaerobic digestion, and ammonia treatment (Albihn and Vinnerås 2007). Animal manure can be treated by conventional composting in a heap due to naturally increased temperatures and pH levels. Composting efficacy in terms of pathogen elimination depends on the aeration and regular mixing of various compost layers as well as the manure type (Heinonen-Tanski et al. 2006). An investigation in Finland determined that vegetables fertilized with composted horse manure did not contain enteric microorganisms, as they did with cattle manure compost. Cattle manure compost was safe only when the manure was 1 year old; however, the storage of manure in large amounts can be very expensive (Holopainen et al. 2002). In another study in Finland about various composting methods of municipal sewage sludge, it was found that methods including thermal drying and closed reactor composting processes sufficiently reduced the level of pathogenic bacteria (*Clostridium perfringens*, *E. coli*, and enterococci); however, the end-products of plants using mesophilic anaerobic digestion contained high levels of *C. perfringens*, and plants using sanitation by quicklime or peat addition were not able to reach adequate microbiological quality standards (Rimhanen-Finne et al. 2004). The addition of wood ash at 6 and 12% to municipal solid waste in windrow composting was able to reduce the presence of pathogenic bacteria (*Listeria*, *Clostridium*) (Fernández-Delgado Juárez et al. 2015). Successful results for *Clostridium* reduction in stored dry digestates (15% of water mass fraction) were obtained in Denmark by pelleting using a pellet press (Pulvirenti et al. 2015).

Precise criteria can be found in EU and national legislations regarding the allowed levels of heavy metals and organic micropollutants in sewage sludge before land spreading. However, very scarce information can be found about national standards for the maximum concentrations of pathogens in sewage sludge among various European countries (Mininni et al. 2015).

6.4 Presence of Pathogenic Bacteria in Vermicompost and the Role of Earthworms as Vectors

In the 1980s, vermicomposting has become a common composting method in Europe (Flack and Hartenstein 1984). Several studies have shown that vermicomposting significantly reduces the amount of pathogenic bacteria or eliminates them completely if vermicomposting is performed for a sufficient time period. For example, during 60-day vermicomposting of sheep manure, no pathogenic bacteria such as *Shigella*, *Salmonella*, or *Enterobacter* could be detected in comparison to the beginning of vermicomposting (Gutiérrez-Miceli et al. 2007). If the initial substrate is less hazardous than sheep manure, such as grass, for example, the elimination of pathogenic bacteria can be reached in 2.5 months, especially when 9 days of thermocomposting has been carried out prior to vermicomposting (Nair et al. 2006). Vermicomposting of at least 2 months has been shown to be sufficient time to reduce pathogenic bacteria in other investigations (Karimi et al. 2017). In comparison, 1-month vermicomposting of human feces mixed with kitchen waste, cattle manure, and molasses in Moldova was appropriate to reduce the number of colony-forming units (CFU) of *Escherichia coli*, *Enterococcus faecalis*, and *Clostridium perfringens* to below the detection limit; however, it was not sufficient to reduce the CFU numbers of coliform bacteria (Andreev et al. 2017).

Domestic waste vermicomposting can be carried out continuously with the regular addition of fresh organic waste. In a 2-year study in Canada, none of the regulated bacterial pathogens were found to be present in vermicomposting bins used for this composting method (Hénault-Ethier et al. 2016). In a similar study in the United States over a 100-day time period, a reduction of *E. coli* numbers was not achieved, likely because turning and thereby aeration of the composted material (fruits and vegetables with shredded paper) was not carried out (Cao et al. 2016).

Several earthworm species have been used in vermicomposting. It has been shown that the earthworm species can alter the composition of microbial populations in the vermicompost if the starting material is the same (Gomez-Brandon et al. 2012). In an artificial inoculation study in North Wales (UK), it was shown that anecic earthworms *Lumbricus terrestris* significantly aided the vertical movement of *E. coli* O157 in soil, whereas epigeic earthworms such as *Dendrobaena veneta* significantly aided the horizontal distribution within compost in laboratory conditions in a 21-day experiment. However, the numbers of *E. coli* O157 were significantly reduced (Williams et al. 2006).

If the starting waste material contains low numbers of pathogenic bacteria, then vermicomposting seems to be a safe waste treatment method. However, if the starting material contains high numbers of pathogenic bacteria, as in the sewage sludge, then quality control measures have to be undertaken in order to assure the maturity and quality of the vermicompost (Grantina-Ievina et al. 2013).

Vermicomposting has been used as a waste treatment and recycling method in a novel "food to waste to food" system that has been carried out in a joint investigation by Norwegian, German and Polish scientists. Other methods used in this system included anaerobic digestion and composting. Various substrates were used in this system, such as source-separated food waste, garden waste, paper waste, wheat straw, and gypsum from house construction plasters. However, the microbiological quality of the digestates and composts used in the digeponics system were not assessed (Stoknes et al. 2016).

6.5 Presence of Antibiotics and Antibiotic-Resistant Bacteria in Various Composted Materials in Temperate Climate Conditions

Antibiotic resistance in microorganisms is a natural phenomenon that is not only caused by the human use of antibiotics and their spread into the environment by feces, manure, sewage sludge, and slaughterhouse wastewater, among others. Antibiotic resistance genes also can be found in biotopes that have never been influenced by humans, such as 30,000-year-old Beringian permafrost sediments (D'Costa et al. 2011) and other natural habitats (Martínez 2008; Davies and Davies 2010). A mixture of naturally occurring antibiotic resistance genes with those coming from medicine and veterinary use can result in new combinations of genes and mobile genetic elements (Wellington et al. 2013).

The consumption rates of antibiotics in Europe vary significantly among different countries. The countries with the highest rates of antibiotic consumption by food-producing animals include Spain, Cyprus, and Italy (Table 6.1). In countries with temperate and northern climates (e.g., Estonia, Latvia, Lithuania, Finland, Sweden, Norway, Iceland), the consumption rates of antimicrobials in animals are significantly lower. Statistically significant associations at the European level have been shown between antimicrobial consumption (AMC) rates and antimicrobial resistance (AMR) in humans and animals—for example, between AMC and AMR for fluoroquinolones in *Escherichia coli* (ECDC/EFSA/EMA 2017).

Table 6.1 Consumption of antibiotics by humans and food-producing animals in 2014 (ECDC/EFSA/EMA 2017)

Country	Consumption (mg/kg)		Country	Consumption (mg/kg)	
	Animals	Humans		Animals	Humans
Spain	418.8	112.6	Slovakia	65.9	140.2
Cyprus	391.5	124.7	United Kingdom	62.1	128.7
Italy	359.9	166.9	Austria	56.3	70.9
Portugal	201.6	116.1	Ireland	48.0	155.6
Hungary	193.1	86.6	Denmark	44.2	143.5
Belgium	158.3	153.4	Luxembourg	40.9	130.2
Germany	149.3	56.9	Romania	39.1	181.7
Poland	140.8	110.7	Latvia	36.7	81.6
Croatia	114.8	128.4	Lithuania	35.5	102.5
France	107.0	174.2	Slovenia	33.4	105.5
Bulgaria	82.9	116.0	Finland	22.3	139.2
Czech Republic	79.5	99.4	Sweden	11.5	119.8
Estonia	77.1	71.7	Iceland	5.2	101.7
Netherlands	68.4	49.9	Norway	3.1	140.1

The amounts of tetracyclines and sulfonamides in pig manure in Germany have been found in maximal concentrations: 53 mg/kg and 38 mg/kg, respectively (Hölzel et al. 2010). The rates for chlortetracycline and sulfadimidine were as high as 46 mg/kg and 20 mg/kg, respectively, in Austria (Martinez-Carballo et al. 2007). Antibiotic compounds also have been detected in soils fertilized with pig manure in the United States, Denmark, and Germany (for a review, see Heuer et al. 2011). A significant increase of clinically relevant antibiotic resistance genes have been found in soil fertilized with diary manure in the United States over 3 years (McKinney et al. 2018). In a study in Finland, where antibiotic use in livestock production is restricted, antibiotic resistance genes and mobile genetic elements were present in fresh cattle and pig manure, and were even increased in stored manure; however, after 6 weeks, several of them were not detectable in the arable soil after manure amendment (Muurinen et al. 2017). In a more targeted study, it was proven that antibiotic use (a standard 3-day course of tetracycline) significantly changed the microbial populations of cow manure (Hammer et al. 2016). In a study in Estonia, land application of cattle slurry and cattle slurry digestate significantly increased the level of antibiotic resistance genes in the soil after three applications during 1 year. Antibiotic resistance genes *SulI*, *intI1*, and *intI2* did not decrease to background levels, even 2 months after the last application (Nõlvak et al. 2016).

Composting reduces more than half of some antibiotics (Kim et al. 2012; Pruden et al. 2013). Antibiotic breakdown is suspected to occur in the thermophilic phase during the first 2 weeks, and its efficiency depends on both the duration of the composting and the temperature (Pruden et al. 2013), as well as on composting methods. For example, the addition of sawdust can increase the reduction of tetracycline and sulfonamide (Kim et al. 2012). In a pilot study in the United States, composting technologies such as watering, aeration, and turning of the compost offered advantages for the degradation of chlortetracycline, monensin, and tylosin; even simple storage of manure resulted in significant antibiotic elimination (Storteboom et al. 2007). The fermentation of livestock waste in biogas plants can effectively degrade antibiotic residues, such as sulfonamides and trimethoprim (Mohring et al. 2009), or fluoroquinolones (ciprofloxacin) in anaerobic digestion of dehydrated sludge (Bloem et al. 2017).

In a comparison of different dairy manure treatment methods (untreated, anaerobically digested, mechanically dewatered, and composted) and subsequent application of the treated manure to the soil, it was observed that fewer antibiotic-resistant bacteria and genes were detected when the waste was simply dewatered and composted. Dewatered composted manure contained significantly fewer CFUs of fecal coliforms, *E. coli*, and *Enterococcus* spp. The amounts of total coliforms and *C. perfringens* were not affected by the treatments. Anaerobic digestion significantly increased the presence of antibiotic resistance genes but composting led to a decrease (Tien et al. 2017). Further examples of the impact of composting and anaerobic digestion on the presence of antibiotics and antibiotic resistance genes are given in Table 6.2. It can be concluded that composting is the most appropriate method for this purpose, as it provided good results in several investigations. In comparison, the results for anaerobic digestion are less effective. These conclusions have been confirmed by experts in the field at the fourth International Symposium on the Environmental Dimension of Antibiotic Resistance (Topp et al. 2018).

In several investigations in Asia, vermicomposting by earthworms or black soldier fly *H. illucens* significantly reduced the abundance of antibiotic resistance genes (Cai et al. 2018; Cui et al. 2018; Huang et al. 2018) and pathogenic bacteria (Cai et al. 2018). This novel composting method that reduced the presence of antibiotic resistance genes has to be investigated in temperate climate conditions in order to evaluate its suitability for this purpose. The necessary quality control measures of produced larva and compost have already been suggested, including the determination of the presence of enthomopathogens (Čičková et al. 2015).

Another novel approach that could be very successful in pathogen elimination has been developed in the United Kingdom. A novel fibrous material obtained from municipal solid waste was used as a feedstock for anaerobic digestion. At the beginning of the fiber extraction process, the municipal solid waste was roto-autoclaved at 140 °C for 45 min at a pressure of 3 bar. The authors did not assess the microbiological quality of the obtained fiber material; however, it can be assumed to be satisfactory after autoclaving (Blake et al. 2017).

Thermal hydrolysis, thermophilic anaerobic digestion, wet oxidation, and pyrolysis have been considered as novel approaches in wastewater sludge treatment in order to reach the EU goal for "end-of-waste" criteria. However, the microbial quality aspects still have to be standardized among EU countries (Mininni et al. 2015).

Table 6.2 Impact of various composting technologies on the antibiotic and antibiotic resistance gene (ARG) degradation

Country	Antibiotic compound/ resistance gene	Composting technology	Effect on antibiotic degradation	References
United States	Florfenicol, sulfadimethoxine, sulfamethazine, tylosin	Aerated and non-aerated dairy manure and wastewater biosolids or ground wood-product (1:3 *v/v*) static pile thermophilic composting; peak temperatures were 66–73 °C, and ≥55 °C was maintained for 6–7 days in the biosolids compost and 17–20 days in the manure compost	≥85, ≥93, and ≥95% antibiotic degradation after 7, 14, and 21 days of composting	Mitchell et al. (2015)
United Kingdom	Chlortetracycline, sulfamethazine, tylosin, pirlimycin, cephapirin	Cattle manure composting up to 42 days under static and turned conditions; thermophilic temperature (>55 °C) maintained for 3 days; alfalfa hay, mulch (pine bark), and sawdust was added to the manure at a C:N ratio of 25–30 and moisture content of 55–65%	Reduction in concentration of chlortetracycline (71–84%) and tetracycline (66–72%) was substantial, while near complete removal of sulfamethazine (97–98%) and pirlimycin (100%) was achieved; tylosin removal was not achieved	Ray et al. (2017)
United States	Sulfonamides, tetracyclines, tetracycline degradation products, and related ARGs	Full-scale advanced anaerobic digester (AAD) receiving continuous manure and antibiotic input, pre-digestion pasteurization (67 °C, 1 h) for hygenization, mesophilic anaerobic digestion with a 22-day hydraulic retention time	Significant reductions in the concentrations of chlortetracycline, oxytetracycline, and tetracycline were observed in manure liquids, concomitant to significant increases in manure solid; sulfonamide resistance genes decreased significantly following AAD, while tetracyclines-resistant genes remained unchanged.	Wallace et al. (2018)

(continued)

Table 6.2 (continued)

Country	Antibiotic compound/ resistance gene	Composting technology	Effect on antibiotic degradation	References
Greece	Resistance of bacterial isolates to ampicillin, norfloxacin, bacitracin, nalidixic acid, ciprofloxacin, ofloxacin, sulfamethaxazole, kanamycin, streptomycin, tetracycline, erythromycin, chloramphenicol etc.	Mixtures of sewage sludge from biological treatment plant and green waste (1:1 and 1:2 v/v), composted for 45 days	Antibiotic resistance was not common among the isolated *E. coli* strains, but was more frequent among the *Enterococcus* spp. isolates	Lasaridi et al. (2018)
Finland	Aminoglycoside, beta-lactam, chloramphenicol, MLSB (macrolide, lincosamide and streptogramin B resistance), multidrug, sulfonamide, tetracycline, vancomycin	Anaerobic digestion of the activated sludge of Helsinki urban wastewater treatment plant followed by drying	All analyzed ARG classes were present in the final product of dried sludge	Karkman et al. (2016)

6.6 Conclusions

The diversity of waste types in northern climatic conditions is high, which reflects the respective sectors of agriculture and industry as well the household and common solutions for the treatment of solid fraction of organic municipal waste. Several waste types may contain high numbers of pathogenic microorganisms, antibiotics, and antibiotic resistance genes. Therefore, various methods exist for waste treatment, including composting, vermicomposting, anaerobic digestion, or combinations of these methods. All listed treatment methods have been shown to be able to reduce (fully or at least partially) the levels of undesired organisms or substances in waste. However, quality control measures have to be undertaken to assure the maturity and quality of the final product (composted waste, digestate, or fertilizer). It has been difficult to reduce the numbers of spore-forming bacteria (*Clostridium* spp. and *Bacillus* spp.) and antibiotic resistance genes, especially in the anaerobic digestion of manure. Drying of the digestate and pelleting have been shown as promising solutions to the problem of spore-forming bacteria, but composting has shown good results for the reduction of antibiotic resistance genes. In several investigations in Asia, vermicomposting by black soldier fly *Hermetia illucens* significantly reduced the abundance of antibiotic resistance genes and pathogenic bacteria; however, this composting method has to be investigated in temperate climate conditions in order to evaluate its suitability for this purpose.

References

Albihn A, Vinnerås B (2007) Biosecurity and arable use of manure and biowaste – treatment alternatives. Livest Sci 112:232–239

Andreev N, Ronteltap M, Boincean B, Lens PNL (2017) Treatment of source-separated human feces via lactic acid fermentation combined with thermophilic composting. Compost Sci Util 25:220–230

Ariunbaatar J, Panico A, Esposito G, Pirozzi F, Lens PNL (2014) Pretreatment methods to enhance anaerobic digestion of organic solid waste. Appl Energ 123:143–156

Atlason RS, Danner RI, Unnthorsson R, Oddsson GV, Sustaeta F, Thorarinsdottir R (2017) Energy return on investment for aquaponics: case studies from Iceland and Spain. Biophys Econ Resour Qual 2. https://doi.org/10.1007/s41247-017-0020-5

Avery LM, Killham K, Jones DL (2005) Survival of *E. coli* O157:H7 in organic wastes destined for land application. J Appl Microbiol 98:814–822

Bagge E, Sahlstrom L, Albihn A (2005) The effect of hygienic treatment on the microbial flora of biowaste at biogas plants. Water Res 39:4879–4886

Blake LI, Halim FA, Gray C, Mair R, Manning DAC, Sallis P, Hutchinson H, Gray ND (2017) Evaluating an anaerobic digestion (AD) feedstock derived from a novel non-source segregated municipal solid waste (MSW) product. Waste Manag 59:149–159

Bloem E, Albihn A, Elving J, Hermann L, Lehmann L, Sarvi M, Schaaf T, Schick J, Turtola E, Ylivainio K (2017) Contamination of organic nutrient sources with potentially toxic elements, antibiotics and pathogen microorganisms in relation to P fertilizer potential and treatment options for the production of sustainable fertilizers: A review. Sci Total Environ 607–608:225–242

Cai M, Ma S, Hu R, Tomberlin JK, Thomashow LS, Zheng L, Li W, Yu Z, Zhang J (2018) Rapidly mitigating antibiotic resistant risks in chicken manure by *Hermetia illucens* bioconversion with intestinal microflora. Environ Microbiol 20:4051–4062

Cao W, Vaddella V, Biswas S, Perkins K, Clay C, Wu T, Zheng Y, Ndegwa P, Pandey P (2016) Assessing the changes in *E. coli* levels and nutrient dynamics during vermicomposting of food waste under lab and field scale conditions. Environ Sci Pollut Res 23:23195–23202

Carballa M, Omil F, Lema JM (2009) Influence of different pretreatments on anaerobically digested sludge characteristics: suitability for final disposal. Water Air Soil Pollut 199:311–321

Cesaro A, Belgiorno V (2014) Pretreatment methods to improve anaerobic biodegradability of organic municipal solid waste fractions. Chem Eng J 240:24–37

Čičková H, Newton GL, Lacy RC, Kozánek M (2015) The use of fly larvae for organic waste treatment. Waste Manag 35:68–80

Coelho JJ, Prieto ML, Dowling S, Hennessy A, Casey I, Woodcock T, Kennedy N (2018) Physical-chemical traits, phytotoxicity and pathogen detection in liquid anaerobic digestates. Waste Manag 78:8–15

Cui G, Li F, Li S, Bhat SA, Ishiguro Y, Wei Y, Yamada T, Fu X, Huang K (2018) Changes of quinolone resistance genes and their relations with microbial profiles during vermicomposting of municipal excess sludge. Sci Total Environ 644:494–502

D'Costa VM, King CE, Kalan L, Morar M, Sung WWL, Schwarz C, Froese D, Zazula G, Calmels F, Debruyne R, Golding GB, Poinar HN, Wright GD (2011) Antibiotic resistance is ancient. Nature 477:457–461

Davies J, Davies D (2010) Origins and evolution of antibiotic resistance. Microbiol Mol Biol Rev 74:417–433

ECDC/EFSA/EMA (2017) ECDC/EFSA/EMA second joint report on the integrated analysis of the consumption of antimicrobial agents and occurrence of antimicrobial resistance in bacteria from humans and food-producing animals. EFSA Journal 15:4872

Faubert P, Barnabé S, Bouchard S, Côté R, Villeneuve C (2016) Pulp and paper mill sludge management practices: What are the challenges to assess the impacts on greenhouse gas emissions? Resour Conserv Recy 108:107–133

Fernández-Delgado Juárez M, Prähauser B, Walter A, Insam H, Franke-Whittle IH (2015) Co-composting of biowaste and wood ash, influence on a microbially driven-process. Waste Manag 46:155–164

Flack FM, Hartenstein R (1984) Growth of the earthworm *Eisenia foetida* on microorganisms and cellulose. Soil Biol Biochem 16:491–495

Franke-Whittle IH, Fernández-Delgado Juárez M, Insam H, Schweizer S, Naef A, Topp A-R, Kelderer M, Rühmer T, Baab G, Henfrey J, Manici LM (2018) Performance evaluation of locally available composts to reduce replant disease in apple orchards of central Europe. Renew Agr Food Syst 34:543–557

Gil A, Siles JA, Martín MA, Chica AF, Estévez-Pastor FS, Toro-Baptista E (2018) Effect of microwave pretreatment on semi-continuous anaerobic digestion of sewage sludge. Renew Energ 115:917–925

Gomez-Brandon M, Lores M, Dominguez J (2012) Species-specific effects of epigeic earthworms on microbial community structure during first stages of decomposition of organic matter. PLoS One 7:e31895

Grantina-Ievina L, Andersone U, Berkolde-Pīre D, Nikolajeva V, Ievinsh G (2013) Critical tests for determination of microbiological quality and biological activity in commercial vermicompost samples of different origins. Appl Microbiol Biotechnol 97:10541–10554

Gutiérrez-Miceli FA, Santiago-Borraz J, Molina JAM, Nafate CC, Abud-Archila M, Llaven MAO, Rincón-Rosales R, Dendooven L (2007) Vermicompost as a soil supplement to improve growth, yield and fruit quality of tomato (*Lycopersicum esculentum*). Bioresour Technol 98:2781–2786

Haimi J, Huhta V (1986) Capacity of various organic residues to support adequate earthworm biomass for vermicomposting. Biol Fert Soils 2:23–27

Hammer TJ, Fierer N, Hardwick B, Simojoki A, Slade E, Taponen J, Viljanen H, Roslin T (2016) Treating cattle with antibiotics affects greenhouse gas emissions, and microbiota in dung and dung beetles. Proc R Soc B 283:20160150

Heinonen-Tanski H, Mohaibes M, Karinen P, Koivunen J (2006) Methods to reduce pathogen microorganisms in manure. Livest Sci 102:248–255

Hénault-Ethier L, Bell TH, Martin VJJ, Gélinas Y (2016) Dynamics of physicochemical variables and cultivable bacteria in vermicompost during steady food waste addition and upon feed interruption. Compost Sci Util 24:117–135

Heuer H, Schmitt H, Smalla K (2011) Antibiotic resistance gene spread due to manure application on agricultural fields. Curr Opin Microbiol 14:236–243

Hill GB, Baldwin SA (2012) Vermicomposting toilets, an alternative to latrine style microbial composting toilets, prove far superior in mass reduction, pathogen destruction, compost quality, and operational cost. Waste Manag 32:1811–1820

Holopainen P, Airaksinen S, Heinonen-Tanski H, Heiskanen M-L (2002) Utilization of composted horse manure with peat bedding in greenhouse and field cultivation. In: Schmilewski G, Rochefort L (eds) Peat in horticulture. Quality and environmental challenges, pp. 154–160. A joined symposium of commissions II (Industrial utilization of peat and peatlands) and V (After-use of cut-over peatlands) of the International Peat Society

Hölzel CS, Harms KS, Küchenhoff H, Kunz A, Müller C, Meyer K, Schwaiger K, Bauer J (2010) Phenotypic and genotypic bacterial antimicrobial resistance in liquid pig manure is variously associated with contents of tetracyclines and sulfonamides. J Appl Microbiol 108:1642–1656

Huang K, Xia H, Wu Y, Chen J, Cui G, Li F, Chen Y, Wu N (2018) Effects of earthworms on the fate of tetracycline and fluoroquinolone resistance genes of sewage sludge during vermicomposting. Bioresour Technol 259:32–39

Insam H, Franke-Whittle IH, Podmirseg SM (2014) Agricultural waste management in Europe, with an emphasis on anaerobic digestion. JIFS 11:3–17

Jain S, Jain S, Wolf IT, Lee J, Tong YW (2015) A comprehensive review on operating parameters and different pretreatment methodologies for anaerobic digestion of municipal solid waste. Renew Sust Energ Rev 52:142–154

Kalnina I, Rugele K, Rubulis J (2018) Digestate management practices in Latvia from nitrogen perspective. Energy Proced 147:368–373
Karimi H, Mokhtari M, Salehi F, Sojoudi S, Ebrahimi A (2017) Changes in microbial pathogen dynamics during vermicomposting mixture of cow manure–organic solid waste and cow manure–sewage sludge. Int J Recycl Org Waste Agric 6:57–61
Karkman A, Johnson TA, Lyra C, Stedtfeld RD, Tamminen M, Tiedje JM, Virta M (2016) High-throughput quantification of antibiotic resistance genes from an urban wastewater treatment plant. FEMS Microbiol Ecol 92:fiw014
Kim K-R, Owens G, Ok YS, Park W-K, Lee DB, Kwon S-I (2012) Decline in extractable antibiotics in manure-based composts during composting. Waste Manag 32:110–116
Kouba A, Lunda R, Hlaváč D, Kuklina I, Hamáčková J, Randák T, Kozák P, Koubová A, Buřič M (2018) Vermicomposting of sludge from recirculating aquaculture system using *Eisenia andrei*: Technological feasibility and quality assessment of end-products. J Clean Prod 177:665–673
Lalander CH, Hill GB, Vinnerås B (2013) Hygienic quality of faeces treated in urine diverting vermicomposting toilets. Waste Manag 33:2204–2210
Lalander C, Diener S, Zurbrügg C, Vinnerås B (2019) Effects of feedstock on larval development and process efficiency in waste treatment with black soldier fly (*Hermetia illucens*). J Clean Prod 208:211–219
Lasaridi K-E, Manios T, Stamatiadis S, Chroni C, Kyriacou A (2018) The evaluation of hazards to man and the environment during the composting of sewage sludge. Sustainability 10:2618
Lazcano C, Gómez-Brandón M, Domínguez J (2008) Comparison of the effectiveness of composting and vermicomposting for the biological stabilization of cattle manure. Chemosphere 72:1013–1019
Lohri CR, Diener S, Zabaleta I, Mertenat A, Zurbrügg C (2017) Treatment technologies for urban solid biowaste to create value products: a review with focus on low- and middle-income settings. Rev Environ Sci Biotechnol 16:81–130
Malińska K, Zabochnicka-Świątek M, Cáceres R, Marfà O (2016) The effect of precomposted sewage sludge mixture amended with biochar on the growth and reproduction of *Eisenia fetida* during laboratory vermicomposting. Ecol Eng 90:35–41
Malińska K, Golańska M, Caceres R, Rorat A, Weisser P, Ślęzak E (2017) Biochar amendment for integrated composting and vermicomposting of sewage sludge – the effect of biochar on the activity of *Eisenia fetida* and the obtained vermicompost. Bioresour Technol 225:206–214
Martínez JL (2008) Antibiotics and antibiotic resistance genes in natural environments. Science 321:365–367
Martinez-Carballo E, Gonzalez-Barreiro C, Scharf S, Gans O (2007) Environmental monitoring study of selected veterinary antibiotics in animal manure and soils in Austria. Environ Pollut 148:570–579
Martinez-Sanchez V, Kromann MA, Astrup TF (2015) Life cycle costing of waste management systems: overview, calculation principles and case studies. Waste Manag 36:343–355
McKinney CW, Dungan RS, Moore A, Leytem AB (2018) Occurrence and abundance of antibiotic resistance genes in agricultural soil receiving dairy manure. FEMS Microbiol Ecol 94:fiy010
Mininni G, Blanch AR, Lucena F, Berselli S (2015) EU policy on sewage sludge utilization and perspectives on new approaches of sludge management. Environ Sci Pollut Res 22:7361–7374
Mitchell A (1997) Production of *Eisenia fetida* and vermicompost from feed-lot cattle manure. Soil Biol Biochem 29:763–766
Mitchell SM, Ullman JL, Bary A, Cogger CG, Teel AL, Watts RJ (2015) Antibiotic degradation during thermophilic composting. Water Air Soil Pollut 226:13
Mohring SAI, Strzysch I, Fernandes MR, Kiffmeyer TK, Tuerk J, Hamscher G (2009) Degradation and elimination of various sulfonamides during anaerobic fermentation: a promising step on the way to sustainable pharmacy? Environ Sci Technol 43:2569–2574
Mondini C, Insam H (2003) Community level physiological profiling as a tool to evaluate compost maturity: a kinetic approach. Eur J Soil Biol 39:141–148

Muurinen J, Stedtfeld R, Karkman A, Pärnänen K, Tiedje J, Virta M (2017) Influence of manure application on the environmental resistome under Finnish agricultural practice with restricted antibiotic use. Environ Sci Technol 51:5989–5999

Nair J, Sekiozoic V, Anda M (2006) Effect of pre-composting on vermicomposting of kitchen waste. Bioresour Technol 97:2091–2095

Ndegwa PM, Thompson SA, Das KC (2000) Effects of stocking density and feeding rate on vermicomposting of biosolids. Bioresour Technol 71:5–12

Nicholson FA, Groves SJ, Chambers BJ (2005) Pathogen survival during livestock manure storage and following land application. Bioresour Technol 96:135–143

Nielfa A, Cano R, Fdz-Polanco M (2015) Theoretical methane production generated by the co-digestion of organic fraction municipal solid waste and biological sludge. Biotechnol Rep 5:14–21

Nigussie A, Kuyper TW, Bruun S, de Neergaard A (2016) Vermicomposting as a technology for reducing nitrogen losses and greenhouse gas emissions from small-scale composting. J Clean Prod 139:429–439

Nõlvak H, Truu M, Kanger K, Tampere M, Espenberg M, Loit E, Raave H, Truu J (2016) Inorganic and organic fertilizers impact the abundance and proportion of antibiotic resistance and integron-integrase genes in agricultural grassland soil. Sci Total Environ 562:678–689

Pruden A, Larsson DG, Amézquita A, Collignon P, Brandt KK, Graham DW, Lazorchak JM, Suzuki S, Silley P, Snape JR, Topp E, Zhang T, Zhu YG (2013) Management options for reducing the release of antibiotics and antibiotic resistance genes to the environment. Environ Health Persp 121:878–885

Pulvirenti A, Ronga D, Zaghi M, Tomasselli AR, Mannella L, Pecchioni N (2015) Pelleting is a successful method to eliminate the presence of *Clostridium* spp. from the digestate of biogas plants. Biomass Bioenergy 81:479–482

Ray P, Chen C, Knowlton KF, Pruden A, Xia K (2017) Fate and effect of antibiotics in beef and dairy manure during static and turned composting. J Environ Qual 46:45–54

Rimhanen-Finne R, Vuorinen A, Marmo S, Malmberg S, Hänninen M-L (2004) Comparative analysis of *Cryptosporidium*, *Giardia* and indicator bacteria during sewage sludge hygienization in various composting processes. Lett Appl Microbiol 38:301–305

Ruiz-Espinoza JE, Méndez-Contreras JM, Alvarado-Lassman A, Martínez-Delgadillo SA (2012) Effect of low temperature thermal pre-treatment on the solubilization of organic matter, pathogen inactivation and mesophilic anaerobic digestion of poultry sludge. J Environ Sci Health A Tox Hazard Subst Environ Eng 47:1795–1802

Ruuskanen M, Muurinen J, Meierjohan A, Pärnänen K, Tamminen M, Lyra C, Kronberg L, Virta M (2016) Fertilizing with animal manure disseminates antibiotic resistance genes to the farm environment. J Environ Qual 45:488–493

Ryckeboer J, Mergaert J, Vaes K, Klammer S, De Clercq D, Coosemans J, Insam H, Swings J (2003) A survey of bacteria and fungi occurring during composting and self-heating processes. Ann Microbiol 53:349–410

Sahlström L, Aspan A, Bagge E, Danielsson-Tham M-L, Albihn A (2004) Bacterial pathogen incidences in sludge from Swedish sewage treatment plants. Water Res 38:1989–1994

Salomone R, Saija G, Mondello G, Giannetto A, Fasulo S, Savastano D (2017) Environmental impact of food waste bioconversion by insects: application of life cycle assessment to process using *Hermetia illucens*. J Clean Prod 140:890–905

Scarlat N, Fahl F, Dallemand J-F, Monforti F, Motola V (2018) A spatial analysis of biogas potential from manure in Europe. Renew Sust Energ Rev 94:915–930

Smith SR, Lang NL, Cheung KHM, Spanoudaki K (2005) Factors controlling pathogen destruction during anaerobic digestion of biowastes. Waste Manag 25:417–425

Spencer JL, Garcia MM (1995) Resistance of chicks and poults fed vermicompost to caecal colonization by *Salmonella*. Avian Pathol 24:157–170

Spencer JL, Chambers JR, Modler HW (1998) Competitive exclusion of *Salmonella typhimurium* in broilers fed with vermicompost and complex carbohydrates. Avian Pathol 27:244–249

Stoknes K, Scholwin F, Krzesiński W, Wojciechowska E, Jasińska A (2016) Efficiency of a novel "food to waste to food" system including anaerobic digestion of food waste and cultivation of vegetables on digestate in a bubble-insulated greenhouse. Waste Manag 56:466–476

Storteboom HN, Kim SC, Doesken KC, Carlson KH, Davis JG, Pruden A (2007) Response of antibiotics and resistance genes to high-intensity and low-intensity manure management. J Environ Qual 36:1695–1703

Sundberg C, Franke-Whittle IH, Kauppi S, Yu D, Romantschuk M, Insam H, Jönsson H (2011) Characterisation of source-separated household waste intended for composting. Bioresour Technol 102:2859–2867

Tien Y-C, Li B, Zhang T, Scott A, Murray R, Sabourin L, Marti R, Topp E (2017) Impact of dairy manure pre-application treatment on manure composition, soil dynamics of antibiotic resistance genes, and abundance of antibiotic-resistance genes on vegetables at harvest. Sci Total Environ 581–582:32–39

Topp E, Larsson DGJ, Miller DN, Van den Eede C, Virta MPJ (2018) Antimicrobial resistance and the environment: assessment of advances, gaps and recommendations for agriculture, aquaculture and pharmaceutical manufacturing. FEMS Microbiol Ecol 94:fix185

Tozzoli R, Di Bartolo I, Gigliucci F, Brambilla G, Monini M, Vignolo E, Caprioli A, Morabito S (2017) Pathogenic *Escherichia coli* and enteric viruses in biosolids and related top soil improvers in Italy. J Appl Microbiol 122:239–247

Tuomela M, Vikman M, Hatakka A, Itävaara M (2000) Biodegradation of lignin in a compost environment: A review. Bioresour Technol 72:169–183

Tyagi VK, Lo SL (2011) Application of physico-chemical pretreatment methods to enhance the sludge disintegration and subsequent anaerobic digestion: an up to date review. Rev Environ Sci Biotechnol 10:215

Vernozy-Rozand C, Montet MP, Lequerrec F, Serillon E, Tilly B, Bavai C, Ray-Gueniot S, Bouvet J, Mazuy-Cruchaudet C, Richard Y (2002) Prevalence of verotoxin-producing *Escherichia coli* (VTEC) in slurry, farmyard manure and sewage sludge in France. J Appl Microbiol 93:473–478

Villar I, Alves D, Pérez-Díaz D, Mato S (2016) Changes in microbial dynamics during vermicomposting of fresh and composted sewage sludge. Waste Manag 48:409–417

Wagner AO, Lackner N, Mutschlechner M, Prem EM, Markt R, Illmer P (2018) Biological pretreatment strategies for second-generation lignocellulosic resources to enhance biogas production. Energies 11:1797

Wallace JS, Garner E, Pruden A, Aga DS (2018) Occurrence and transformation of veterinary antibiotics and antibiotic resistance genes in dairy manure treated by advanced anaerobic digestion and conventional treatment methods. Environ Pollut 236:764–772

Watcharasukarn M, Kaparaju P, Steyer JP, Krogfelt KA, Angelidaki I (2009) Screening *Escherichia coli*, *Enterococcus faecalis*, and *Clostridium perfringens* as indicator organisms in evaluating pathogen-reducing capacity in biogas plants. Microbial Ecol 58:221–230

Wellington EMH, Boxall ABA, Cross P, Feil EJ, Gaze WH, Hawkey PM, Johnson-Rollings AS, Jones DL, Lee NM, Otten W, Thomas CM, Williams AP (2013) The role of the natural environment in the emergence of antibiotic resistance in Gram-negative bacteria. Lancet Infect Dis 13:155–165

Williams AP, Roberts P, Avery LM, Killham K, Jones DL (2006) Earthworms as vectors of *Escherichia coli* O157:H7 in soil and vermicomposts. FEMS Microbiol Ecol 58:54–64

Part II
Modern Tools and Techniques for Composting Research

Chapter 7
Molecular Tools and Techniques for Understanding the Microbial Community Dynamics of Vermicomposting

Mukesh K. Meghvansi, K. K. Chaudhary, Mohammad Haneef Khan, Sazada Siddiqui, and Ajit Varma

Abstract Earthworms have the remarkable capability of decomposing the organic matter through a biochemical oxidative process popularly called as vermicomposting. As this process significantly alters the physical and biochemical attributes of organic matter through solubilization of minerals, the resultant product called vermicompost is highly rich in nutrients that are very useful for improving the soil fertility. The organic matter decomposition is mainly carried out by microorganisms, while the earthworms provide a milieu favorable for activity of microorganisms, through ingestion and churning of the organic matter. Hence, investigation on the pattern of changes taking place in the community of microorganisms during vermicomposting is very important as it has a direct bearing on vermicompost quality. With the advancement in tools and techniques for scientific research on vermicomposting, our understanding about the entire process has tremendously increased over past few decades. Of particular relevance, here are the molecular tools and techniques that have greatly enhanced our knowledge in understanding the microbial community dynamics of vermicomposting. This chapter provides an

M. K. Meghvansi (✉)
Defence Research Laboratory, Tezpur, Assam, India

Present Address: Bioprocess Technology Division, Defence Research and Development Establishment, Gwalior, Madhya Pradesh, India

K. K. Chaudhary
Department of Plant Protection, Hamelmalo Agricultural College, Hamelmalo, Eritrea

M. H. Khan
Professor Joy Michelle Bergelson Lab, Department of Ecology and Evolution, University of Chicago, Chicago, IL, USA

S. Siddiqui
Department of Biology, College of Science, King Khalid University, Abha, Saudi Arabia

A. Varma
Amity Institute of Microbial Technology, Amity University Uttar Pradesh, Noida, Delhi-NCR, India
e-mail: ajitvarma@amity.edu

M. K. Meghvansi, A. Varma (eds.), *Biology of Composts*, Soil Biology 58,
https://doi.org/10.1007/978-3-030-39173-7_7

overview of molecular tools and techniques being used for unravelling the mechanisms of interactions of microorganisms with various biotic and abiotic factors under a unique micro-ecosystem where the earthworms are crucial drivers of the process owing to their active role in the stimulation of microbial populations.

Keywords Molecular tools and techniques · Microbial community dynamics · Vermicomposting

7.1 Introduction

Earthworms are soft-bodied and metamerically segmented invertebrates belonging to the class Oligochaeta of phylum Annelida. They are present in a wide variety of environments and soil types wherever there is an adequate moisture and organic matter available. They ingest the organic substrate that is degraded effectively when it passes through their digestive tract, and then excreted as castings (Fig. 7.1). Hence, Aristotle had called them "the intestine of the earth." Earthworm burrowing results into physicochemical changes in the soil including improved aggregation, stability, porosity, and organic matter dynamics in terms of quality and quantity, nutrient cycling, chemical forms of nutrient in soil and their availability to plants (Meghvansi et al. 2011). In fact, earthworms have the capability to influence a given area structurally to a very significant extent. Hence, they have been referred to as "ecosystem engineers" by many researchers (Ojha and Devkota 2014; Le Bayon et al. 2017).

The ability of earthworms to act upon the diverse organic substrates and turn them into a nutritive "organic fertilizer" has been known to the mankind since many centuries. In fact, the perennial fertility of Nile river valley of Egypt was considered

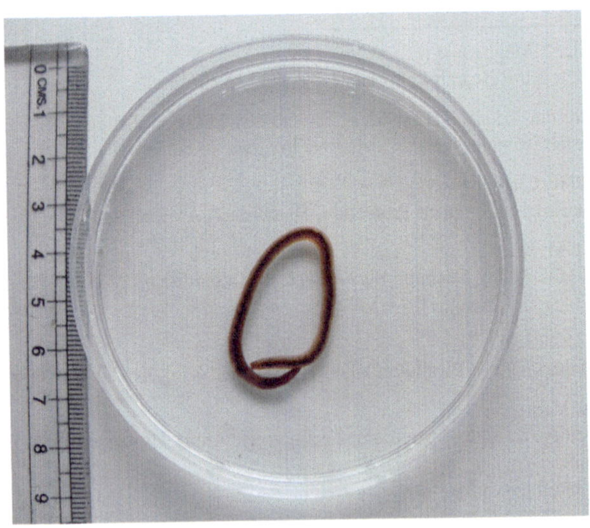

Fig. 7.1 Ingested organic matter visible through the earthworm body

to be associated with intense earthworm activity, and ancient farmers were well aware of the earthworm's contribution to their prosperity. As mentioned by Minnich (1977) in *The Earthworm* Book, Cleopatra (69–30 BC) had declared earthworms as sacred, and Egyptian were not allowed to remove even a single earthworm from the land. Egyptian farmers were not even allowed to touch the earthworms for fear of offending the god of fertility (Minnich 1977). This subject also generated great interest to Charles Darwin (1881) who devoted his last scientific book exclusively to earthworms discussing the link of earthworm activity to the physical soil processes. Darwin estimated that an acre of earthworms brought some 16 tonnes of subterranean material to the surface annually. His book served as a foundation for what today is referred to as bioturbation research (Wilkinson et al. 2009). Interestingly, Darwin's book, *The Formation of Vegetable Mould Through the Action of Worms, With Observations on Their Habits*, published in 1881, sold even better than *On the Origin of Species* during Darwin's lifetime. The modern researchers have identified the environmental conditions optimal for earthworm activity and successfully utilized it for solid waste management practices (Meghvansi and Veer 2014; Meghvansi et al. 2016). In view of the increasing environmental and economic problems associated with disposal of organic wastes from domestic, agricultural, and industrial sources, several developed as well as developing countries have already adopted vermicomposting technology at domestic as well as industrial scale for the organic waste management (Domínguez and Edwards 2004; Meghvansi et al. 2015).

Researchers worldwide have generated a significant body of evidence on the strong and complex interplay of earthworm activity and microbial community dynamics of drillosphere and rhizosphere. The microbial population in the worm casts has been found to be greater than in surrounding soil (Parthasarathi and Ranganathan 1998). It is also frequently observed that the bacterial count is considerably enhanced in the soil when vermicompost is applied, as compared to untreated soil. According to Pizl and Novakova (2003), microorganisms form an inevitable constituent of earthworm's natural diet. Microbial hotspots created in the rhizosphere and detritusphere have been investigated by many researchers. Based on the knowledge accumulated so far, the vermicomposting has been considered as a biphasic process. Whereas the first phase primarily deals with modification of the physical attributes and microbial composition of organic waste (Lores et al. 2006), the second phase invariably includes movement of the earthworms toward newer layers of substrate, during which microbes take over the waste decomposition processed by the earthworms (Aira et al. 2007). Considering the importance of microbial community dynamics for vermicomposting, a great deal of researchers have made in-depth studies under varied conditions to unravel the underlying mechanisms of tripartite interactions of earthworm–microorganism–soil factors. This has become possible mainly due to the advancements made in the field of molecular tools and techniques. In this chapter, we provide an overview of various technological developments that happened leading to state-of-the-art molecular tools and techniques that are currently being used for understanding the microbial community dynamics of vermicomposting. We also discuss the advantages and limitations of various molecular techniques through a critical meta-analysis of the literature that may become the basis for making further optimization and improvements for better utilization to meet the desired goals related to waste decomposition.

7.2 Microbial Community Dynamics of Vermicomposting

Microbial community dynamics of vermicomposting may be referred typically to the process of change and development taking place in microbial communities during the course of earthworm-mediated biodegradation of organic substrate. This approach involves mainly two aspects: structural and functional community dynamics. Whereas structural dynamics deals with the change in the types and numbers of microbial species happening during the vermicomposting process, the functional dynamics throw the light on the change in role or functions played by the microorganisms over time. Combined investigation on structural and functional dynamics would make a significant contribution toward a more comprehensive understanding of the vermicomposting process. Considerable efforts have been made to understand the various aspects of microbial community dynamics of vermicomposting. It has been observed that there exists a significant variation in the pattern of microbial community dynamics thereby necessitating further investigations. For instance, in a study on bacterial community dynamics during coconut leaf vermicompost production and in earthworm gut contents using the 16S rRNA gene-based analyses, it was revealed that the bacterial communities changed considerably during vermicompost production (Gopal et al. 2017). More specifically, the information on the most abundantly distributed operational taxonomic units (OTUs) indicated that a unique microbiome represents an individual stage of the vermicomposting process. This study further concluded that bacterial communities keep on changing continuously during the entire course of vermicomposting and that some of the OTUs corresponding to a particular stage may be a suitable option for making further improvements in the process (Fig. 7.2; Gopal et al. 2017). In a study, microbial community structure and function were investigated under the influence of *E. fetida* in a continuous feeding reactor having sequential addition of fresh layers of pig slurry, for the purpose of making age gradient. Earthworm activity considerably reduced the bacterial and fungal biomass and microbial diversity as compared to control (Gómez-Brandón et al. 2011). In a similar study that evaluated, through microbial dynamics, the use of earthworm species *E. andrei* for maturation of pre-composted pig manure in comparison with maturation under static conditions and with vermicomposting of fresh pig manure, a direct effect of microbiota evolution on the degradative processes could be registered (Villar et al. 2017). It implies that the outcome of interactions of microorganisms with earthworms is governed by the earthworm species and the quality and/or substrate availability (Gómez-Brandón et al. 2012). Similarly, an experiment with vermicomposting for excess sludge stabilization indicated a considerable increase in Shannon index in vermicomposting, and decrease in common composting with time, suggesting that earthworm activity could contribute to enhanced diversity of microbial community (Zhao et al. 2018).

Fig. 7.2 Change in bacterial communities associated with each stage of coconut leaf vermicomposting process. Change in the measured property mentioned on the left is indicated by the change in height of horizontal boxes from left to right across the figure (Source: Gopal et al. 2017; reprinted by permission from Springer Nature)

7.3 Why Is It Important to Study Microbial Community Dynamics of Vermicomposting?

In the process of vermicomposting, earthworms and associated microorganisms jointly act on the organic waste, although the mechanism of earthworm–microbe interaction is very complex and yet to be deciphered in its entirety. In fact, the earthworm–microbe interaction starts at a very early life (cocoon) stage of earthworms. In a recent study on cocoon microbiome of the earthworms *E. andrei* and *E. fetida*, 275 and 176 species of bacteria were reported, respectively. Vertically transmitted symbionts, Microbacteriaceae, *Verminephrobacter* and *Ca. Nephrothrix* dominated contributing for 88% and 66% of the sequences, respectively. Interestingly, *Verminephrobacter* and Ca. Nephrothrix displayed a significant rate of variation in a sequence indicating the possibility of their biparental origin acquired during mating (Aira et al. 2018). Therefore, microbial activity and population could provide valuable insights not only for stabilization of organic materials but also have implications for the effective utilization of the finished product for improving soil fertility. Another important aspect is that the earthworm activity is known to increase the population of plant growth-promoting bacteria (Sinha et al. 2010) thereby potentially influencing plant growth. Since the soils characterized by low organic matter and microbial activity are prone to plant root diseases (Stone et al. 2004), the use of vermicompost and vermiwash as organic amendments have the potential for plant disease suppression under such a scenario (Meghvansi 2011; Khan et al. 2015). Although the mechanisms of plant disease suppression under the influence of vermicompost application are very complex, yet microbial antagonism through their capability to compete for space and nutrients, has been subscribed as one of the possible reasons for the disease suppression (Meghvansi et al. 2011; Pathma and Sakthivel 2012). In view of this, the microbial repertoire of vermicompost assumes greater significance for field application. Deciphering the microbial community dynamics could prove very useful for ensuring the quality of vermicompost.

7.4 Molecular Tools and Techniques for Understanding the Microbial Community Dynamics of Vermicomposting

For the purpose of understanding the microbial community structure and dynamics of vermicomposting, both cultivation-dependent and cultivation-independent molecular methods have been employed. Various studies have established that the culture-based techniques are capable of recovering only 1–10% of the true microbial diversity within an environment (Hugenholtz et al. 1998; Muyzer 1999). A great deal of research evidence is now available wherein new classes, phylotypes, and

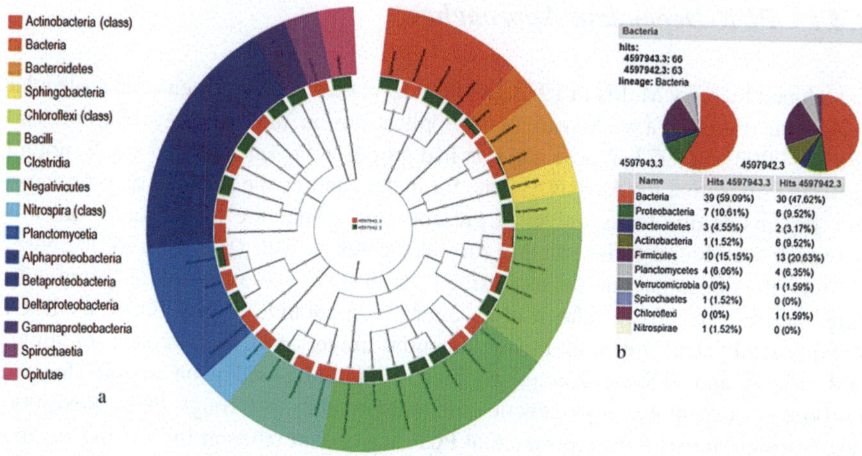

Fig.7.3 (**a**) Phylogenetic tree of the metagenomic sequences from *E. foetida* (mg-rast id 4597942.3) and *P. excavatus* (mg-rast id 4597943.3), created on the basis of sequence hits on RDP database. (**b**) Pie chart illustrating the distribution of bacterial phylum in the metagenomes, based on the hits per sequence using RDP database. (Source: Singh et al. 2015; reprinted with permission from Elsevier GmbH)

families have been reported using cultivation-independent molecular methods (Fig. 7.3). Hence, remarkable variation is encountered in the results of molecular surveys of bacterial communities when compared with traditional culture-based approaches. In fact, the studies on microbial diversity through cultivation-independent molecular methods have generated plethora of information that has not only led to unearthing a significant amount of gene pool but has also highlighted the challenges associated with deciphering their phylogenetic attributes as well as ecological significance (Agrawal et al. 2015; Singh et al. 2015). To circumvent the limitations of cultivation-dependent molecular methods that are time consuming and often do not represent true diversity, research focus has shifted to cultivation-independent methods so much so that these techniques have been regarded among the most important developments of environmental microbiology since the 1980s (Keller and Zengler 2004). This chapter primarily discusses the various cultivation-independent molecular methods that follow the approach of extracting the nucleic acids, amplifying the rDNA/rRNA and their subsequent analysis using advanced tools and techniques. Other approaches such as community-level physiological profiling, carbon utilization pattern, and phospholipid fatty acid profiling although widely used are beyond the scope of this chapter. A significant emphasis has been laid on next-generation and third-generation sequencing technologies that have been deployed to study the microbial community dynamics of compost and vermicompost.

7.4.1 PCR-Dependent Approaches

Developed by Kary Mullis in 1983, the PCR has become an indispensable molecular technique used for a wide variety of applications in microbiology (Bartlett and Stirling 2003). In 1977, Woese and Fox had proposed the suitability of small subunit (SSU) of ribosomal RNA (rRNA) gene for understanding the phylogeny of prokaryotic organisms due to its universal presence in cellular life forms. Later on, this approach was found to be suitable for eukaryotes as well (Woese, 1987). Subsequently, the ribosomal internal transcribed spacer (ITS) was explored for community fingerprinting. In 2012, Schoch et al. after a comparative evaluation of six DNA regions established utility of ITS as potential universal DNA barcodes for fungi. SSU rRNA and ITS are among the most commonly used gene regions for the purpose of community phylogenetic composition analysis through PCR. Nevertheless, various studies have reported that PCR has certain inherent limitations leading to biases in the template to product ratios of target sequences amplified from environmental DNA (Polz and Cavanaugh 1998), which tend to increase with increasing numbers of amplification cycles. These PCR limitations pose challenges for estimating the abundance of individual genes present in environmental samples. To overcome this, extraction of microbial nucleic acid from environmental matrices and subsequent amplification of structural and functional gene markers through PCR in combination with other approaches has been attempted in analysis of microbial community in vermicompost and compost. Some of these techniques used commonly are discussed here.

7.4.1.1 Random Amplification of Polymorphic DNA (RAPD)

In RAPD, the segments of DNA are amplified randomly using short primers of 8–12 nucleotides. These primers bind at uncertain place on a large template of genomic DNA. Unlike conventional PCR analysis, this technique requires no specific knowledge of the DNA sequence of the target organism. Moreover, it is a simple, cost effective yet powerful technique that can be used for differentiating between genetically distinct individuals. Malik et al. (1994) used RAPD fingerprinting to comparing the structures of a pilot-scale compost community at various four different times throughout its development. Results revealed that with RAPD fingerprinting, it was possible to distinguish community structures within the reactor at the different sampling times. These results were further validated using another approach community DNA cross hybridization (Malik et al. 1994). The availability of a large number of molecular markers requiring meager quantity of DNA without the need for cumbersome and time-consuming cloning and sequencing made the RAPD technique very popular (Bardakci 2001). In spite of its simplicity, RAPD technique has inherent limitations. It has less resolving power than the species-specific DNA comparison methods. Moreover, RAPD products are dependent on the thermostable DNA polymerase (Schierwater and Ender 1993) and on the concentrations of primer and template (Muralidharan and Wakeland 1993) used in the PCR. Therefore, experimental reproducibility of the results is of great concern particularly for the weak bands. Another limitation with the RAPD technique is that most of RAPD markers have dominant character thereby making it very difficult to distinguish homozygotes and heterozygotes. In view of these limitations, it is difficult to ascertain as to what extent RAPD markers could be useful for carrying out phylogenetic analysis of species with varying degrees of relatedness (van de Zande and Bijlsma 1995).

7.4.1.2 Amplified Ribosomal DNA Restriction Analysis

Using restriction fragment length polymorphism (RFLP) technique, DNA sample is fragmented with the help of a restriction enzyme which selectively cleaves the DNA molecule at the place where a short specific sequence is recognized. Earlier, RFLP analysis was an important technique for distinguishing individuals, populations, or species because of its property of utilizing variations in homologous DNA sequences. Amplified Ribosomal DNA Restriction Analysis (ARDRA) is a further modification of RFLP involving an enzymatic amplification using primers directed at the conserved regions at the ends of the 16S gene. Then, PCR amplified 16S rDNA sequences are digested using tetracutter restriction enzymes such as *Alu I* and *Hae II*. Digested fragments are then resolved on agarose or polyacrylamide gels. The restriction digestion pattern obtained through this approach is considered to be representative of the species analyzed and can be employed for the phylogenetic characterization of the community (Sklarz et al. 2009). ARDRA was successfully used to monitor the changes in the microbial community structure during composting and to compare the differences in the structures between different composting processes while degradation of polyhydroxyalkanoates (Uchiyama et al. 2002). Similarly, other researchers have used the ARDRA for investigating changes in the genetic structure of contaminated soil bacterial communities following different amendments (Ntougias et al. 2004; Pérez-de-Mora et al. 2006). The ARDRA has certain advantages in terms of the method simplicity, the universal availability of PCR primers, reproducibility, amenability to computer database analysis, and comparison of fingerprints using a database of ARDRA patterns (Alves et al. 2005). ARDRA can be deployed to identify the unique clones and to estimate the OTUs in libraries based on restriction profiles (Rastogi and Sani 2011). Nevertheless, this technique also shares the limitations of PCR biases. Moreover, sometimes, such profiles generated from complex microbial communities are difficult to separate on the agarose or polyacrylamide gel (Rastogi and Sani 2011).

7.4.1.3 Terminal Restriction Fragment Length Polymorphism

Somewhat similar to the methods described above, Terminal Restriction Fragment Length Polymorphism (T-RFLP) is also based on PCR amplification of a target gene. However, in this technique which was used for the first time by Liu et al. (1997) for the amplification of the 16S rDNA target gene from the DNA of several isolated bacteria as well as environmental samples, a fluorescently labeled oligonucleotide primer is used for amplification of rRNA gene fragments and the digestion of the PCR products with one or more restriction enzymes (Edel-Hermann et al. 2004). The most commonly used dyes are 6-carboxyfluorescein (6-FAM), ROX, carboxytetramethylrhodamine (TAMRA), and hexachlorofluorescein (HEX). The results can be visualized as a graph (electropherogram) where the x-axis represents the size of fragments while the y-axis denotes fluorescence intensity of each fragment. The band of electrophoretic gel is akin to peak on the electropherogram representing the total fluorescence. Therefore, T-RFLP is distinct from ARDRA and RFLP in the sense that here only the terminal fragments are read while in the case of ARDRA and RFLP, all the restriction fragments are read. Important developments in T-RFLP methodology to distinguish the total microbial diversity and community composition in the diverse ecosystems have been reviewed by Chauhan et al. (2011). Székely et al. (2009) investigated the microbial community succession in summer and winter cycles of mushroom compost using 16S ribosomal deoxyribonucleic acid (rDNA)-based denaturing gradient gel electrophoresis (DGGE) and T-RFLP. More recently, bacterial and fungal diversity of vermicompost leachate produced from six diverse substrates at four- and eight-week duration was studied by Donohoe (2018) using T-RFLP analysis. T-RFLP can provide highly reproducible data when used with an automated sequencer. However, being a PCR-dependent technique, T-RFLP is also not immune from inherent biases associated with DNA extraction methodologies and PCR reagents/conditions (Brooks et al. 2015). Moreover, since in T-RFLP, only terminal fragments are used which affect the results in terms of underrepresentation of the diversity because any two distinct sequences sharing a terminal restriction site would yield only one peak on the graph thereby rendering them inseparable. To minimize the errors arising due to this, different corrective approaches are used in T-RFLP which further complicates the analysis.

7.4.1.4 Ribosomal RNA (rRNA) Intergenic Spacer Analysis (RISA)

This technique deals with the PCR amplification of the intergenic region between the small (16S) and large (23S) subunit rRNA in the rRNA operon employing oligonucleotide primers targeted to conserved regions in 16S and 23S genes. The PCR product is then electrophoresed in a polyacrylamide gel. DNA banding pattern is then visualized by silver staining. Here, each DNA band is regarded as corresponding to at least one organism in the original assemblage (Fisher and Triplett 1999). This technique focusses on the length heterogeneity of the ISR, which has been shown to range between 150 and 1500 bp with the majority of the ISR lengths being between 150 and 500 bp (Sigler et al. 2002). Apart from being time consuming and cumbersome, RISA too is fraught with inherent limitations of silver staining. To overcome these limitations, an improved method called automated RISA (ARISA) has been developed that involves PCR amplification of the target DNA fragment using a fluorescence-tagged oligonucleotide primer. Subsequently, electrophoresis is performed on an automated system using laser detection of fluorescent DNA fragments (Fisher and Triplett 1999). This technique has the capability of generating DNA fragments up to 1400 bp in length (Borneman and Triplett 1997). ARISA was used during the initial stages of composting for determining temporal changes of bacterial community by Schloss et al. (2003). The results revealed a remarkable shift in bacterial community dynamics from *Bacillus* to lactic acid sequences. Later on, it reverted to *Bacillus*-type sequences again. Similarly, bacterial community structure and diversity of agricultural soil treated with a varied quantity of municipal solid waste compost and other fertilizers were studied using DGGE and ARISA by Cherif et al. (2008). Based on ARISA profiles, it was possible to assign dominant populations to low and high GC-Gram-positive bacteria, Cyanobacteria, Spirochetes, and Cytophagales. Cherif et al. (2008) also observed that DGGE was more accurate for bacterial identification. On the other hand, ARISA had more practical utility in terms of handling as well as quicker estimation of dominant bacteria.

7.4.1.5 Denaturing Gradient Gel Electrophoresis (DGGE), Temperature Gradient Gel Electrophoresis (TGGE) and Single-Strand Conformation Polymorphism (SSCP)

Primarily developed for single point mutation detection in certain diseases (Fischer and Lerman 1979), DGGE was first described by Muyzer et al. (1995) for the analysis of bacterial diversity in complex communities. This technique is based on the principle that single-stranded DNA molecule tends to migrate more slowly as compared to the double-stranded molecule during electrophoresis, owing to increased interaction of the unbonded nucleotides in the single-stranded molecule with the gel matrix (Strathdee and Free 2013). It involves the separation of PCR products on polyacrylamide gels containing an increasing gradient of chemical denaturants (usually urea and formamide) through which the DNA molecules pass by electrophoresis. At a given concentration of denaturant, double-stranded DNA gets denatured and retarded making a branched structure depending upon its quantity of GC and the arrangement of base sequences, that can be visualized in gel. In order to prevent complete meltdown of double-stranded DNA, this technique uses specific PCR primers that have a 35–40 nt GC-rich tail called a GC clamp (Abrams et al. 1990). When a gradient of temperature is used as denaturant instead of chemicals, it is called Temperature Gradient Gel Electrophoresis (TGGE). One of the major advantages of DGGE is that bands can be excised from the gel and subsequently sequenced to unearth the phylogenetic relatedness of microbial communities. Huang et al. (2013) investigated changes of microbial community during vermicomposting of vegetable wastes by hatchling, juvenile, and adult *E. fetida* through analysis of the extracted bacterial 16S rDNA and fungal 18S rDNA with quantitative polymerase chain reaction (PCR), denaturing gradient gel electrophoresis (DGGE), and sequencing. Similarly, Zhen et al. (2014), based on DGGE fingerprinting analysis established that application of manure compost plus bacterial fertilizers could immediately improve the microbial community structure and diversity of degraded cropland soils. Limitations of DGGE include separation of relatively short (ca. 500 bp) DNA fragments. Theoretically, a single DGGE band corresponds to a distinct DNA sequence but in practice, often, it is very difficult to get separate fragments from different sequences. Sometimes, due to co-migration of DNA fragments, a single DGGE band might actually contain multiple sequences thereby further compromising the accuracy of results (Gafan and Spratt 2005). Therefore, the retardation of the fragment in the gel matrix may not necessarily indicate phylogenetic relatedness at high resolution, like the species level (Kisand and Wikner 2003). Some researchers have highlighted the challenges of result reproducibility associated with the use of GC clamp. Rettedal et al. (2010) have hypothesized that repeat syntheses of identical 40-base GC-clamp primers lead to different DGGE profiles. Hence, it becomes very crucial to take into consideration these limitations of DGGE while interpreting the final results.

Similar to DGGE, SSCP is also based on change in the migrational behavior of the fragments under specified conditions representing conformational polymorphism on a single strand. Nucleic acid chains have a strong tendency for base pairing but in absence of complementary strand, RNA and single-stranded DNA pair with themselves yielding highly complex conformations depending upon the sequences (Mülhardt and Beese 2007). SSCP was originally developed for detecting known or novel polymorphisms/point mutations in DNA (Orita et al. 1989). This technique is most commonly used for screening patients for mutations associated with specific diseases (Konstantinos et al. 2008). However, later on, it has found applications for studying microbial community dynamics of various environmental samples. SSCP methodology includes PCR amplification of a random DNA fragment of approximately 250 bp, its denaturation and then application of the probe to nondenatured gel in which DNA is resolved electrophoretically. Figure 7.4 shows the Cipher mutation detection system (CBS Scientific, USA) used for running DGGE/TGGE/SSCP. Unlike DGGE/TGGE, this technique obviates the need for GC base pairs. Here, the temperature plays crucial role in obtaining polymorphisms. Therefore, many gels are run at various temperatures and compared for getting optimal resolution. Structural divergence of bacterial communities of compost and vermicompost has been studied using SSCP by many workers (Fracchia et al. 2006; Sen et al. 2008; Partanen et al. 2010).

7.4.2 PCR-Independent Approaches

As discussed in preceding section, PCR-dependent approaches have several limitations primarily with regard to accuracy and reproducibility arising due to potential PCR bias issues. In view of this, large number of PCR-independent techniques has been developed that are in vogue for studying microbial community dynamics. Here, we discuss three important techniques which are commonly used.

Fig. 7.4 Cipher Genetic Analysis System (CBS Scientific, USA) used for detecting mutation and polymorphism through DGGE, TGGE, and SSCP. Gel assembly is placed in the tank filled with buffer

7.4.2.1 DNA Microarray

DNA microarray technology exploits the fundamental properties of DNA that is hybridization to complementary sequences. In DNA microarray, thousands of nucleic acids samples are immobilized as miniaturized spots on a solid support, typically a glass slide or nylon chip. Each spot has a definite position and contains a known DNA sequence or gene called probe. Such slides are popularly called DNA chips or biochips. The exact location and sequence of each spot is registered in a computer database. The technique typically involves extraction of mRNA from the experimental samples and their conversion into cDNA. The samples are then labeled with a fluorescent dye and allowed to bind to probe through hybridization. The array is then washed to remove unbound sample and scanned to measure fluorescence signal at each spot on the array. The data through comparative analysis of various samples in microarrays can be used to create gene expression profiles representing simultaneous changes in the expression of genes under a given condition. Historically, the seminal work of Grunstein and Hogness (1975) with colony hybridization method using nitrocellulose filters made the foundation of DNA microarray technology, which was subsequently improved and automated over the years. With advancement in the printing technologies, oligo synthesis chemistries, slide surface chemistries, and methods for fluorescence detection this technique, microarray technology has now become one of the indispensable tools for studying genome-wide expression levels of genes in a given organism and different conditions. The in-depth microbial community analysis of the compost under varying conditions and at different stages has been investigated extensively using COMPOCHIP microarray (Danon et al. 2008; Franke-Whittle et al. 2009, 2014). More recently, performance of locally available composts to reduce replant disease in apple orchards was assessed by Franke-Whittle et al. (2018).In this study, researchers investigated the bacterial communities of the composts using the COMPOCHIP microarray, and observed that differences in plant growth response were mainly attributed to the microbial changes introduced into the soil through composts rather than to changes in soil chemical and biological parameters (Franke-Whittle et al. 2018). Despite its incredible utility, only a handful of researchers have used DNA microarray technology for elucidating functional capabilities of microbial community of composting and vermicomposting processes. This technique also has some limitations. For instance, microarray primarily is an indirect measure of relative concentration through a signal generated at a given position on the microchip. These signals often may not be linearly proportional to the concentration of the species hybridizing to the array (Bumgarner 2013). Moreover, it could detect the only sequences for which complimentary sequence are available on the microchip. Hence, array designing for a specific group of genes (reference chip) itself is a very daunting task for deriving meaningful conclusions from the data obtained. Various challenges and possible solutions in the development and application of DNA microarrays in microbial ecology are discussed by Wagner et al. (2007).

7.4.2.2 Next-Generation Sequencing (NGS) Technologies

Although the efforts on nucleic acid sequencing started in as early as the 1960s, yet it was only in 1977 when, Sanger was able to sequence the first DNA genome, that of bacteriophage ϕX174. This classical approach employed the plus and minus technique for DNA polymerase to synthesize from a primer with the use of radiolabeled nucleotides. Subsequently, the position of nucleotides in the covered sequence could be discerned by running the products on a polyacrylamide gel and comparing between the eight lanes (Sanger 1977). This is considered as the beginning of first-generation sequencing technologies. Further, a significant improvement in the technique was made by Sanger and Nicklen (1977) through the development of a dideoxy technique that used chemical analogues of the deoxyribonucleotides (dNTPs). In the same year, Maxam and Gilbert described a new method (chemical cleavage technique) that employed the treatment of radiolabeled DNA with chemicals that break the chain at specific bases (Maxam and Gilbert 1977). In 1987, first automated capillary electrophoresis-based sequencing equipment AB370 was introduced in the market by Applied Biosystems. In 1998, AB3730xl, a higher version of this equipment was launched. These two instruments were primarily used for Human Genome projects (Collins et al. 2003). In parallel, serious research efforts continued for improvements in the sequencing technologies. Consequently, massively parallel high-throughput sequencing techniques have been

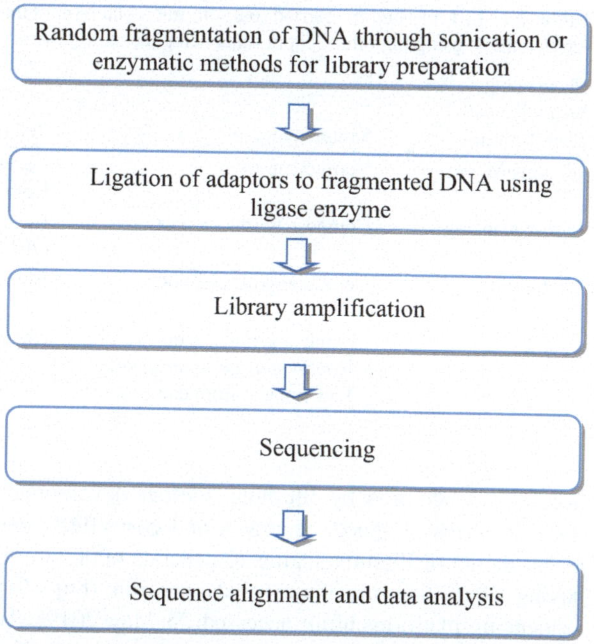

Fig. 7.5 A typical workflow of the NGS platform

developed that has in recent times revolutionized the way the genetic information is obtained. These next-generation sequencing platforms are not only quick and accurate but cost effective also. For instance, the HiSeq X® TenSystem, launched in 2014, has the capability of sequencing over 45 human genomes in a single day for approximately $1000 (Anonymous 2014). On the contrary, the first human genome mapping took more than a decade at the cost of about three billion dollars. A typical workflow of NGS platform includes random fragmentation of DNA for library preparation, ligation of adaptors (oligos bound to the 5′and 3′ end of each DNA fragment), amplification of library using clonal amplification method/PCR, and then sequencing on automated sequencer (Fig. 7.5). In this section, we discuss some of the most commonly used next-generation sequencing technologies and platforms. Examples of studies carried out using NGS platform for microbial community analysis of compost and vermicompost are provided in Table 7.1.

Illumina Sequencing

In 1990s, Balasubramanian and Klenerman who were working on the use of fluorescently labeled nucleotides to observe the mobility of a polymerase at the single-molecule level as it synthesized DNA immobilized to a surface, started a company called Solexa with the help of Abingworth Management, a venture capital firm. In 2006, the first genome analyzer was launched by Solexa having the capability to sequence 1 gigabase (Gb) of data in a single run. The very next year,

Table 7.1 List of studies carried out on microbial community dynamics of compost and vermicompost using high throughput sequencing technologies

Sequencing technologies/ platform used	Sample type	References
Illumina/MiSeq	Vermicompost	Liu et al. (2018), Cai et al. (2018), Kolbe et al. (2019)
Pyrosequencing	Compost/Vermicompost	Lv et al. (2015), Galitskaya et al. (2017), Gopal et al. (2017)
Ion torrent	Vermicompost; compost	Blomström et al. (2016), Blaya et al. (2016), Mickan et al. (2018)
PacBio	Complete genome sequence of *Pseudomonas aeruginosa* strain YL84 isolated from compost	Chan et al. (2014)

Solexa was acquired by Illumina. Further developments in the technology were made in terms of speed, accuracy, and cost effectiveness. NovaSeq sequencer, a recent model of Illumina claims to generate 6000 Gb of output in ~44 h of the run having 20 billion maximum reads per run (https://sapac.illumina.com/systems/sequencing-platforms.html; accessed 25 May 2019). Today, Illumina is the most successful sequencing system with a claimed >70% dominance of the market, particularly with the HiSeq and MiSeq platforms (Kulski 2016). Illumina technology basically relies on the use of bridge amplification for polony generation and sequencing-by-synthesis (SBS) approach. It uses reversible dye terminators which makes it possible to identify single bases as they are incorporated into DNA strands. Enzymatic processes and imaging steps of the Illumina technology take place in a flow cell having multiple separate lanes depending upon the specific model of the instrument. While sequencing, the polonies on the flow cell are read, through fluorescently labeled dNTPs incorporated into the growing DNA chain, one nucleotide at a time in repetitive cycles (Buermans and den Dunnen 2014).

Ion Torrent Sequencing

Ion Torrent Sequencing or Ion semiconductor sequencing is also based on sequencing by synthesis approach. Whereas Illumina being an optics-based sequencer uses visible light-collecting photons from arrays and forming a visual image to detect how bases are incorporated into the genome, Ion Torrent does not use fluorescence or chemiluminescence instead it is based on detection of H^+ ions released during the polymerization of DNA. When dNTP is incorporated into a growing DNA, a covalent bond is formed with a release of pyrophosphate and an H^+ ion. As the dNTP gets incorporated only in case of being complimentary to the unpaired template nucleotide, Ion Torrent sequencer determines whether or not a hydrogen ion is released upon providing a single species of dNTP in the reaction. Ion Torrent workflow includes placing the DNA template with DNA polymerase in designated microwells. Under the layer of microwells, there is an ion-sensitive layer attached with by ion-sensitive field-effect transistor (ISFET) housed in CMOS semiconductor chip. Upon biochemical reaction, the release of H^+ ion changes the pH of the solution which is detected by ISFET. The unattached dNTP molecules are washed out before the next cycle when a different dNTP species is introduced (Pennisi 2010). As the Ion Torrent obviates the need for expensive optics-based gadgets and modified nucleotides or special enzymes, the upfront and operating costs are also less as compared to that of Illumina. Performance of Illumina and Ion Torrent sequencing has been compared using diverse samples by the researchers convincing that both the platforms have similar capacity (Lahens et al. 2017; Speranskaya et al. 2018) but sometimes results might also differ. Plausible reasons for variation in results include the inability to generate full read lengths for specific organisms, and differences in rates of sequence error which affect classification of certain species. This indicates that the selection of sequencing platforms alone can also create differential bias in results pertaining to community profiles (Salipante et al. 2014).

454 Pyrosequencing

Like Illumina, pyrosequencing is also based on the sequencing-by-synthesis approach, in which the sequencing is performed through the detection of the nucleotide incorporated by a DNA polymerase. However, when the fragmented DNA is incubated with DNA polymerase, ATP sulfurylase, apyrase enzymes, adenosine 5′ phosphosulfate, and luciferin substrates, it releases pyrophosphate which is converted into ATP using ATP sulfurylase. ATP in turn participates in the conversion of luciferin to oxyluciferin with the assistance of luciferase emitting light in proportion to the amount of ATP participating in the conversion, and is detected by the detector for determining the number and type of nucleotides added. Unutilized nucleotides and ATP are removed from the system through degradation by apyrase thereby allowing the reaction to begin again with other nucleotides (Greenwood 2018). The processing time of pyrosequencing is relatively fast (~20 min/96-well plate) and the cost of the reaction is comparable to other medium throughput technologies (Marsh 2007). A Swedish company named Pyrosequencing AB used to manufacture the Pyrosequencers. This company was renamed as Biotage in 2003 that was acquired by Qiagen in 2008. Another company named 454 Life technologies (Subsidiary of CuraGen) subsequently bought the license of this technology which was further acquired by Roche in 2007. This technique became very popular alternative to Sanger sequencing until the advent of more efficient sequencing technologies such as Illumina and Ion Torrent. In 2013, Roche announced its plan to discontinue this product by 2016 (Hollmer 2013).

7.4.2.3 Third-Generation Sequencing (TGS) Technologies

The NGS technologies as discussed in the preceding sections come under the class of Second-Generation technologies. In parallel, the work had already begun for developing the technologies which obviates the need for cumbersome steps of library making, amplification and sequencing. TGS or long-read sequencing is fundamentally different from clonal-based sequencing methods and is typically characterized by reading the sequences at the level of single-molecule and requires no PCR amplification. There are several technical challenges associated with TGS technologies primarily related to lower signal quality due to longer read-length (Treangen and Salzberg 2011) on which active research work is still going on. Nevertheless, a few of such technologies like PacBio SMRT and Oxford Nanopore have made inroads. Here we discuss these two promising technologies.

PacBio SMRT Sequencing

In a way, PacBio SMRT sequencing is also a sequencing by synthesis type of reaction. However, it differs from previously described technologies as it performs the sequencing at single-molecule level in real time. Core of this technology is zero-mode waveguides (ZMW) that consists of holes in a metal film that are of sub-wavelength size. With this method, it is possible to study the single-molecule dynamics at micromolar concentrations with microsecond temporal resolution (Levene et al. 2003). The library preparation includes ligating the hairpin adapters onto DNA molecules, thereby circularizing them into a construct termed a SMRTbell. Then, primer and polymerase are annealed to the adapter and library is put on a SMRT cell that contains ZMWs. The polymerase in a ZMW incorporates fluorescently labeled nucleotides thereby emitting a fluorescent signal that is recorded by a highly sensitive camera (Rhoads and Au 2015). In 2011, PacBio introduced its first commercialized SMRT sequencer having a read length of around 1100 bases. PacBio has continuously made incremental improvements in the reagent chemistry and equipment. Its RS II system with P6-C4 chemistry has an average read length of 10,000–15,000 bases. In 2015, a new system called Sequel was launched. In March 2018, an upgrade in Sequal system was announced claiming to achieve up to 10 Gb per SMRT Cell for de novo genome assembly, effectively doubling the throughput when using ultra-long inserts (>40 kb) (Anonymous 2018). More recently during annual AGBT Conference, PacBio has presented Sequel II System, a new SMRT Cell (8 M) with eight million ZMW's, increasing the expected throughput per SMRT Cell by a factor of eight (Anonymous 2019).

Oxford Nanopore Sequencing

Another major breakthrough in the TGS technologies is the Oxford Nanopore sequencing which relies on protein nanopores embedded into a synthetic membrane present in an electrophysiological solution. An ionic current is allowed to pass through the nanopores. When DNA or RNA molecules pass through the nanopores, they cause interruption in current which is recorded in real time to determine the sequence of bases. The utility of an electric field capable of driving single-stranded nucleic acid molecules through a nanometer-sized ion channel in a lipid membrane for providing direct, high-speed detection of the base sequence was described in as early as 1996 by Kasianowicz and co-workers. Nevertheless, the first nanopore sequencer MINION became commercially available only in 2014. It has a small form factor with pocket-sized USB-interfaced device that is capable of generating extra-long read from libraries. In spite of higher error rates at base level, MINION has generated considerable attention of the scientific community worldwide due to its portability, low upfront and running cost, longer read length, and greater speed in data generation. It is an emerging and promising sequencing solution that could potentially surpass all currently available high-throughput sequencing technologies (Marx 2015; Laver et al. 2015).

7.5 Conclusions and Future Perspectives

In order to provide food to the burgeoning population of the world, it is essential to maintain soil fertility for getting higher productivity of the crops. In view of the drawbacks associated with chemical fertilizers in terms of environmental pollution degradation, the use of organic fertilizers appears more prudent. Earthworms are considered as farmers' friend. However, in order to fully utilize the potential of earthworms, it is imperative to understand the complex microbial community dynamics of vermicomposting. In the past few decades, advancement in instrumentation has improved our understanding about the microbial community dynamics of biologically degrading environmental samples like vermicompost. Now it is possible to record the minutest changes taking place in community structure, within unprecedented time. Enormous amounts of data generated with the help of high throughput sequencers have allowed us to dive deep into the complex interactions of microorganisms and earthworms. Today, data on whole-genome assemblies of many of the bacterial and fungal species associated with vermicomposting are available. Even though the output and error rate appears to have scope for substantial improvements, the NGS and TGS technologies have opened up new vistas not only for characterizing the microbial communities for finding out the evolutionary linkages but also for utilizing the data for devising strategies for more efficient vermicomposting in order to produce greater quality of the product. In near future, it is expected to formulate vermicompost quality attributes utilizing the signatures of microbial community dynamics generated from third-generation sequencing technologies. For this to achieve, portable hand-held sequencing platforms like that of Oxford Nanopore hold immense potential.

References

Abrams ES, Murdaugh SE, Lerman LS (1990) Comprehensive detection of single base changes in human genomic DNA using denaturing gradient gel-electrophoresis and a GC clamp. Genomics 7:463–475

Agrawal PK, Agrawal S, Shrivastava R (2015) Modern molecular approaches for analyzing microbial diversity from mushroom compost ecosystem. 3 Biotech 5:853–866

Aira M, Monroy F, Domínguez J (2007) Microbial biomass governs enzyme activity decay during aging of worm-worked substrates through vermicomposting. J Environ Qual 36:448–452

Aira M, Pérez-Losada M, Domínguez J (2018) Diversity, structure and sources of bacterial communities in earthworm cocoons. Sci Rep 8:6632

Alves A, Phillips AJL, Henriques I, Correia A (2005) Evaluation of amplified ribosomal DNA restriction analysis as a method for the identification of Botryosphaeria species. FEMS Microbiol Lett 245:221–229

Anonymous (2014) Illumina. HiSeq X Ten Series of Sequencing Systems. www.illumina.com/documents/products/datasheets/datasheet-hiseq-x-ten.pdf. Accessed 25 May 2019

Anonymous (2018) New Software, Polymerase for Sequel System Boost Throughput and Affordability. https://www.pacb.com/blog/new-software-polymerase-sequel-system-boost-throughput-affordability.

Anonymous (2019) Latest PacBio Sequencing Advancements Including Preview of Sequel II System to be Featured at Annual AGBT Conference. https://www.globenewswire.com/news-release/2019/02/26/1742290/0/en.

Bardakci F (2001) Random amplified polymorphic DNA (RAPD) markers. Turk J Biol 25:185–196

Bartlett JMS, Stirling D (2003) A short history of the polymerase chain reaction. PCR protocols, methods in molecular biology 226 (2nd ed), pp 3–6

Blaya J, Marhuenda FC, Pascual JA, Ros M (2016) Microbiota characterization of compost using omics approaches opens new perspectives for Phytophthora root rot control. PLoS One 11: e0158048

Blomström AL, Lalander C, Komakech AJ, Vinnerås B, Boqvist S (2016) A metagenomic analysis displays the diverse microbial community of a vermicomposting system in Uganda. Infect Ecol Epidemiol 6:32453

Borneman J, Triplett EW (1997) Molecular microbial diversity in soils from Eastern Amazonia: evidence for unusual microorganisms and population shifts associated with deforestation. Appl Environ Microbiol 63:2647–2653

Brooks JP, Edwards DJ, Harwich MD, Rivera MC, Fettweis JM, Serrano MG, Reris RA, Sheth NU, Huang B (2015) The truth about metagenomics: quantifying and counteracting bias in 16S rRNA studies. BMC Microbiol 15:66. https://doi.org/10.1186/s12866-015-0351-6

Buermans HPJ, den Dunnen JT (2014) Next generation sequencing technology: advances and applications. Biochim Biophys Acta Mol basis Dis 1842:1932–1941

Bumgarner R (2013) Overview of DNA microarrays: types, applications, and their future. Current protocols in molecular biology, Unit–221. https://doi.org/10.1002/0471142727.mb2201s101

Cai L, Gong X, Sun X, Li S, Yu X (2018) Comparison of chemical and microbiological changes during the aerobic composting and vermicomposting of green waste. PLoS One 13:e0207494

Chan K-G, Yin W-F, Lim YL (2014) Complete genome sequence of *Pseudomonas aeruginosa* strain YL84, a quorum-sensing strain isolated from compost. Genome Announc 2:e00246–e00214

Chauhan PS, Shagol CC, Yim W, Tipayno SC, Kim CG, Sa T (2011) Use of terminal restriction length polymorphism (T-RFLP) analysis to evaluate uncultivable microbial community structure of soil. Korean J Soil Sci Fert 44:127–145

Cherif H, Ouzari H, Marzorati M, Brusetti L, Jedidi N, Hassen A, Daffonchio D (2008) Bacterial community diversity assessment in municipal solid waste compost amended soil using DGGE and ARISA fingerprinting methods. World J Microbiol Biotechnol 24:1159. https://doi.org/10.1007/s11274-007-9588-z

Collins FS, Morgan M, Patrinos A (2003) The human genome project: lessons from large-scale biology. Science 300:286–290
Danon M, Franke-Whittle IH, Insam H, Chen Y, Hadar Y (2008) Molecular analysis of bacterial community succession during prolonged compost curing. FEMS Microbiol Ecol 65:133–144
Darwin C (1881) The formation of vegetable mould, through the action of worms, with observations on their habits. John Murray, London
Domínguez J, Edwards CA (2004) Vermicomposting organic wastes: a review. In: Shakir Hana SH, Mikhail WZA (eds) Soil zoology for sustainable development in the 21st century. Self-Publisher, Cairo, pp 369–395
Donohoe K (2018) Chemical and microbial characteristics of vermicompost leachate and their effect on plant growth. PhD Thesis, University of Sydney
Edel-Hermann V, Dreumont C, Pérez-Piqueres A, Steinberg C (2004) Terminal restriction fragment length polymorphism analysis of ribosomal RNA genes to assess changes in fungal community structure in soils. FEMS Microbiol Ecol 47:397–404
Fischer SG, Lerman LS (1979) Length independent separation of DNA restriction fragments in 2-dimensional gel electrophoresis. Cell 16:191–200
Fisher MM, Triplett EW (1999) Automated approach for ribosomal intergenic spacer analysis of microbial diversity and its application to freshwater bacterial communities. Appl Environ Microbiol 65:4630–4636
Fracchia L, Dohrmann AB, Martinotti MG, Tebbe CC (2006) Bacterial diversity in a finished compost and vermicompost: differences revealed by cultivation-independent analyses of PCR-amplified 16S rRNA genes. Appl Microbiol Biotechnol 71:942–952
Franke-Whittle IH, Knapp BA, Fuchs J, Kaufmann R, Insam H (2009) Application of COMPOCHIP microarray to investigate the bacterial communities of different composts. Microbial Ecol 57:510–521
Franke-Whittle IH, Confalonieri A, Insam H, Schlegelmilch M, Korner I (2014) Changes in the microbial communities during co-composting of digestates. Waste Manag 34:632–641
Franke-Whittle IH, Juárez M, Insam F-D, Schweizer H, Naef S, Topp A et al (2018) Performance evaluation of locally available composts to reduce replant disease in apple orchards of Central Europe. Renewable Agric Food Syst 34:543. https://doi.org/10.1017/S1742170518000091
Gafan GP, Spratt DA (2005) Denaturing gradient gel electrophoresis gel expansion (DGGEGE)-an attempt to resolve the limitations of co-migration in the DGGE of complex polymicrobial communities. FEMS Microbiol Lett 253:303–307
Galitskaya P, Biktasheva L, Saveliev A, Grigoryeva T, Boulygina E, Selivanovskaya S (2017) Fungal and bacterial successions in the process of co-composting of organic wastes as revealed by 454 pyrosequencing. PLoS One 12:e0186051
Gómez-Brandón M, Aira M, Lores M, Domínguez J (2011) Changes in microbial community structure and function during vermicomposting of pig slurry. Bioresour Technol 102:4171–4178
Gómez-Brandón M, Lores M, Domínguez J (2012) Species-specific effects of Epigeic earthworms on microbial community structure during first stages of decomposition of organic matter. PLoS One 7:e31895
Gopal M, Bhute SS, Gupta A, Prabhu SR, Thomas GV, Whitman WB, Jangid K (2017) Changes in structure and function of bacterial communities during coconut leaf vermicomposting. Antonie Van Leeuwenhoek 110:1339–1355
Greenwood M (2018) What is pyrosequencing?. News-Medical. https://www.news-medical.net/life-sciences/What-is-Pyrosequencing.aspx.
Grunstein M, Hogness DS (1975) Colony hybridization: a method for the isolation of cloned DNAs that contain a specific gene. Proc Natl Acad Sci U S A 72:3961–3965
Hollmer M (2013) Roche to close 454 Life Sciences as it reduces gene sequencing focus. Fierce Biotech. https://www.fiercebiotech.com/medical-devices/roche-to-close-454-life-sciences-as-it-reduces-gene-sequencing-focus. Accessed 25 May 2019

Huang K, Li F, Wei Y, Chen X, Fu X (2013) Changes of bacterial and fungal community compositions during vermicomposting of vegetable wastes by *Eisenia foetida*. Bioresour Technol 150:235–241

Hugenholtz P, Goebel BM, Pace NR (1998) Impact of culture-independent studies on the emerging phylogenetic view of bacterial diversity. J Bacteriol 180:4765–4774

Kasianowicz JJ, Brandin E, Branton D, Deamer DW (1996) Proc Natl Acad Sci U S A 93:13770–13773

Keller M, Zengler K (2004) Tapping into microbial diversity. Nat Rev Microbiol 2:141–150

Khan MH, Meghvansi MK, Gupta R, Chaudhary KK, Prasad K, Siddiqui S, Veer V, Varma A (2015) Combining application of vermiwash and arbuscular mycorrhizal fungi for effective plant disease suppression. In: Meghvansi MK, Varma A (eds) Organic amendments and soil suppressiveness, Soil Biology, vol 46. Springer International Publishing, Switzerland, pp 479–493

Kisand V, Wikner J (2003) Limited resolution of 16S rDNA DGGE caused by melting properties and closely related DNA sequences. J Microbiol Methods 54:183–191

Kolbe AR, Aira M, Gómez-Brandón, Pérez-Losada M, Domínguez J (2019) Bacterial succession and functional diversity during vermicomposting of the white grape marc *Vitis vinifera* v. Albariño. Sci Rep 9:7472.doi:https://doi.org/10.1038/s41598-019-43907-y

Konstantinos KV, Panagiotis P, Antonios VT, Agelos P, Argiris NV (2008) PCR–SSCP: a method for the molecular analysis of genetic diseases. Mol Biotechnol 38:155–163

Kulski JK (2016) Next-generation sequencing — an overview of the history, tools, and "Omic" applications. In: Kulski JK (ed) Next generation sequencing advances, applications and challenges. IntechOpen Limited, London, pp 1–60

Lahens NF, Ricciotti E, Smirnova O, Toorens E, Kim EJ, Baruzzo G, Hayer KE, Ganguly T, Schug J, Grant GR (2017) A comparison of Illumina and Ion Torrent sequencing platforms in the context of differential gene expression. BMC Genomics 18:602. https://doi.org/10.1186/s12864-017-4011-0

Laver T, Harrison J, O'Neill PA, Moore K, Farbos A, Paszkiewicz K, Studholme DJ (2015) Assessing the performance of the Oxford Nanopore technologies MinION. Biomol Detect Quantif 3:1–8

Le Bayon R-C, Bullinger-Weber G, Schomburg A, Turberg P, Schlaepfer R, Guenat C (2017) Earthworms as ecosystem engineers: a review. In: Horton CG (ed) Earthworms: types, roles and research. Nova Science Publishers, New York, pp 129–177

Levene MJ, Korlach J, Turner SW, Foquet M, Craighead HG, Webb WW (2003) Zero-mode waveguides for single-molecule analysis at high concentrations. Science 299(5607):682–686

Liu W, Marsh T, Cheng H, Forney L (1997) Characterization of microbial diversity by determining terminal restriction fragment length polymorphisms of genes encoding 16S rRNA. Appl Environ Microbiol 63:4516–4522

Liu M, Wang C, Wang F, Xie Y (2018) Vermicompost and humic fertilizer improve coastal saline soil by regulating soil aggregates and the bacterial community. Arch Agron Soil Sci 65:281–293

Lores M, Gómez-Brandón M, Pérez-Díaz D, Domínguez J (2006) Using FAME profiles for the characterization of animal wastes and vermicomposts. Soil Biol Biochem 38:2993–2996

Lv B, Xing M, Yang J, Zhang L (2015) Pyrosequencing reveals bacterial community differences in composting and vermicomposting on the stabilization of mixed sewage sludge and cattle dung. Appl Microbiol Biotechnol 99:10703–10712

Malik M, Kain J, Pettigrew C, Ogram A (1994) Purification and molecular analysis of microbial DNA from compost. J Microbiol Methods 20:183–196

Marsh S (2007) Pyrosequencing applications. In: Marsh S (ed) Pyrosequencing protocols, methods in molecular biology. Humana, Totowa, NJ, pp 15–24

Marx V (2015) Nanopores: a sequencer in your backpack. Nat Methods 12(11):1015–1018

Maxam AM, Gilbert W (1977) A new method for sequencing DNA. Proc Natl Acad Sci U S A 74:560–564

Meghvansi MK (2011) Contribution of Earthworm bioturbation to soil suppressiveness. DRDO Science Spectrum, DESIDOC, Defence R & D Organisation, New Delhi, p 167–172
Meghvansi MK, Veer V (2014) Vermibiotechnology for solid waste management. Kashi Block Industry (Press), Tezpur, India
Meghvansi MK, Singh L, Srivastava RB, Varma A (2011) Assessing the role of earthworms in biocontrol of soilborne plant fungal diseases. In: Karaca A (ed) Biology of earthworms, Soil Biology, vol 24. Springer, Berlin, Heidelberg, pp 173–190
Meghvansi MK, Khan MH, Gupta R, Chaudhary KK, Siddiqui S, Veer V (2015) Vermibiotechnology: relevance, challenges and future prospects for India. South Asian J Exp Biol 5:222–228
Meghvansi MK, Khan MH, Gupta R, Chaudhary KK, Veer V (2016) Comparative evaluation of epigeic earthworm species suggests better nutrient mineralization efficiency of vegetable market solid waste by *Eisenia fetida*. South Asian J Exp Biol 6:234–240
Mickan BS, Abbott LK, Fan J, Har MM, Siddique KHM, Solaiman ZM, Jenkins SN (2018) Application of compost and clay under water-stressed conditions influences functional diversity of rhizosphere bacteria. Biol Fertil Soils 54:55–70
Minnich J (1977) The earthworm book. Rodale Press, Emmaus, PA
Mülhardt C, Beese EW (2007) DNA Analysis. Molecular biology and genomics. Academic, Cambridge, pp 151–168
Muralidharan K, Wakeland EK (1993) Concentration of primer and template qualitatively affects products in random-amplified polymorphic DNA PCR. BioTechniques 14:362–364
Muyzer G (1999) Genetic fingerprinting of microbial communities: present status and future perspective. In: Bell CR, Brylinsky M, Johnson-Green P (eds) Proceedings of the 8th International Symposium on Microbial Ecology, Atlantic Canada Society for Microbial Ecology, Halifax, NovaScotia, pp 1–10
Muyzer G, Teske A, Wirsen CO, Jannasch HW (1995) Phylogenetic relationships of Thiomicrospira species and their identification in deep-sea hydrothermal vent samples by denaturing gradient gel electrophoresis of 16S rDNA fragments. Arch Microbiol 164:165–172
Ntougias S, Zerva kis GI, Kavroulakis N, Ehaliotis C, Papadopoulou KK (2004) Bacterial diversity in spent mushroom compost assessed by amplified rDNA restriction analysis and sequencing of cultivated isolates. Syst Appl Microbiol 27:746–754
Ojha RB, Devkota D (2014) Earthworms: 'Soil and ecosystem Engineers'—a review. World J Agric Res 2:257–260
Orita M, Suzuki Y, Sekiya T, Kaayashi K (1989) Rapid and sensitive detection of point mutations and DNA polymorphisms using the polymerase chain reaction. Genomics 5:874–879
Partanen P, Hultman J, Paulin L, Auvinen P, Romantschuk M (2010) Bacterial diversity at different stages of the composting process. BMC Microbiol 10:94. https://doi.org/10.1186/1471-2180-10-94
Parthasarathi K, Ranganathan LS (1998) Pressmud vermicast are hot spots of fungi and bacteria. Ecol Environ Cons 4:81–86
Pathma J, Sakthivel N (2012) Microbial diversity of vermicompost bacteria that exhibit useful agricultural traits and waste management potential. Springerplus 1:26. https://doi.org/10.1186/2193-1801-1-26
Pennisi E (2010) Semiconductors inspire new sequencing technologies. Science 327(5970):1190. https://doi.org/10.1126/science.327.5970.1190
Pérez-de-Mora A, Burgos P, Madejón E, Cabrera F, Jaeckel P, Schloter M (2006) Microbial community structure and function in a soil contaminated by heavy metals: effects of plant growth and different amendments. Soil Biol Biochem 38:327–341
Pizl V, Novakova A (2003) Interactions between microfungi and *Eisenia andrei* (Oligochaeta) during cattle manure vermicomposting. Pedobiologia 47:895–899
Polz MF, Cavanaugh CM (1998) Bias in template-to-product ratios in multi-template PCR. Appl Environ Microbiol 64:3724–3730

Rastogi G, Sani RK (2011) Molecular techniques to assess microbial community structure, function, and dynamics in the environment. In: Ahmad I (ed) Microbes and microbial technology: agricultural and environmental applications. Springer Nature, Singapore, pp 29–57
Rettedal EA, Clay S, Brozel VS (2010) GC-clamp primer batches yield 16S rRNA gene amplicon pools with variable GC clamps, affecting denaturing gradient gel electrophoresis profiles. FEMS Microbial Lett 312:55–62
Rhoads A, Au KF (2015) PacBio sequencing and its applications. Genomics Proteomics Bioinformatics 13:278–289
Salipante SJ, Kawashima T, Rosenthal C, Hoogestraat DR, Cummings LA, Sengupta DJ et al (2014) Performance comparison of illumina and ion torrent next generation sequencing platforms for 16S rRNA-based bacterial community profiling. Appl Environ Microbiol 80:7583–7591
Sanger F (1977) Nucleotide sequence of bacteriophage phi X174 DNA. Nature 265:687–695
Sanger F, Nicklen ARC (1977) DNA sequencing with chain-terminating. Proc Natl Acad Sci U S A 74:5463–5467
Schierwater B, Ender A (1993) Different thermostable DNA polymerases may amplify different RAPD products. Nucleic Acids Res 21:4647–4648
Schloss PD, Hay AG, Wilson DB, Walker LP (2003) Quantitative and qualitative changes in microbial community structure during the early stages of composting using ARISA. FEMS Microbiol Ecol 46:1–9
Schoch CL, Seifert KA, Huhndorf S, Robert V, Spouge JL, Levesque CA, Chen W, Fungal Barcoding Consortium (2012) Nuclear ribosomal internal transcribed spacer (ITS) region as a universal DNA barcode marker for Fungi. Proc Natl Acad Sci U S A 109:6241–6246
Sen B, Hamelin J, Bru-Adan V, Godon JJ, Chandra TS (2008) Structural divergence of bacterial communities from functionally similar laboratory-scale vermicomposts assessed by PCR-CE-SSCP. J Appl Microbiol 105:2123–2132
Sigler WV, Crivii S, Zeyer J (2002) Bacterial succession in glacial forefield soils characterized by community structure, activity and opportunistic growth dynamics. Microb Ecol 44:306–316
Singh A, Singh DP, Tiwari R, Kumar K, Singh RV, Singh S, Prasanna R, Saxena AK, Nain L (2015) Taxonomic and functional annotation of gut bacterial communities of Eisenia foetida and Perionyx excavates. Microbiol Res 175:48–56
Sinha RK, Agarwal S, Chauhan K, Valani D (2010) The wonders of earthworms and its vermicompost in farm production: Charles Darwin's 'friends of farmers', with potential to replace destructive chemical fertilizers from agriculture. Agric Sci 1:76–94
Sklarz M, Angel R, Gillor O, Soares MIM (2009) Evaluating amplified rDNA restriction analysis assay for identification of bacterial communities. Antonie Van Leeuwenhoek 96:659–664
Speranskaya AS, Khafizov K, Ayginin AA, Krinitsina AA, Omelchenko DO, Nilova MV, Severova EE, Samokhina EN, Shipulin GA, Logacheva MD (2018) Comparative analysis of illumina and ion torrent high-throughput sequencing platforms for identification of plant components in herbal teas. Food Control 93:315–324
Stone A, Steven JS, Heather D (2004) Suppression of soilborne diseases in field agricultural systems: organic matter management, cover cropping, and other cultural practices. In: Magdoff F, Weil RR (eds) Soil organic matter in sustainable agriculture. CRC, Boca Ratan, pp 132–179
Strathdee F, Free A (2013) Denaturing gradient gel electrophoresis (DGGE). In: Makovets S (ed) DNA electrophoresis: methods and protocols, methods in molecular biology, vol 1054. Springer, New York, pp 145–157
Székely AJ, Sipos R, Berta B, Vajna B, Hajdú C, Márialigeti K (2009) DGGE and T-RFLP analysis of bacterial succession during mushroom compost production and sequence-aided T-RFLP profile of mature compost. Microb Ecol 57:522–533
Treangen TJ, Salzberg SL (2011) Repetitive DNA and next-generation sequencing: computational challenges and solutions. Nat Rev Genet 13(1):36–46. [published correction appeared in Nat Rev Genet 13:146]

Uchiyama K, Suzuki T, Tatsumi H, Kanetake H, Shioya S (2002) Amplified 16S ribosomal DNA restriction analysis of microbial community structure during rapid degradation of a biopolymer, PHA, by composting. In: Insam H, Riddech N, Klammer S (eds) Microbiology of composting. Springer, Berlin, Heidelberg, pp 83–98

van de Zande L, Bijlsma R (1995) Limitations of the RAPD technique in phylogeny reconstruction in drosophila. J Evol Biol 8:645–656

Villar I, Alves D, Mato S (2017) Product quality and microbial dynamics during vermicomposting and maturation of compost from pig manure. Waste Manag 69:498–507

Wagner M, Smidt H, Loy A, Zhou J (2007) Unravelling microbial communities with DNA-microarrays: challenges and future directions. Microbial Ecol 53:498–506

Wilkinson MT, Richards PJ, Humphreys GS (2009) Breaking ground: pedological, geological, and ecological implications of soil bioturbation. Earth-Sci Rev 97:257–272

Woese CR (1987) Bacterial evolution. Microbiol Rev 51:221–271

Woese CR, Fox GE (1977) Phylogenetic structure of the prokaryotic domain: the primary kingdoms. Proc Natl Acad Sci U S A 74:5088–5090

Zhao C, Wang Y, Wang Y, Wu F, Zhang J, Cui R, Wang L, Mu H (2018) Insights into the role of earthworms on the optimization of microbial community structure during vermicomposting of sewage sludge by PLFA analysis. Waste Manag 79:700–708

Zhen Z, Liu H, Wang N, Guo L, Meng J et al (2014) Effects of manure compost application on soil microbial community diversity and soil microenvironments in a temperate cropland in China. PLoS One 9:e108555

Chapter 8
Facile Monitoring of the Stability and Maturity of Compost Through Fast Analytical Instrumental Techniques

Pablo Martín-Ramos and Jesús Martín-Gil

Abstract This chapter delivers a panorama of the latest progress on characterization techniques to study the biodegradation and transformation of organic residues and structural agents during composting. Although biological tests can provide valuable information on compost stability and maturity, further insight into the transformation by composting can be easily gained through the monitoring of a set of chemical parameters obtained from other analytical instrumental techniques. Spectroscopic and thermal analysis methods allow the facile tracking of physicochemical parameters such as the C/N ratio, the water-soluble organic carbon, the mineralization rate, etc. and supply data associated with kinetic and thermodynamic processes. The different sections covered herein explain and review how they may be used for the rapid identification of fulvic and humic acids or of the different lignocellulosic constituents; for the determination of aromaticity or of carbon contents (both soluble and in solid-phase); for the tracking of dehydration, decarboxylation reactions, degree of humification, C/N ratio evolution, indirect microbial activity, etc. Examples of the biodegradation of different heterogeneous raw materials are discussed and achievements reported in the last few years are described.

Keywords Composting · Humification · Maturity · Monitoring · Spectroscopy · Stability · Thermal analysis

P. Martín-Ramos (✉)
Department of Agricultural and Environmental Sciences, EPS, Instituto Universitario de Investigación en Ciencias Ambientales de Aragón (IUCA), Universidad de Zaragoza, Huesca, Spain
e-mail: pmr@unizar.es

J. Martín-Gil
Agriculture and Forestry Engineering Department, ETSIIAA, Universidad de Valladolid, Palencia, Spain
e-mail: mgil@iaf.uva.es

M. K. Meghvansi, A. Varma (eds.), *Biology of Composts*, Soil Biology 58,
https://doi.org/10.1007/978-3-030-39173-7_8

8.1 Introduction

Composting is a widespread technology to convert residual organic matter into a sanitized, stabilized, and mature product, with a substantial lessening of weight and volume during the process, making it easier to store and facilitating its commercialization as a fertilizer or as a plant soil amendment.

The main compost comprehensive properties, which are indicative of the degree of organic matter decomposition and of its potential phytotoxicity, are maturity and stability. Nevertheless, compost stability and compost maturity are not synonymous (Iannotti et al. 1994). The first generally refers to degree of completion of the composting process, while the second is representative either of the degree of humification of the material or of the microbial biomass level of activity (Butler et al. 2001). As composting is a microbial-based process, both compost maturity and stability are the outcome of microbial activities. Thus, the chemical composition of the compost and its status in terms of organic matter decomposition play a major role in the determination of the microbial status (Wu and Ma 2002).

An appropriate index to evaluate the maturity of a compost should have an overall trend that can be described by a monotonic function, so that its measurement yields unequivocal results. Moreover, the stabilization of the index should not be reached too early, that is, prior to the end of the thermophilic period (De Nobili and Petrussi 1988). The degree of compost maturity at the end of the process may be regarded as crucial in terms of product marketability (Butler et al. 2001).

8.2 Chemical and Biochemical Parameters to Track the Compost Process

The difficulty of compost maturity evaluation is still one of the main problems faced by composting programs. Different methods—either biological, chemical, or physicochemical—have been assessed to characterize the maturity of composts and their agrochemical properties (Barberis and Nappi 1996; Senesi and Brunetti 1996). The tests used for this purpose are usually based on the study of one (or several) of the major changes that organic matter eventually undergoes during composting before its stabilization, e.g., changes in cation exchange capacity (CEC), evolved carbon dioxide-carbon (CO_2-C), C/N ratio, ammonium-N, nitrate-N, total organic carbon (TOC), dissolved organic carbon (DOC), total Kjeldahl nitrogen (TN), potassium (TK), phosphorus (TP), fulvic acid (FA), humic acid (HA), degree of polymerization (DP), CO_2, temperature, pH, or odor.

It is known that temperature rapidly increases from the mesophilic stage to the thermophilic stage, and then progressively decreases near the maturation phase; and that a similar behavior occurs for pH, which becomes slightly alkaline when maturity is reached. In the composting process, the concentration of total carbon, NH_4^+-N, NH_4/NO_3 ratio, C/N ratio, and CO_2 decrease, while TK, TP, HA, CEC, DP, and TN increase. Among these parameters, the diminution of the C/N ratio has been frequently chosen as the main indicator of the maturity of a compost (Iqbal et al. 2010).

At present there is consensus on the impossibility of using a particular parameter to determine the maturity, stability, and/or quality of compost by means of a single method, because of the high variability observed in composts, provided that their characteristics have a strong dependence on the raw materials and on the chosen composting system (Bustamante et al. 2010). This fact is particularly notable for materials such as winery-distillery residues, which show an atypical trend of humification parameters (e.g., humification index and humification rate[1]) during the development of the composting process (Bustamante et al. 2008). In these cases, an increase of humified to non-humified carbon ratio must be used to monitor compost maturity.

One of the best solutions to problems on compost maturity evaluation could be the monitoring of the evolution of organic matter (Castaldi et al. 2005). According to the literature, the amount and the chemical structure of humic substances, mainly HAs formed during the composting process, constitute a very relevant index for the assessment of the maturity and stability of a compost (Tomati et al. 2000). The percentage of HAs is calculated from the C_{ha}/C_{ext} quotient multiplied by 100, where C_{ha} and C_{ext} stand for HA-like carbon and extractable organic carbon, respectively (Torres-Climent et al. 2015).

Biochemical parameters to track the compost process refer to the plant bioassays related to phytotoxicity (Iwegbue et al. 2006; Katayama and Kubota 1995) and respirometric techniques (Sadaka et al. 2006).

Phytotoxicity appraisal involves germination and plant growth tests, and is affected both by the contents of toxic substances (e.g. DOC, volatile fatty acids (VFAs)) and by the nutrients content (e.g., P, N, K, humus). The germination index (GI) has frequently been used to appraise phytotoxicity and hence the maturity of a compost (Antil et al. 2013; Matei et al. 2016): it generally increases in a rapid manner during the first 7 days, and then it presents a slight increase in the final stage of the mesophilic period, as a result of the decomposition of the phytotoxic organic compounds.

One of the most important approaches for determining stability is the use of respirometric methods. These methods are based on the higher rates of O_2 consumption and CO_2 production featured by immature compost material due to the rapid development of microorganisms in easily biodegradable compounds (Barrena Gómez et al. 2005). Thompson et al. (2001) proposed a compost stability classification based on respiration rates (evolved CO_2-C/volatile solids (VS) content in the sample) with the following categories: very stable (1 mg CO_2-C/(g VS·d)), stable (2–4 mg CO_2-C/(g VS·d)), moderately unstable (5–7 mg CO_2-C/(g VS·d)), unstable

[1]The humification ratio (HR) is calculated by multiplying by 100 the extractable organic carbon/total carbon quotient; and the humification index (HI) is determined by multiplying by 100 the humid acid-like carbon/total carbon rate.

raw compost (8–9 mg CO_2-C/(g VS·d)), raw compost (10–11 mg CO_2-C/(g VS·d)), and raw feedstock (>11 mg CO_2-C/(g VS·d)).

The reader interested in conventional methods for bio-stability assessment is referred to the recent and complete survey by Lü et al. (2018).

8.3 Overview of Analytical Instrumental Techniques

Since aforementioned methods are laborious and habitually empirical, direct and rigorous analytical tests may be deemed as necessary. In particular, spectroscopic methods, including Fourier-Transform Infrared spectroscopy (FTIR), Raman spectroscopy, UV–Vis spectroscopy, Fluorescence EEM spectroscopy (and Fluorescence Reflectance Imaging, FRI), and Nuclear Magnetic Resonance (NMR) have been directly used on raw samples of compost without any previous extraction (Chefetz et al. 1996). Thermal analysis methods, mainly thermogravimetry (TG) and differential scanning calorimetry (DSC), have been used to characterize soil organic matter and its humic fractions (Provenzano and Senesi 1999), and thus constitute a simple, rapid, and reliable approach that can be applied to the characterization of whole compost samples (Dell'Abate et al. 1997). The following sections provide details on each of these techniques.

8.3.1 *Vibrational Spectroscopy Techniques*

8.3.1.1 Fourier-Transform Infrared spectroscopy (FTIR) and Diffuse Reflectance Infrared Fourier-Transform Spectroscopy (DRIFT)

Common series of FTIR or DRIFT bands for compost are: a broad band at around 3400 cm^{-1} (H-bonded OH groups); sharp peaks at 2925 and 2850 cm^{-1} (ascribed to C–H stretching vibration of alkyls); a small shoulder at 1791–1733 cm^{-1} (associated with nonconjugated carboxylic C=O), which disappears when pH reaches neutrality; a strong broad peak at around 1630 cm^{-1} (aromatic C=C, C=O and/or C=O of bonded conjugated ketones, quinone, C=O stretching of amide I); a peak at 1460 cm^{-1} (C-H deformation of CH_2 or CH_3 groups); a peak at 1385 cm^{-1} (COOH and CH_3 groups); peaks in the 1200–1280 cm^{-1} range (C–O stretching of aryl ether, C-O and OH of COOH groups, amide III); a strong peak at around 1420 cm^{-1} (aromatic C=C); and a band at 1040 cm^{-1} (C-O stretching of aromatic ether and carbohydrates, Si-O of silicates) (Ouatmane et al. 2000).

To clarify whether the peaks observed at 1420 cm^{-1} and at 1044 cm^{-1} are associated with mineral forms or with organic matter, the infrared spectra of residual ashes should be recorded. Those spectra may be similar to rock phosphate spectrum and exhibit peaks attributable to calcite (at 1420 and 874 cm^{-1}) and/or to phosphate minerals or silicates (1100–950 cm^{-1}, 620 cm^{-1}), thus indicating that the infrequent peaks observed in the compost spectrum actually arise from phosphate rock residues in which the ashes are rich.

Nonetheless, it is worth noting that composts differ from each other in some particular bands. Specifically, the peak at 1540 cm^{-1} (arising from amide II structures), present in domestic solid wastes, agrees with the high nitrogen content in which this type of compost is rich. Besides, freshly collected farmyard manure and sawdust of wood composts show a peak at 1511 cm^{-1}, due to aromatic skeletal vibrations, in agreement with the high lignocellulosic content of these materials.

Differences During Composting The intensity of certain peaks can vary in the FTIR spectra, leveling off or down during the composting process. In particular, for urban and municipal wastes (Chefetz et al. 1996), a decrease in the intensity of the band at 1740 cm^{-1} can be observed, ascribed to organic matter decarboxylation during the decomposition process. Further, as composting time increases, the peak at 1540 cm^{-1} decreases, probably because of the degradation of peptide structures. Upon completion of the composting process, the characteristic peaks are more flattened and aromatic ring stretching vibration peaks at 665 and 671 cm^{-1} significantly increase, indicating that the composting process of easily biodegradable compounds (proteins, carbohydrates, and fatty compounds) is on decline. The increase in the level of aromatization evidences the curing reaction during decomposition.

Materials that are not composted can be identified by bands indicative of reactivity, such as the one at 1320 cm^{-1}.

Semiquantitative changes in the IR spectra can be monitored by calculating the ratios between the intensities of major peaks (Fig. 8.1): the peaks at 2930, 2850,

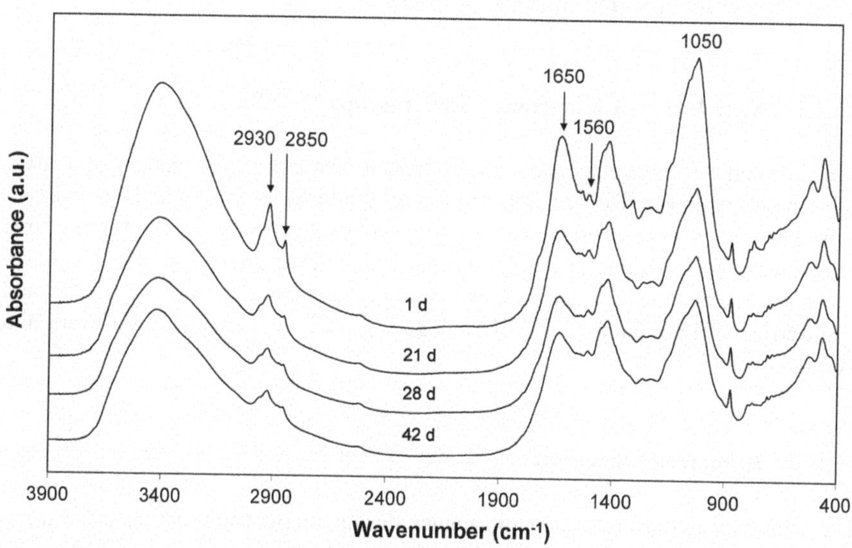

Fig. 8.1 DRIFT spectra at five stages of a composting process for yard/kitchen waste: raw, and after 1, 21, 28, and 42 days. Adapted from Böhm (2009)

1650, 1560, and 1050 cm^{-1} are the ones typically considered (Chefetz et al. 1996; Inbar et al. 1990). The aromatic C/aliphatic C ratio (1650/2930 cm^{-1}) has been reported to increase from 0.88 to 1.10; the aromatic C/aliphatic C ratio (1650/2850 cm^{-1}) from 0.79 to 1.54; the aromatic C/polysaccharides ratio (1650/1050 cm^{-1}) from 2.39 to 2.80; and the aromatic C/amide II bond ratio (1650/1560 cm^{-1}) from 0.94 to 1.5. These changes are indicative of a decrease in the level of polysaccharides, amide, and aliphatic components, accompanied by an increase in aromatic structures in the mature compost. Such relative enhancement in aromatic structures results from aliphatic moieties decomposition (Chefetz et al. 1996).

The linear correlation between the 1650/2930 cm^{-1} (aromatic to aliphatic) ratio and the C/N ratio, calculated according to $y = 1.263 - 0.014x$ formula, exhibited a $R^2 = 0.936$, indicating that FTIR can be deemed as a useful and reliable index for compost maturity evaluation. The highest increase was observed for domestic urban samples, followed by farmyard manure, while sawdust wood samples gave little changes. An increase in 1511/2930 cm^{-1} ratio was also found in the spectra of farmyard manure. In contrast, a noticeable increase in the intensity of two peaks (at 1420 and 1044 cm^{-1}) was found in all the spectra. This trend would agree with an increase in the ash content.

By the end of the composting process, except for sawdust wood, the spectra usually exhibit less complex features than those found in fresh composts. Regardless of the differences in the nature of the raw materials, as the maturity of the compost increases, the spectra are more similar to each other and to those reported in the literature (Chefetz et al. 1996; Inbar et al. 1990) for mature composts. Hence, such similarity between composts may be regarded as an indicator of maturity from a practical point of view (Ouatmane et al. 2000).

8.3.1.2 Near Infrared Reflectance Spectroscopy (NIRS)

Near infrared reflectance spectroscopy (NIRS) is also a practical method of analyzing manure composts (Reeves and Van Kessel 2000; Ye et al. 2005). Once calibrations have been developed, it can provide concentration estimates for multiple components: for instance, available nitrogen can be estimated from uric acid nitrogen in poultry manure compost with an extremely high correlation ($r = 0.99999$) and small errors of prediction (SE $= 0.61$ mg·g^{-1}), as demonstrated by Fujiwara and Murakami (2007).

8.3.1.3 Raman Spectroscopy

The structural evolution of organic matter during thermal maturation (viz. aforementioned increase in aromaticity and reduction in heteroatoms and aliphatic content) is also notably revealed on Raman spectra. Two major peaks, known as *G* and

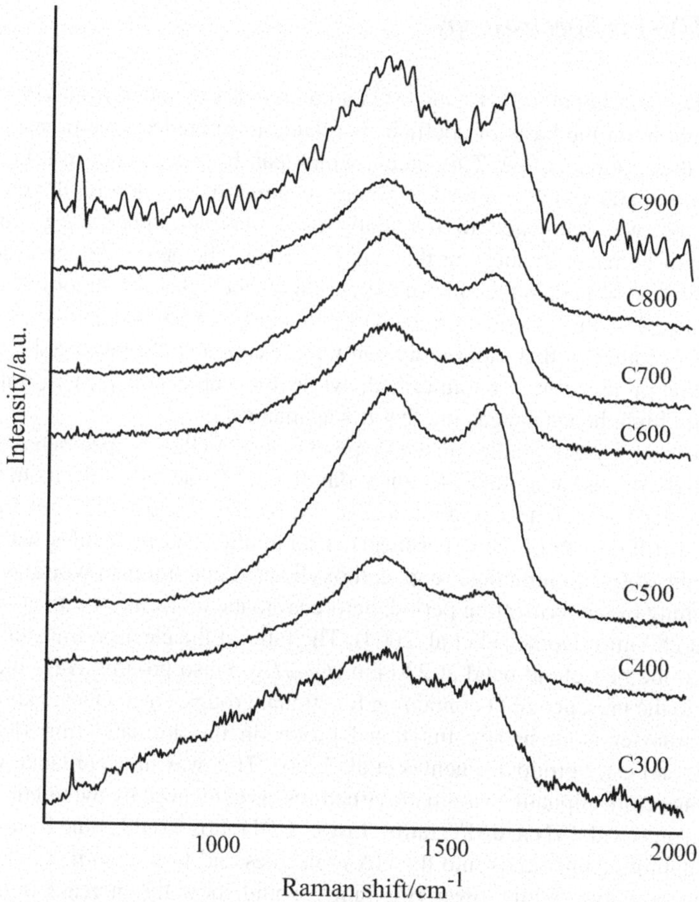

Fig. 8.2 Raman spectra of the sludge biochar samples obtained at various temperatures (Zhang et al. 2015)

D bands, are detectable at Raman shifts of 1358 and 1580 cm^{-1}, respectively (Fig. 8.2). During maturation, many of the hydrogenated and oxygenated chemical groups in the organic precursors rich in heteroatoms are lost, and the number of attachments that separate from aromatic carbons increases. Aforementioned *D* band (at 1350–1367 cm^{-1}) refers to disorder in the atoms, resulting from the Raman-active A_{1g} symmetry, which is related to discontinuities of the sp^2 carbon network-like heteroatoms and in-plane lattice defects. Therefore, the *D* band may be used as an indication of the attachments that separate from organic matter alongside with the generation of hydrocarbons (Khatabi et al. 2018).

8.3.2 UV-Vis Spectroscopy

In 2004, Domeizel et al. (2004) suggested a novel index obtained from UV spectral deconvolution, on the basis of coefficients related to reference spectra and, specifically, to their ratios (a_1/a_2). This index, which can be determined in a facile and quick manner, and which is aimed at giving accurate information on the quality of the compost, was compared to frequently used maturity indices, e.g., the total extraction of humic substances or the C_{HA}/C_{FA} ratio. The new index showed more representative results than the ratio of extraction (as the values of abnormal concentrations caused by this classical protocol of extraction can be disregarded), and than the C_{HA}/C_{FA} ratio (for that supposedly non-mature compost, the proposed UV index showed that it was not yet humidified, while the value of C_{HA}/C_{FA} ratio was abnormally high, hence suggesting that it was mature).

In more recent years, studies on the shape of compost UV–vis spectra have shown that the highest band at around 210 nm at day 0 would gradually shift its maximum toward longer wavelengths, peaking at 228 nm on day 7 and at 232 nm from day 35 onward (till day 49 of the experiment). This implies that molecules with absorbance in the 200–220 nm range (e.g., carboxylic acid and nitrates) would appear at the beginning of the fermenting period, before aromatic molecules (with absorbance at around 280 nm) (Domeizel et al. 2004). The ratio of the electron transfer band at 253 nm to the benzenoid band at 230 nm (E_{253}/E_{230}) also progressively increased, pointing to the presence of O-containing functional groups, since O is a major active electron transfer atom in any functional group on the aromatic ring (hydroxyl, carbonyl, and ester groups) (Fuentes et al. 2006). This was in accordance with the shift change from aliphatic to aromatic structures, as evidenced by the relationship of E_{253}/E_{230} ratio with C/N and H/C ratios. Lower C/N ratios would come from binding of N-containing compounds into the HA molecules and loss of some C via microorganisms activities, while lower H/C ratios would show the increase in aromatic structures (Sanmanee et al. 2011). The constant C/N and H/C ratios in the final stage would suggest the appearance of resistant forms of HA. As it could be discerned by UV spectral deconvolution, after day 49, the shapes of spectra remained the same, with a small decay after 3 months. Thus, it would be advisable to use the compost at day 49, in order to profit from the highest HA quality, and to avoid keeping it longer than 3 months.

To evaluate the maturity of compost, a widely used alternative as an indicator is the E_4/E_6 ratio of the optical densities of HA (at 465 nm) and FA (at 665 nm), determined from the absorbance of 0.05 N $NaHCO_3$ solutions containing 0.02% (w/v) compost (Chanyasak et al. 1982; Chen and Inbar 1993). E_4/E_6 ratios decrease along the composting process, which implies that high molecular weight compounds increase, as a result of increased condensation and aromatization processes that produce more polycondensed HA during composting. Thus, E_4/E_6 may prove useful as an index to assess the molecular weight of humic substances.

Other parameter which reflects the UV–Vis absorbance of humus is the SUVA254, that is, the specific UV absorbance measured at 254 nm and normalized by the dissolved organic carbon (DOC) concentration. Shao et al. (2009) reported that the SUVA254 value increased from 0.04 to 3.13 L/mg/m during the active stage of aerobic bio-stabilization, but after day 30 it remained steady at approximately 3.00 L/mg/m, indicating the stabilization of the municipal solid waste (MSW).

8.3.3 Fluorescence Excitation–Emission Matrix (EEM) Spectroscopy

In 2001, Provenzano et al. (2001) demonstrated that compost maturity could be assessed by fluorescence excitation–emission matrix (EEM) spectroscopy with high instrumental sensitivity. Excitation (Ex)/emission (Em) wavelength pairs (at 330/425 nm) were proposed in relation to HAs in domestic solid waste after 30 days of composting.

In the beginning of the composting process (day 0, Fig. 8.3), fluorescence EEM spectra showed two peaks: *A*, characterized by a 220/340 nm Ex/Em wavelength pair, which falls in the region related to typical aromatic proteins (e.g., tryptophan); and *B*, represented by a 280/328 nm Ex/Em wavelength pair and associated with soluble microbial byproduct-like compounds (Chen et al. 2003; Lv et al. 2013). During composting, the *A* and *B* peaks would show a decreasing trend, and both peaks would finally disappear in the final stage of the experiment, which can be explained by organic matter biodegradation due to microbial activities (Marhuenda-Egea et al. 2007). This result is in good agreement with that reported by Xing et al. (2012). On the 7th day, a third peak would emerge (peak *C*, corresponding to the 270/436 nm Ex/Em wavelength pair, associated with FA-like substances) (Chen et al. 2003). The presence of this peak indicates FAs formation. Marhuenda-Egea et al. (2007) found that various molecular compounds derived from lignin and other degraded plant materials could contribute to FAs fluorescence. Simultaneously, another FA-like compounds-related peak (peak *D*) was also detected, with an Ex/Em wavelength pair at 230/410 nm. This peak allowed the monitoring of changes in FA-like materials during the composting of MSW (He et al. 2011; Saidpullicino et al. 2007). However, as the composting proceeded, peak *D* vanished after day 21. As reported by Marhuenda-Egea et al. (2007), the FA-like components would be synthesized in the humic macromolecules during their formation. The last peak (peak *E*) was detected on the 60th day, at longer excitation and emission wavelengths (Ex/Em wavelength pair at 320/416 nm), and was related to HA-like organics (Antízar-Ladislao et al. 2006). Its intensity gradually increased as composting progressed. Slightly different Ex/Em wavelength pairs were reported by He et al. (2011) (at 313/406 nm) for HAs in extracts from composted MSW.

Yu et al. (2010) using fluorescence EEM in combination with parallel factor (PARAFAC) analysis concluded that components 1 [Ex/Em wavelengths = (230, 330)/410, attributable to humic-like substances] and 3 [Ex/Em

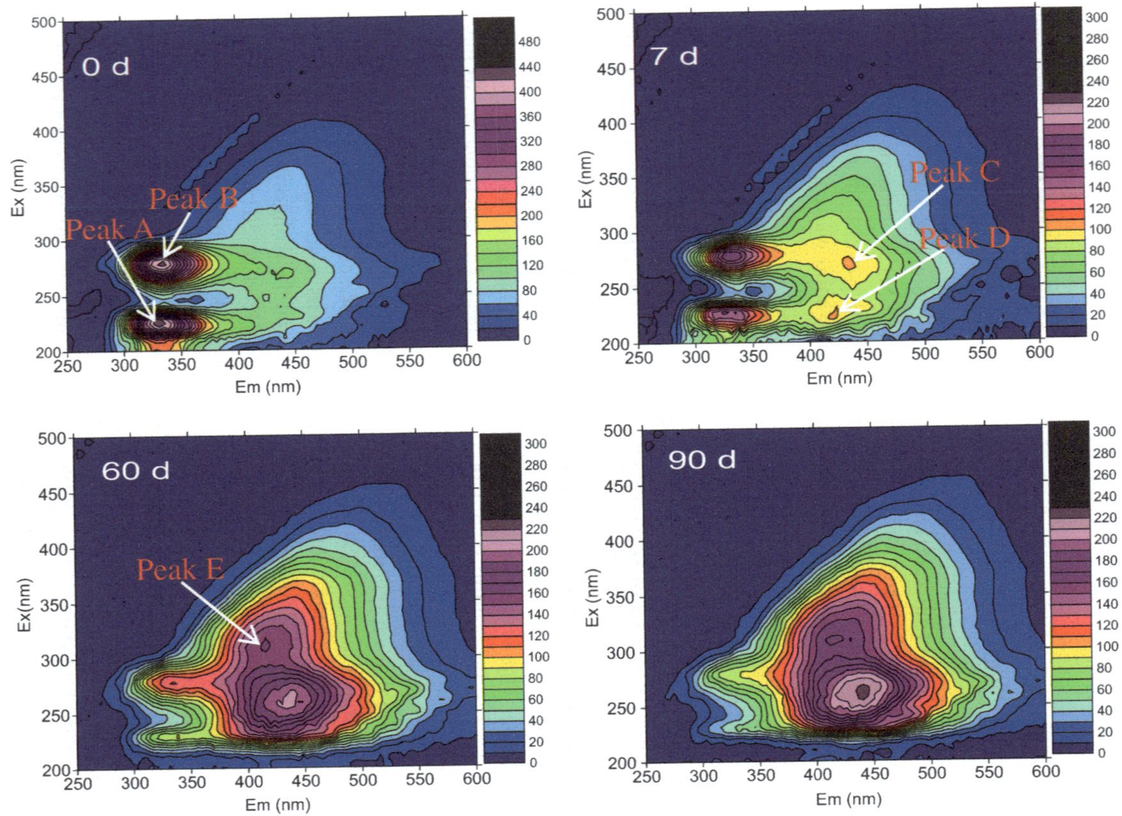

Fig. 8.3 Fluorescence excitation–emission matrix spectra of water extractable organic matter at different stages of vermicomposting. Note: normalized to 1 mg/L DOC by dividing corresponding DOC (Lv et al. 2013)

wavelengths = (220, 280)/340, belonging to protein-like substances] would be the most suitable to assess compost maturity.

Working on composting of animal manures, Yu et al. (2011) found that the EEM contours of water-extract organic matter from immature composts featured four peaks (at Ex/Em wavelengths of 220/340, 280/340, 220/410, and 330/410 nm), whereas those from mature composts only exhibited two (at Ex/Em wavelengths of 230/420 and 330/420 nm).

Fluorescence regional integration (FRI), proposed by Chen's group, is an analytical approach that can assess the heterogeneity of a compost scrutinizing all the wavelength-dependent fluorescence intensity data from EEM spectra (Chen et al. 2003; Tang et al. 2011). The EEM spectrum may be divided into five excitation–emission regions: region I associated with tyrosine-like organic compounds, region II related to tryptophan-like organic compounds, region III corresponding to FA-like materials, region IV connected to soluble microbial byproduct-like materials, and region V related to HA-like materials. By normalizing the cumulative excitation–emission area volumes to relative regional areas, the percent fluorescence responses ($P_{i,n}$) could be determined. The $P_{i,n}$ of regions I and II exhibited a decreasing trend during the composting process, indicating a decrease in simple aromatic proteins. On the other hand, the $P_{i,n}$ of regions III and V increased as the composting progressed, revealing an increase in HA-like and FA-like materials (Marhuenda-Egea et al. 2007). Moreover, no obvious change in the $P_{V,n}/P_{III,n}$ ratio was observed in the first 21 days, while it sharply rose afterwards, pointing to humic substances formation (Tang et al. 2011). Busato et al. (2012) reported a similar phenomenon: HAs greatly increased after day 30 during vermicompost maturation. As explained by He et al. (2011), a higher humification degree is associated with condensed aromatic structures more resistant to biodegradation. In addition, a decrease in tyrosine- and tryptophan-like materials and an increase in humic- and fulvic-like substances characterized the transformation during composting, finally resulting in an enhanced stability of the end product.

8.3.4 ^{13}C-NMR Spectroscopy

According to Chen and Inbar (1993), this is a highly promising physical method for the reliable characterization of organic matter and it can provide valuable information on organic matter transformation during composting.

Solid-state ^{13}C NMR spectroscopy with cross-polarization and magic angle spinning (^{13}C-CPMAS-NMR) spectra of bulk compost samples with different degrees of maturity are shown in Fig. 8.4, while Table 8.1 summarizes the relative distribution of signal areas (Spaccini and Piccolo 2007).

Methylene groups (aliphatic rings and chain) and alkyl groups (terminal methyl) presence is evidenced by the two distinct resonances that appear at around δ 21 and 33 ppm, respectively; and the presence of methoxy C and N-alkyl C from lignin and protein residues, respectively, is demonstrated by the resonances in the δ 46–60 ppm

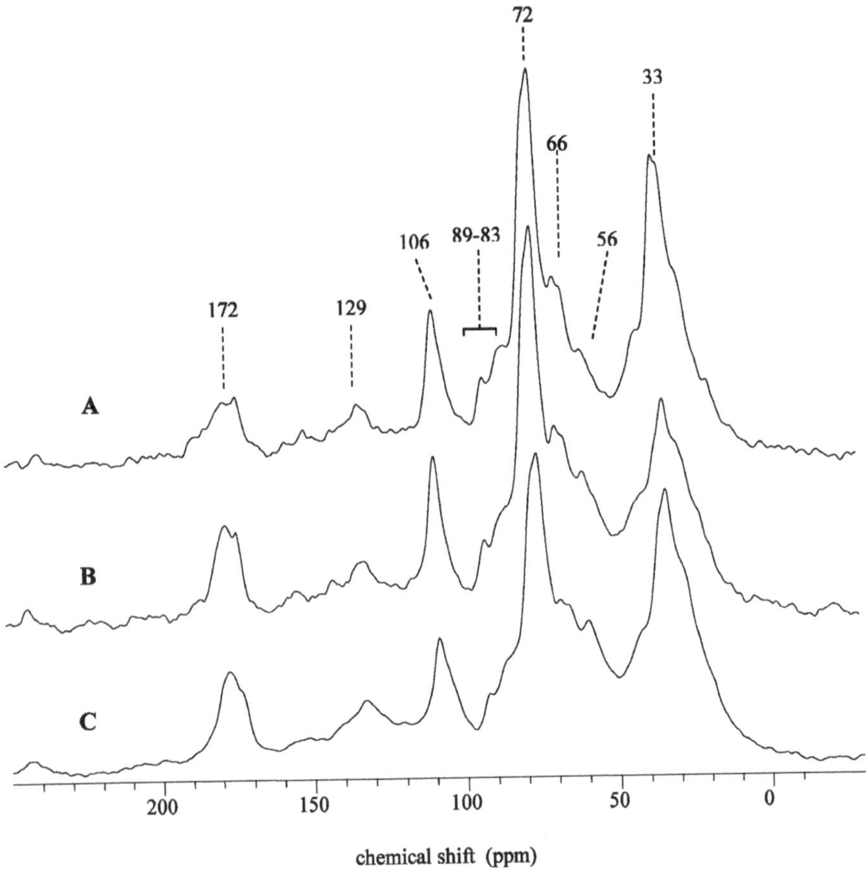

Fig. 8.4 ^{13}C-CPMAS-NMR spectra of bulk compost samples at increasing maturity stages: (**a**) compost 60; (**b**) compost 90; (**c**) compost 150. Reprinted with permission from Spaccini and Piccolo (2007). Copyright American Chemical Society

Table 8.1 Relative distribution (percent) of signal area over chemical shift regions (ppm) in ^{13}C-CPMAS-NMR spectra of compost samples

Sample	0–60	60–110	110–160	160–200	HB/HI[a]
Compost 60	37.6	50.8	7.2	4.3	0.81
Compost 90	30.6	56.1	7.3	6.1	0.61
Compost 150	45.3	37.6	9.6	7.4	1.22

Reprinted with permission from Spaccini and Piccolo (2007). Copyright American Chemical Society
[a]Hydrophobic carbons/hydrophilic carbons

range. The δ 60–110 ppm chemical shift region is associated with polysaccharides (e.g., the peak appearing at around 72 ppm corresponds to C4 of cellulose). Signals at δ 106 and 129 ppm are due to anomeric carbons of polysaccharides and C-substituted aromatic C (Chefetz et al. 1996), while the strong peak at δ 143 ppm can be ascribed to methoxy-attached aromatic C. The chemical shift region between δ 160 and 220 ppm includes C=O groups from ester, amino acids, and quinones.

Distinctive changes in the spectrum resulting from organic matter decomposition can be observed, such as those reported by Chefetz et al. (1996), which include a decrease in the 31 ppm peak; the appearance of a peak at 56 ppm as a shoulder in the mature compost; and an increase in peaks in the aromatic region (at 130 and 150 ppm). Those noted by Iwegbue et al. (2006) were the following: (*i*) a 14.4% decrease in total aliphatic C (0–112 ppm), from 78.4% in the raw material to 64.0% in the mature compost; a 17.0% decrease in polysaccharides (60–112 ppm), from 42.3% to 35.1%; (*iii*) an 11.0% decrease in alkyl groups (0–50 ppm); (*iv*) an increase of aromatic components, including phenolic and aromatic carbons, from 15.9% up to 24.5% of the total C; (*v*) an increase in total aromaticity from 17.5% to 27.7%; (*vi*) a 9.5% increase in carboxyl C; and (*vii*) an increase from 1.9% to 33% in C=O carbonyl carbon. Thus, the compost maturation process was characterized by a gradual decrease in alkyl components, while more resistant cellulose polysaccharides began their transformation at later composting stages.

The distinct changes observed can be summarized through the hydrophobic carbons/hydrophilic carbons ratio, HB/HI, calculated as: $[(0 - 60) + (110 - 160)/(60 - 110) + 160/200])$. From the spectra of samples reported by Spaccini and Piccolo (2007), HB/HI values raised to 1.22 for mature compost (150 days).

8.3.5 *Pyrolysis-GC/MS, TMAH Thermochemolysis-GC/MS and HPLC*

Pyrolysis gas chromatography–mass spectrometry (Py-GC/MS) has been used for the characterization of compost components. Pyrograms obtained by double shot Py–GC/MS (Fig. 8.5) allow to further separate the various compounds present in the sample, setting apart labile fractions (usually released during the first desorption stage at 300 °C) from refractory fractions (usually released during the second pyrolysis stage at 500 °C) (González-Vila et al. 2009). Pyrolysates of compost are mainly constituted of nitrogen-containing compounds, terpenoids, lignin moieties and carbohydrates, mono- and diacids, and methyl esters (Som et al. 2009).

The branched to linear acids and the aliphatic mono to diacids ratios have been reported to be useful to monitor compost stability (Som et al. 2009).

Offline pyrolysis with tetramethylammonium hydroxide followed by GC/MS (Pyr-TMAH-GC-MS) chromatograms of less mature compost (60 d) and mature compost (150 d) are depicted in Fig. 8.6. A list of the compounds identified in the compost samples is presented in Table 8.2 (Spaccini and Piccolo 2007).

Zhu et al. (2016) used a headspace, solid-phase micro-extraction method (HS-SPME) followed by GC/MS detection to study changes in volatile organic compound contents in compost samples, and observed that most of the volatile components could only be detected before day 22. Conversely, alkanes, alkenes,

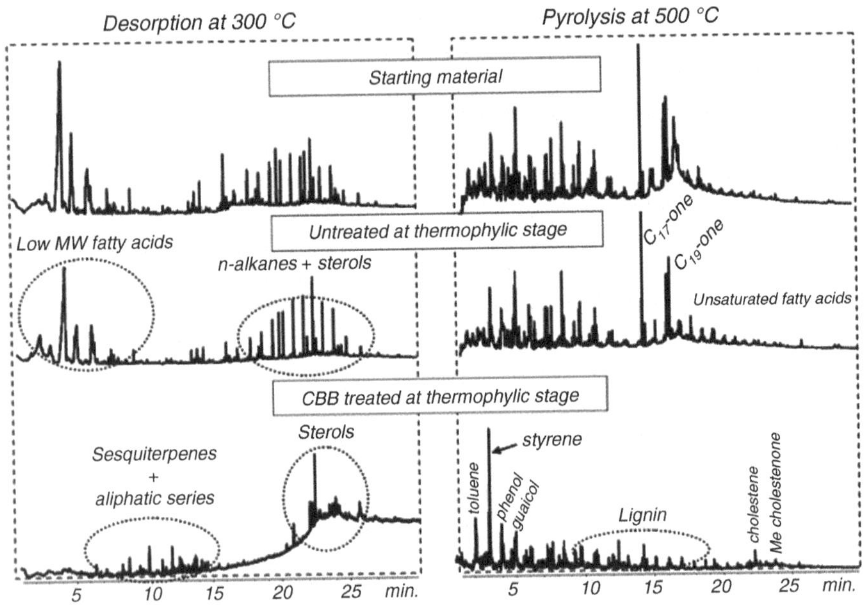

Fig. 8.5 Pyrograms obtained by double shot Py-GC/MS at 300 °C (left) and 500 °C (right) of urban waste compost from microbial bio-accelerator (CBB)-treated and untreated compost piles (González-Vila et al. 2009)

and benzenes increased and eventually stabilized after day 22. Important factors for compost quality such as phenol and acid substances were virtually undetectable on day 39 and on day 13 in natural compost samples and maggot-treated compost samples, respectively. During the summer, the authors concluded that the appropriate composting times for natural compost and maggot-treated compost samples were approximately 40 days and 22 days, respectively.

As regards high performance liquid chromatography (HPLC), the molecular weight of the humic fraction can be used to evaluate the suitability for transformation of the starting materials, the degree of humification and the efficiency of a composting process, as demonstrated by Tomati et al. (2000).

Low molecular weight fractions decrease and their conversion into high molecular weight fractions has been described by Chanyasak et al. (1982) and Roletto et al. (1985). These authors noted that, over composting time, the 5–10 kDa molecular weight fraction disappeared, while those with higher molecular weight increased. The study by Tomati et al. (2000) of this latter region allowed observe two separated fractions: one with molecular weights in the 100–200 kDa range (A_1), and another one with molecular weights >200 kDa (A_2). During composting, the A_2/A_1 ratio (humic acid evolution index, HAEI) tends to reach a constant value, which is indicative of the maximum possible degree of HA polymerization (Tomati et al. 2000). Although it varies as a function of the material composted and the composting process, it can be deemed as a reliable index to evaluate the maturity of a compost.

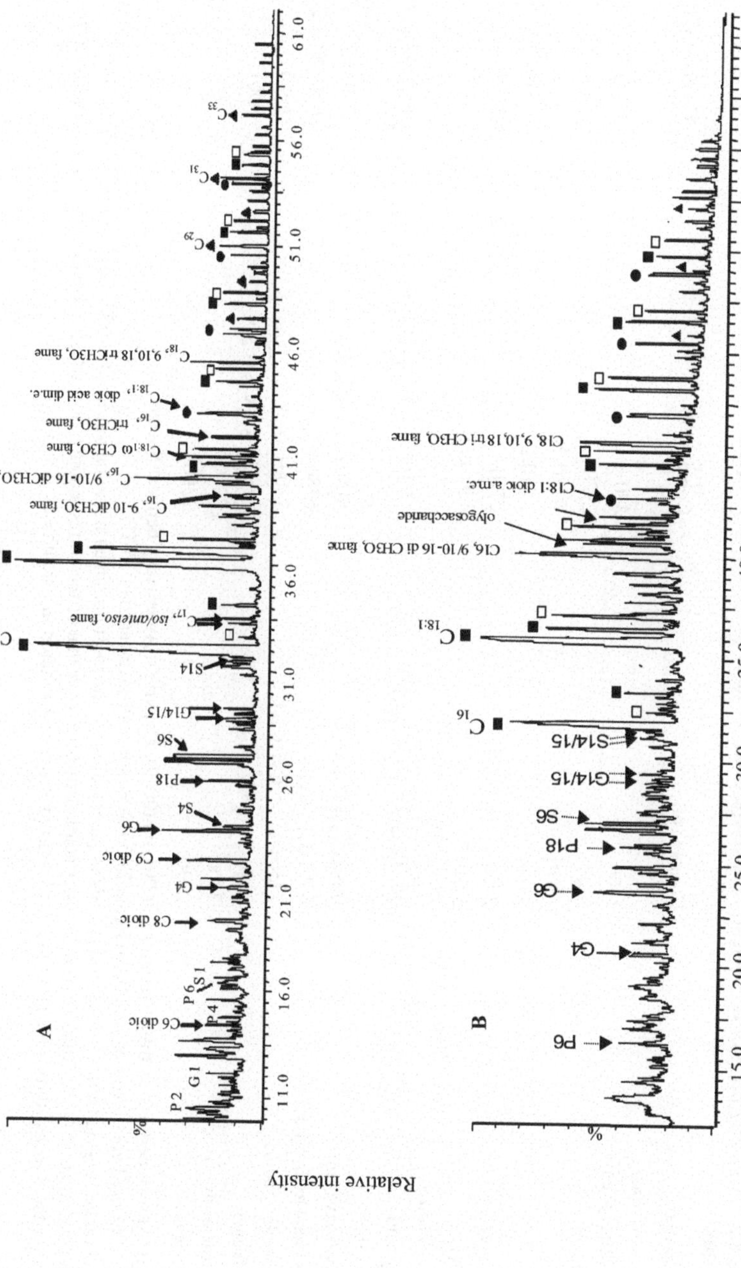

Fig. 8.6 Total ion chromatograms of: (**a**) compost 60; (**b**) compost 150; (filled square) FAME; (open square) ω-hydroxy-FAME; (filled circle) alkanedioic acid DIME; (filled triangle) *n*-alkanes. Reprinted with permission from Spaccini and Piccolo (2007). Copyright American Chemical Society

Table 8.2 Yields (μg per g of dry weight)[a] and composition[b] of main thermochemolysis products released from compost materials at different compost maturities

Compound	Compost 60	Compost 90	Compost 150
Lignin products			
p-hydroxyphenyl	1480	870	650
Guaiacyl	3240	2920	2830
Ad/Al_G	3.8 (0.3)	4.1 (0.2)	4.3 (0.3)
Γ_G[c]	2.9 (0.2)	2.8 (0.2)	3.0 (0.1)
Syringyl	2950	2710	2730
Ad/Al_S[c]	5.0 (0.2)	4.9 (0.3)	5.1 (0.2)
Alkyl products			
Fatty acids	36,100 $C_{12} \div C_{30}$ ($C_{18:1}$)	15,100 $C_{12} \div C_{30}$ ($C_{18:1}$)	13,250 $C_{12} \div C_{30}$ ($C_{18:1}$)
$>C_{20}$ (%)	6.1	14.3	31.2
ω-Hydroxy acids	9850 $C_{14} \div C_{26}$ (C_{18})	8150 $C_{14} \div C_{26}$ (C_{16})	7300 $C_{14} \div C_{26}$ (C_{16})
Mid-chain hydroxy acids	8350 (C_{16}, C_{18})	6780 (C_{16}, C_{18})	6200 (C_{16}, C_{18})
Alkanedioic acids	8150 $C_{18:1} \div C_{24}$ (C_{18})	7300 $C_{18:1} \div C_{24}$ (C_{18})	5900 $C_{18:1} \div C_{24}$ (C_{20})
n-Alkanes	3140 $C_{25} \div C_{33}$ (C_{29})	1390 $C_{25} \div C_{33}$ (C_{29})	900 $C_{25} \div C_{33}$ (C_{29})
Diterpenoids	2200	2210	2140
Triterpenoids	2230	2300	2280

[a] $n = 3$. Overall coefficient of variations: lignin lower than 10%; alkyl lower than 15%

[b] Total range varying from C_i to C_j; compounds in parentheses are the most dominant homologues; number after colon refers to double bond

[c] Structural indices: Ad/Al (ratio of peak areas of acidic structures over that of the corresponding aldehydes) = G6/G4, S6/S4); and Γ_G (over the sum of peak areas for the threo/erythro isomers) = G6/(G14 + G15); standard deviation in parentheses

8.3.6 Elemental Analysis (High-T CHNS Combustion Analysis) for C/N Ratios

Brown or woody vegetal samples, such as wood chips or dried leaves, tend to have a lower proportion of N than grass clippings and other green vegetation (and therefore a higher C/N ratio). Generally, high C to N ratios result in a slower decomposition, since the microorganisms suffer from N-deficiency and therefore the population growth is slow. In contrast, a low C to N ratio usually involves an excess of N in relation to energy, and it can result in high N_2 losses. Compost will not heat up if it is too low in N, but if the N proportion is too high, it may become too hot, killing the microorganisms, or it may go anaerobic, which would result in foul odor nuisance. Indiscriminate or dogmatic application of C/N ratio to fresh compost formulas may result in an over calculation of the needed carbon.

The total N, C, and S contents can be determined using a CHNS analyzer. Freeze-dried and ground samples (5–10 mg) are mixed with an oxidizer (V_2O_5) in a tin capsule, which is then heated at 1000 °C in a reactor. The sample and container melt and tin promotes flash combustion in a temporarily enriched O_2 atmosphere. A constant flow of He carries the combustion products (viz. CO_2, SO_2, and NO_2) to a glass column packed with an oxidation catalyst (WO_3) and a reducer (Cu), in which NO_2 is reduced to N_2. Subsequently, CO_2, SO_2, and N_2 are transported by the carrier gas to, and separated by, a column, and are finally quantified with a thermal conductivity detector (TCD). The chromatographic responses are calibrated against standards, and the obtained CHNS elemental contents are expressed in weight percent.

Compost may be finished anywhere around a C/N ratio of 17 or lower (in some cases, as low as 10:1, the ratio found in bacterial cells and stable soil humus) and this is dependent on lab sieving techniques. It is also worth noting that the loss of nitrogen during the composting process may make the total C/N appear to level off (or even increase).

For some authors, a product is not considered to be a compost unless the C:N is less than 20:1 (Fourti 2013; Iglesias Jiménez and Pérez García 1991). For instance, in the particular case of the composting process of MSW, it generally begins with a higher C/N ratio in windrows free of sludge (i.e., constituted with 100% of MSW; W1 in Fig. 8.7) than when mixed with sewage sludge (W2): 32 and 28.5, respectively. Subsequently, a net decrease is observed, reaching values of 18.6 (W1) and 14.6 (W2) at the end of the process.

8.3.7 Thermal Analysis

According to the TG analysis results, composts of different origin show two consecutive weight losses, between 240 °C and 365 °C, during the composting process. In the case of date palm waste and date palm–couch grass mixture composts

Fig. 8.7 Progress of C/N ratio over time in windrows W1 and W2 (Fourti 2013)

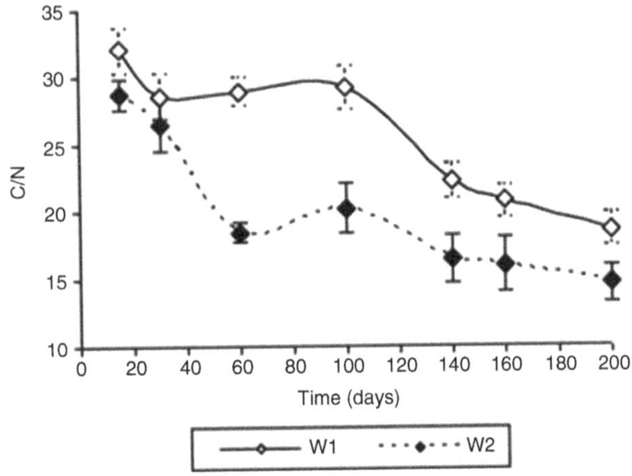

reported by Zahra El Ouaqoudi et al. (2015), such losses were sensitized by two exothermic peaks (Fig. 8.8). The gradual reduction in the mass loss related to the first peak at 200–350 °C indicated a progressive degradation of carbohydrates, aliphatic compounds, and some easily biodegradable aromatic structures. The intensity of the mass loss related to the second peak at 400–500 °C, associated with more complex aromatic structures with high molecular weights, decreased at the beginning of the composting process, and then increased after 6 months for both composts. The decrease in the intensity of this latter peak may be ascribed to degradation of aromatic compounds (e.g., lignin), while its increase may be correlated to the release of aromatic structures after the deterioration of lignocellulose (i.e., to the condensation of these structures). When the R_{TG} ratio (calculated as the weight losses associated with the second phenomenon divided by those associated with the first one) reaches 0.8, the compost may be regarded as mature (Som et al. 2009).

The DSC technique has also been applied to characterize samples derived from composts of different organic waste materials (Ouatmane et al. 2000; Provenzano and Senesi 1999). In the thermogram of a typical soil humic acid, a strong endotherm in the low-temperature region (at ca. 140 °C), and an exotherm in the high-temperature region (at ca. 500 °C) can be observed, which can be ascribed to dehydration and/or to loss of peripheral polysaccharide chains and to polycondensation and oxidation of aromatic nuclei of the molecules, respectively. In addition to the aforementioned effects, composted materials show a typical exotherm in the medium temperature region (at ca. 360 °C), attributable to peptide structures loss. The relative intensity of this exotherm seems to be associated with the humification degree of the sample, and it decreases as composting time increases.

Thermograms of sawdust wood samples (day 0 to day 365) show an exotherm at ca. 380 °C, whereas the high-temperature exotherm does not appear. Thermograms of both urban solids and farm manure samples exhibit the characteristic exotherm in the medium temperature region (at 350–360 °C) and show a similar trend with increasing composting time: the intensity of this peak decreases as composting time

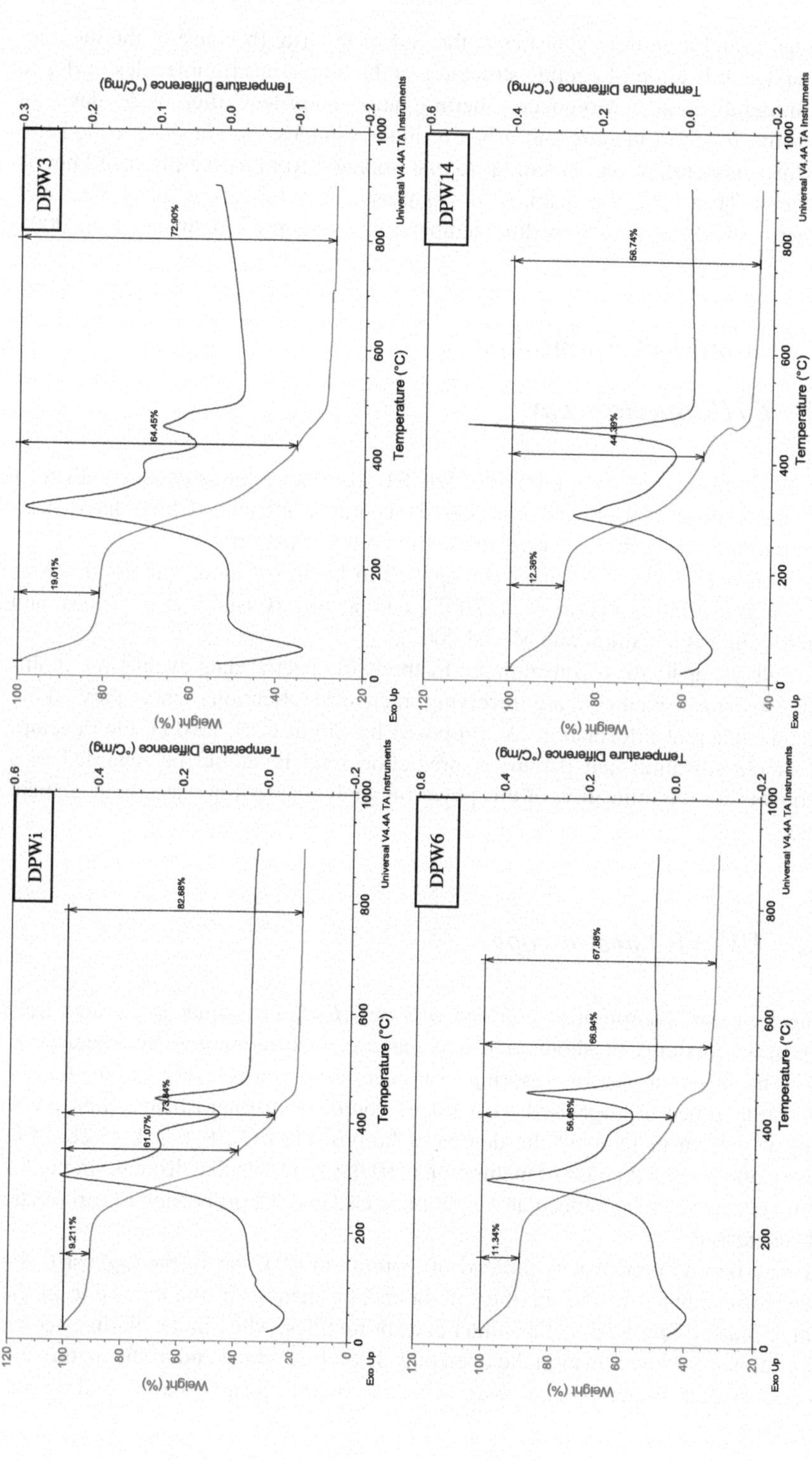

Fig. 8.8 TG-DTA of date palm waste in an oxidizing atmosphere (Zahra El Ouaqoudi et al. 2015)

is increased and it entirely vanishes at the end of the trial (because of the incorporation and stabilization of peptide structures in the humic macromolecules or due to their mineralization). Moreover, thermograms recorded after 365 days of composting for farm manure and urban residues samples show a sharp endotherm in the low-temperature region similar to that obtained from native dissolved humic substances. Therefore, the maturity of compost can be assessed using the latter endotherm, in addition to the medium temperature exotherm (Ouatmane et al. 2000).

8.4 Examples of Application

8.4.1 FTIR Spectroscopy

Since 1993 (Inbar et al. 1993), FTIR or DRIFT absorbance ratios of several distinct peaks for sewage sludge and composted separated manure (CSM) have been reported to hold a significant correlation with compost maturity.

In the last 15 years, FTIR has been applied to landfilled municipal solid wastes during in situ aeration (Tesar et al. 2007), composting (Castaldi et al. 2005), and anaerobic digestion (Smidt and Meissl 2007).

Nowadays, apart from conventional methods of spectral data evaluation, multivariate statistical methods are receiving increased attention, since they allow extracting inherent information. As proposed by Smidt et al. (2011), the development of classification and parameter prediction models should be regarded as a prerequisite for the utilization of this powerful analytical tool in waste management practice.

8.4.2 UV–Vis Spectroscopy

Changes in the absorption at 465 and 665 nm (E_4/E_6) of aqueous extracts from composting mixtures of shoots and hen manure were monitored by Matei et al. (2017) in different seasons (spring, summer, and winter) (Fig. 8.9). As the composting process progressed, two stages could be distinguished: a first phase during which an increase in the degree of humification (E_4/E_6) took place up to 60 days; and a second phase of maturation (250 days), in which a decrease in the E_4/E_6 ratio occurred as the molecular weight increased and the proportion of carboxylic acids decreased.

It may be observed that at the end of maturation (250 days), the highest E_4/E_6 values corresponded to the mixture of shoots/hen manure in the open pile in the summer season, followed by the winter and spring piles, while in the biodigester the E_4/E_6 ratios were lower than in the open pile. Therefore, at the end of the maturation process, greater mineralization was achieved in the spring season and in the biodigester.

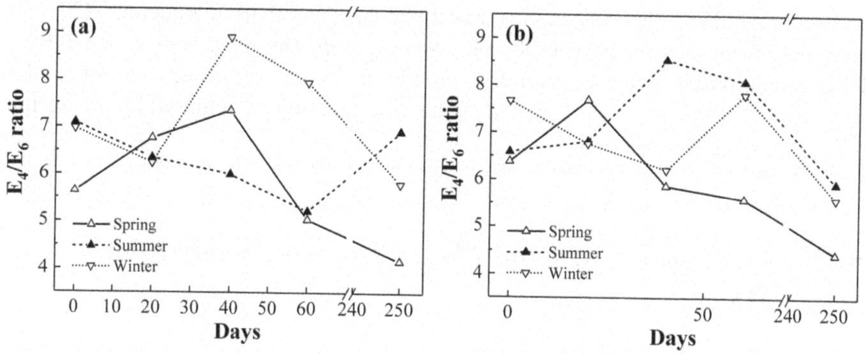

Fig. 8.9 Seasonal variation of the absorbance ratio corresponding to 465 and 665 nm (E_4/E_6) as a function of time for composting processes involving mixtures of grapevine shoots with poultry manure in: (**a**) an open pile; (**b**) a batch closed biodigester. Adapted from Matei et al. (2017)

8.4.3 EEM Spectroscopy

As indicated in Sect. 8.3.3, EEM data can be explained by fluorescent peaks, by FRI, or by PARAFAC. The comparison of these techniques to monitor the biodegradation process allowed to state that FRI is not suitable for proteins and humus, provided that they can feature multi-peak fluorophores characteristics. Conversely, PARAFAC can resolve the EEM signals of the unknown samples from those of overlapping and uncalibrated interferents, which makes it more suitable for a quantitative tracking of the composting process (Zheng et al. 2014). A recent contribution on the usefulness of EEM to analyze compost was reported by Zhao et al. (2017), who assessed the humification degree in FAs from different composts, with a view to revealing their roles after soil amending.

8.4.4 ^{13}C-NMR

Over the past few years, some studies have chosen ^{13}C-CPMAS-NMR technique to monitor the stabilization process during the composting of various organic waste materials, analyzing either the complete sample or the extracted HAs (Caricasole et al. 2011; Martinez-Sabater et al. 2009; Torres-Climent et al. 2015). For instance, swine, cattle and chicken manures and composts were monitored during a 70-day composting process without the addition of bulking agents (Mao et al. 2017), observing that cattle and chicken composts were relatively stable after 36 and 56 days, respectively, whereas swine manure composts required 70 days to be mature.

NMR also showed powerful capabilities to gain insight into the evolution of C pools during composting, in particular when recalcitrant C versus labile C pools were compared (alkyl/O-alkyl ratio) (Martin-Mata et al. 2016). In general, ^{13}C-NMR analyses on various types of bulk composts revealed (in agreement with FTIR or DRIFT results) an increase in alkyl, aromatic, and carboxyl group contents and a decrease in carbohydrates, together with the accumulation of modified lignin, during the composting process.

8.4.5 Pyrolysis-GC/MS and TMAH Thermochemolysis-GC/MS

Even if pyrolysis and TMAH thermochemolysis techniques have been extensively used in organic matter research in the past decade, their application to compost research has been more limited. A recent application has been the study of the biotransformation of lignin components during co-composting of sewage sludge activated with palm tree waste using Py-GC-MS (El Fels et al. 2014). Two groups of compounds were identified: one (1-ethyl-2-methylbenzene, styrene, toluene, 2,4-dimethylbenzene, ethylbenzene, 2-methylnaphthalene, and 4-methylphenol) for which a reduction in concentration occurred, due to metabolization and biotransformation into other compounds; and another one (benzofuran, phenol, ethylmethoxyphenol, and dimethoxyphenol) for which concentrations increased with co-composting time. The authors suggested that these lignin constituents would possibly be released in parallel with lignin degradation and would become incorporated into humic substances.

8.4.6 C/N Ratios

The water-soluble (WS) C/N ratio is an alternative to total C/N quotients discussed above, and this soluble C to soluble N ratio may be higher or lower than the total C/N ratio. A WS C/N ratio lower than 10 is a good sign, indicating that the material will not immobilize nitrogen upon soil addition. Conversely, if the WS C/N ratio is higher than 10 or larger than the total C/N ratio, it suggests that the material would not be fully decomposed (Woods End Laboratories Inc. 2013).

8.4.7 Thermal Analysis

Although according to the TG-DTG analysis results, compost shows its largest weight loss in the 240 °C to 350 °C temperature range, the subsequent weight loss that occurs between 350 and 500 °C (sensitized by an exothermic at 360 °C) is the one which is currently receiving more interest. Ali et al. (2012) attributed the weight loss in this latter range to the degradation of complex aromatic structures and peptides.

8.5 Concluding Remarks

As reported above, fast analytical instrumental techniques allow a facile monitoring of the stability and maturity of compost. The end of the composting process can be detected:

- By FTIR vibrational spectroscopy, when the 1650/2930, 1650/2850, 1650/1050, and 1650/1560 cm^{-1} band intensity ratios are larger than 1.10, 1.54, 2.80, and 1.50, respectively; or from Raman signals, by analyzing the *D band* at around 1358 cm^{-1}.
- From UV–vis spectra, by the appearance of a shoulder at 235 nm, upon the increase of the E_{253}/E_{230} ratio, upon the decrease of the E_{465}/E_{665} ratio or when the SUVA254 index exceeds 3.13 L/mg/m.
- By EEM spectroscopy, upon the increase of the $\lambda_{Ex}/\lambda_{Em}$ pair at 330/420 nm or by the increase of $P_{V,n}/P_{III,n}$ ratio.
- From the relative distribution of signal area over chemical shift regions in ^{13}C-NMR spectra, when the hydrophobic carbons/hydrophilic carbons ratio (HB/HI) increases over 1.22.
- From Py-GC/MS pyrograms, by the peak areas of acidic structures over those of the corresponding aldehydes (Ad/Al ratio).
- By HPLC, by the HAEI index.
- From CHNS analyses, by the attaining of C/N ratios of around 17 and WS C/N ratios above 10.
- From thermal analysis, by a R_{TG} value of 0.8 and a decrease in the intensity of the DSC exothermic effect at around 360 °C.

Acknowledgments PMR gratefully acknowledges the financial support of Santander Universidades through the "Becas Iberoamérica Jóvenes Profesores e Investigadores, España" scholarship program.

References

Ali M, Bhatia A, Kazmi AA, Ahmed N (2012) Characterization of high rate composting of vegetable market waste using Fourier transform-infrared (FT-IR) and thermal studies in three different seasons. Biodegradation 23:231–242. https://doi.org/10.1007/s10532-011-9502-0

Antil RS, Raj D, Narwal RP, Singh JP (2013) Evaluation of maturity and stability parameters of composts prepared from organic wastes and their response to wheat. Waste Biomass Valor 4:95–104. https://doi.org/10.1007/s12649-012-9141-7

Antízar-Ladislao B, Lopez-Real J, Beck AJ (2006) Investigation of organic matter dynamics during in-vessel composting of an aged coal–tar contaminated soil using fluorescence excitation–emission spectroscopy. Chemosphere 64:839–847. https://doi.org/10.1016/j.chemosphere.2005.10.036

Barberis R, Nappi P (1996) Evaluation of compost stability. In: de Bertoldi M, Sequi P, Lemmes B, Papi T (eds) The science of composting. Springer, Dordrecht, pp 175–184. https://doi.org/10.1007/978-94-009-1569-5_18

Barrena Gómez R, Vázquez Lima F, Gordillo Bolasell MA, Gea T, Sánchez Ferrer A (2005) Respirometric assays at fixed and process temperatures to monitor composting process. Bioresour Technol 96:1153–1159. https://doi.org/10.1016/j.biortech.2004.09.026

Böhm K (2009) Compost quality determination using infrared spectroscopy and multivariate data analysis. University of Natural Resources and Applied Life Sciences, Vienna

Busato JG, Lima LS, Aguiar NO, Canellas LP, Olivares FL (2012) Changes in labile phosphorus forms during maturation of vermicompost enriched with phosphorus-solubilizing and diazotrophic bacteria. Bioresour Technol 110:390–395. https://doi.org/10.1016/j.biortech.2012.01.126

Bustamante MA, Paredes C, Marhuenda-Egea FC, Pérez-Espinosa A, Bernal MP, Moral R (2008) Co-composting of distillery wastes with animal manures: carbon and nitrogen transformations in the evaluation of compost stability. Chemosphere 72:551–557. https://doi.org/10.1016/j.chemosphere.2008.03.030

Bustamante MA, Suárez-Estrella F, Torrecillas C, Paredes C, Moral R, Moreno J (2010) Use of chemometrics in the chemical and microbiological characterization of composts from agroindustrial wastes. Bioresour Technol 101:4068–4074. https://doi.org/10.1016/j.biortech.2010.01.085

Butler TA, Sikora LJ, Steinhilber PM, Douglass LW (2001) Compost age and sample storage effects on maturity indicators of biosolids compost. J Environ Qual 30:2141–2148. https://doi.org/10.2134/jeq2001.2141

Caricasole P, Provenzano MR, Hatcher PG, Senesi N (2011) Evolution of organic matter during composting of different organic wastes assessed by CPMAS ^{13}C NMR spectroscopy. Waste Manag 31:411–415. https://doi.org/10.1016/j.wasman.2010.09.020

Castaldi P, Alberti G, Merella R, Melis P (2005) Study of the organic matter evolution during municipal solid waste composting aimed at identifying suitable parameters for the evaluation of compost maturity. Waste Manag 25:209–213. https://doi.org/10.1016/j.wasman.2004.12.011

Chanyasak V, Hirai M, Kubota H (1982) Changes of chemical components and nitrogen transformation in water extracts during composting of garbage. J Ferment Technol 60:439–446

Chefetz B, Hatcher PG, Hadar Y, Chen Y (1996) Chemical and biological characterization of organic matter during composting of municipal solid waste. J Environ Qual 25. https://doi.org/10.2134/jeq1996.00472425002500040018x

Chen Y, Inbar Y (1993) Chemical and spectroscopic analysis of organic matter transformation during composting in relation to compost maturity. In: Hoitink HAJ, Keener HM (eds) Science and engineering of composting: design, environmental, microbiological and utilization aspect. Renaissance Publications, Worthington, OH, pp 551–600

Chen W, Westerhoff P, Leenheer JA, Booksh K (2003) Fluorescence excitation−emission matrix regional integration to quantify spectra for dissolved organic matter. Environ Sci Technol 37:5701–5710. https://doi.org/10.1021/es034354c

De Nobili M, Petrussi F (1988) Humification index (HI) as evaluation of the stabilization degree during composting. J Ferment Technol 66:577–583. https://doi.org/10.1016/0385-6380(88)90091-x

Dell'Abate MT, S. Canali, Trinchera A, Benedetti A, Sequi P (1997, September 9–14) Evaluation of the stability of organic matter in composts by means of humification parameters and thermal analysis. In: Drozd J, Gonet SS, Senesi N, Weber J (eds) Proceedings of the 8th Meeting of the International Humic Substances Society. Polish Society of Humic Substances, Wroclaw, pp 842–846

Domeizel M, Khalil A, Prudent P (2004) UV spectroscopy: a tool for monitoring humification and for proposing an index of the maturity of compost. Bioresour Technol 94:177–184. https://doi.org/10.1016/j.biortech.2003.11.026

El Fels L, Lemee L, Ambles A, Hafidi M (2014) Identification and biotransformation of lignin compounds during co-composting of sewage sludge-palm tree waste using pyrolysis-GC/MS. Int Biodeterior Biodegrad 92:26–35. https://doi.org/10.1016/j.ibiod.2014.04.001

Fourti O (2013) The maturity tests during the composting of municipal solid wastes. Resour Conserv Recycl 72:43–49. https://doi.org/10.1016/j.resconrec.2012.12.001

Fuentes M, González-Gaitano G, García-Mina JM (2006) The usefulness of UV–visible and fluorescence spectroscopies to study the chemical nature of humic substances from soils and composts. Org Geochem 37:1949–1959. https://doi.org/10.1016/j.orggeochem.2006.07.024

Fujiwara T, Murakami K (2007) Application of near infrared spectroscopy for estimating available nitrogen in poultry manure compost. Soil Sci Plant Nutr 53:102–107. https://doi.org/10.1111/j.1747-0765.2007.00109.x

González-Vila FJ, González-Pérez JA, Akdi K, Gómis MD, Pérez-Barrera F, Verdejo T (2009) Assessing the efficiency of urban waste biocomposting by analytical pyrolysis (Py–GC/MS). Bioresour Technol 100:1304–1309. https://doi.org/10.1016/j.biortech.2008.06.067

He X, Xi B, Wei Z, Guo X, Li M, An D, Liu H (2011) Spectroscopic characterization of water extractable organic matter during composting of municipal solid waste. Chemosphere 82:541–548. https://doi.org/10.1016/j.chemosphere.2010.10.057

Iannotti DA, Grebus ME, Toth BL, Madden LV, Hoitink HAJ (1994) Oxygen respirometry to assess stability and maturity of composted municipal solid waste. J Environ Qual 23. https://doi.org/10.2134/jeq1994.00472425002300060007x

Iglesias Jiménez E, Pérez García V (1991) Composting of domestic refuse and sewage sludge. I. Evolution of temperature, pH, C/N ratio and cation-exchange capacity. Resour Conserv Recycl 6:45–60. https://doi.org/10.1016/0921-3449(91)90005-9

Inbar Y, Chen Y, Hadar Y (1990) Humic substances formed during the composting of organic matter. Soil Sci Soc Am J 54. https://doi.org/10.2136/sssaj1990.03615995005400050019x

Inbar Y, Hadar Y, Chen Y (1993) Recycling of cattle manure: the composting process and characterization of maturity. J Environ Qual 22. https://doi.org/10.2134/jeq1993.00472425002200040032x

Iqbal MK, Shafiq T, Ahmed K (2010) Effect of different techniques of composting on stability and maturity of municipal solid waste compost. Environ Technol 31:205–214. https://doi.org/10.1080/09593330903426703

Iwegbue CMA, Egun AC, Emuh FN, Isirimah NO (2006) Compost maturity evaluation and its significance to agriculture. Pak J Biol Sci 9:2933–2944. https://doi.org/10.3923/pjbs.2006.2933.2944

Katayama A, Kubota H (1995) Compost testing method for the compatibility of plants. In: Proceedings of the first international symposium. University Essen and Hamburg, pp 363–384

Khatabi S, Ostadhassan M, Aghajanpour A (2018) Raman spectroscopy: an analytical tool for evaluating organic matter. J Oil Gas Petrochem Sci 1:28–33. https://doi.org/10.30881/jogps.00007

Lü F, Shao L-M, Zhang H, Fu W-D, Feng S-J, Zhan L-T, Chen Y-M, He P-J (2018) Application of advanced techniques for the assessment of bio-stability of biowaste-derived residues: a minireview. Bioresour Technol 248:122–133. https://doi.org/10.1016/j.biortech.2017.06.045

Lv B, Xing M, Yang J, Qi W, Lu Y (2013) Chemical and spectroscopic characterization of water extractable organic matter during vermicomposting of cattle dung. Bioresour Technol 132:320–326. https://doi.org/10.1016/j.biortech.2013.01.006

Mao J, Huang J, Yu Z, Gao H, Yan X, Chang J, Wang C, Hu J, Zhang L (2017) Chemical structures and characteristics of animal manures and composts during composting and assessment of maturity indices. PLoS One 12:e0178110. https://doi.org/10.1371/journal.pone.0178110

Marhuenda-Egea FC, Martínez-Sabater E, Jordá J, Moral R, Bustamante MA, Paredes C, Pérez-Murcia MD (2007) Dissolved organic matter fractions formed during composting of winery and distillery residues: evaluation of the process by fluorescence excitation–emission matrix. Chemosphere 68:301–309. https://doi.org/10.1016/j.chemosphere.2006.12.075

Martin-Mata J, Lahoz-Ramos C, Bustamante MA, Marhuenda-Egea FC, Moral R, Santos A, Saez JA, Bernal MP (2016) Thermal and spectroscopic analysis of organic matter degradation and humification during composting of pig slurry in different scenarios. Environ Sci Pollut Res Int 23:17357–17369. https://doi.org/10.1007/s11356-016-6838-3

Martinez-Sabater E, Bustamante MA, Marhuenda-Egea FC, El-Khattabi M, Moral R, Lorenzo E, Paredes C, Galvez LN, Jorda JD (2009) Study of the evolution of organic matter during composting of winery and distillery residues by classical and chemometric analysis. J Agric Food Chem 57:9613–9623. https://doi.org/10.1021/jf901027v

Matei PM, Sanchez-Bascones M, Bravo-Sanchez CT, Martin-Ramos P, Martin-Villullas MT, Garcia-Gonzalez MC, Hernandez-Navarro S, Navas-Gracia LM, Martin-Gil J (2016) Hygienization and control of *Diplodia seriata* fungus in vine pruning waste composting and its seasonal variability in open and closed systems. Waste Manag 58:126–134. https://doi.org/10.1016/j.wasman.2016.08.002

Matei PM, Báscones MS, Antolín JM, Bravo CT, Sánchez Sastre LF, Silva Castro I, Correa Guimaraes A, Martín-Ramos P, Martín-Gil J (2017) Estudio de la mineralización de la materia orgánica en el proceso de compostaje de sarmientos de vid mediante espectroscopía IRTF-RTA y UV/VIS. In: López Núñez R, Cabrera Capitán F (eds) Reciclando los residuos para mejorar los suelos y el medioambiente. V Jornadas de la Red Española de Compostaje. Red Española de Compostaje, Sevilla, Spain, pp 43–47

Ouatmane A, Provenzano MR, Hafidi M, Senesi N (2000) Compost maturity assessment using calorimetry, spectroscopy and chemical analysis. Compost Sci Util 8:124–134. https://doi.org/10.1080/1065657x.2000.10701758

Provenzano MR, Senesi N (1999) Thermal properties of standard and reference humic substances by differential scanning calorimetry J Therm Anal Calorim 57:517–526. https://doi.org/10.1023/a:1010176326691

Provenzano MR, de Oliveira SC, Silva MR, Senesi N (2001) Assessment of maturity degree of composts from domestic solid wastes by fluorescence and Fourier transform infrared spectroscopies. J Agric Food Chem 49:5874–5879. https://doi.org/10.1021/jf0106796

Reeves JB, Van Kessel JS (2000) Near-infrared spectroscopic determination of carbon, Total nitrogen, and ammonium-N in dairy manures. J Dairy Sci 83:1829–1836. https://doi.org/10.3168/jds.S0022-0302(00)75053-3

Roletto E, Chiono R, Barberis E (1985) Investigation on humic matter from decomposing poplar bark. Agric Wastes 12:261–272. https://doi.org/10.1016/0141-4607(85)90025-3

Sadaka SS, Richard TL, Loecke TD, Liebman M (2006) Determination of compost respiration rates using pressure sensors. Compost Sci Util 14:124–131. https://doi.org/10.1080/1065657x.2006.10702273

Saidpullicino D, Erriquens F, Gigliotti G (2007) Changes in the chemical characteristics of water-extractable organic matter during composting and their influence on compost stability and maturity. Bioresour Technol 98:1822–1831. https://doi.org/10.1016/j.biortech.2006.06.018

Sanmanee N, Panishkan K, Obsuwan K, Dharmvanij S (2011) Study of compost maturity during humification process using UV-spectroscopy. World Acad Sci Eng Technol 80:403–405. https://doi.org/10.5281/zenodo.1079774

Senesi N, Brunetti G (1996) Chemical and physico-chemical parameters for quality evaluation of humic substances produced during composting. In: de Bertoldi M, Sequi P, Lemmes B, Papi T (eds) The science of composting. Springer, Dordrecht, pp 195–212. doi:https://doi.org/10.1007/978-94-009-1569-5_20

Shao ZH, He PJ, Zhang DQ, Shao LM (2009) Characterization of water-extractable organic matter during the biostabilization of municipal solid waste. J Hazard Mater 164:1191–1197. https://doi.org/10.1016/j.jhazmat.2008.09.035

Smidt E, Meissl K (2007) The applicability of Fourier transform infrared (FT-IR) spectroscopy in waste management. Waste Manag 27:268–276. https://doi.org/10.1016/j.wasman.2006.01.016

Smidt E, Bohm K, Schwanninger M (2011) The application of FT-IR spectroscopy in waste management. In: Nikolic G (ed) Fourier transforms - new analytical approaches and FTIR strategies. IntechOpen Limited, London, pp 405–430. https://doi.org/10.5772/15998

Som MP, Lemée L, Amblès A (2009) Stability and maturity of a green waste and biowaste compost assessed on the basis of a molecular study using spectroscopy, thermal analysis, thermodesorption and thermochemolysis. Bioresour Technol 100:4404–4416. https://doi.org/10.1016/j.biortech.2009.04.019

Spaccini R, Piccolo A (2007) Molecular characterization of compost at increasing stages of maturity. 2. Thermochemolysis–GC-MS and ^{13}C-CPMAS-NMR spectroscopy. J Agric Food Chem 55:2303–2311. https://doi.org/10.1021/jf0625407

Tang Z, Yu G, Liu D, Xu D, Shen Q (2011) Different analysis techniques for fluorescence excitation–emission matrix spectroscopy to assess compost maturity. Chemosphere 82:1202–1208. https://doi.org/10.1016/j.chemosphere.2010.11.032

Tesar M, Prantl R, Lechner P (2007) Application of FT-IR for assessment of the biological stability of landfilled municipal solid waste (MSW) during *in situ* aeration. J Environ Monit 9:110–118. https://doi.org/10.1039/b614002e

Thompson W, Leege P, Millner P, Watson M (2001) Test methods for the examination of composting and compost. The United States Composting Council Research and Education Foundation. The United States Department of Agriculture, Reston, VA

Tomati U, Madejon E, Galli E (2000) Evolution of humic acid molecular weight as an index of compost stability. Compost Sci Util 8:108–115. https://doi.org/10.1080/1065657x.2000.10701756

Torres-Climent A, Gomis P, Martín-Mata J, Bustamante MA, Marhuenda-Egea FC, Pérez-Murcia MD, Pérez-Espinosa A, Paredes C, Moral R (2015) Chemical, thermal and spectroscopic methods to assess biodegradation of winery-distillery wastes during composting. PLoS One 10:e0138925. https://doi.org/10.1371/journal.pone.0138925

Woods End Laboratories Inc. (2013) Laboratory test interpretation v.10, vol 2018. Mt Vernon, ME

Wu L, Ma LQ (2002) Relationship between compost stability and extractable organic carbon. J Environ Qual 31:1323–1328. https://doi.org/10.2134/jeq2002.1323

Xing M, Li X, Yang J, Huang Z, Lu Y (2012) Changes in the chemical characteristics of water-extracted organic matter from vermicomposting of sewage sludge and cow dung. J Hazard Mater 205–206:24–31. https://doi.org/10.1016/j.jhazmat.2011.11.070

Ye W, Lorimor JC, Hurburgh CRJ, Zhang H, Hattey J (2005) Application of near-infrared reflectance spectroscopy for determination of nutrient contents in liquid and solid manures. Trans ASAE 48:1911–1918. https://lib.dr.iastate.edu/abe_eng_pubs/408

Yu G-H, Luo Y-H, Wu M-J, Tang Z, Liu D-Y, Yang X-M, Shen Q-R (2010) PARAFAC modeling of fluorescence excitation–emission spectra for rapid assessment of compost maturity. Bioresour Technol 101:8244–8251. https://doi.org/10.1016/j.biortech.2010.06.007

Yu G-H, Wu M-J, Luo Y-H, Yang X-M, Ran W, Shen Q-R (2011) Fluorescence excitation–emission spectroscopy with regional integration analysis for assessment of compost maturity. Waste Manag 31:1729–1736. https://doi.org/10.1016/j.wasman.2010.10.031

Zahra El Ouaqoudi F, El Fels L, Lemée L, Amblès A, Hafidi M (2015) Evaluation of lignocelullose compost stability and maturity using spectroscopic (FTIR) and thermal (TGA/TDA) analysis. Ecol Eng 75:217–222. https://doi.org/10.1016/j.ecoleng.2014.12.004

Zhang J, Lü F, Zhang H, Shao L, Chen D, He P (2015) Multiscale visualization of the structural and characteristic changes of sewage sludge biochar oriented towards potential agronomic and environmental implication. Sci Rep 5. https://doi.org/10.1038/srep09406

Zhao Y, Wei Y, Zhang Y, Wen X, Xi B, Zhao X, Zhang X, Wei Z (2017) Roles of composts in soil based on the assessment of humification degree of fulvic acids. Ecol Indic 72:473–480. https://doi.org/10.1016/j.ecolind.2016.08.051

Zheng W, Lü F, Phoungthong K, He P (2014) Relationship between anaerobic digestion of biodegradable solid waste and spectral characteristics of the derived liquid digestate. Bioresour Technol 161:69–77. https://doi.org/10.1016/j.biortech.2014.03.016

Zhu F, Pan Z, Hong C, Wang W, Chen X, Xue Z, Yao Y (2016) Analysis of volatile organic compounds in compost samples: a potential tool to determine appropriate composting time. Waste Manag 58:98–106. https://doi.org/10.1016/j.wasman.2016.06.021

Chapter 9
Recent Advances in Assessing the Maturity and Stability of Compost

Sazada Siddiqui, Saad Alamri, Suliman Al Rumman, Mohammed A. Al-Kahtani, Mukesh K. Meghvansi, Mouna Jeridi, Tanveer Shumail, and Mahmood Moustafa

Abstract Safe and sound use of compost in plant agronomy involves the application of mature and stable compost, which have many prospective benefits for soil improvement. Though, to be measured useful, a compost should be of high quality. It should be nontoxic to plants, humans, ecosystem, and valuable for soil. In addition, it must be exempt free from pathogenic bodies, hold only nominal quantities of external resources, have tolerable amounts of organic pollutants and free from trace elements and be satisfactorily mature and stable because immature and unstable compost has a number of problems. The present review chapter deals with the methods most commonly used for estimating maturity and stability of compost. Based on in-depth literature analysis, a case is proposed for the usage of a battery of tests owing to the fact that a solitary and stand-alone test for both compost maturity and stability has not yet been agreed upon by the researchers. In addition, we provide the guidelines and regulations for compost stability and maturity all around the world.

Keywords Compost maturity · Compost stability · Maturity indicators

S. Siddiqui (✉) · S. Alamri · S. Al Rumman · M. A. Al-Kahtani · M. Jeridi · T. Shumail · M. Moustafa
Department of Biology, College of Science, King Khalid University, Abha, Saudi Arabia
e-mail: sasdeky@kku.edu.sa; salrumman@kku.edu.sa; mjoridi@kku.edu.sa; mfmostfa@kku.edu.sa

M. K. Meghvansi
Defence Research Laboratory, Tezpur, Assam, India

Present Address: Bioprocess Technology Division, Defence Research and Development Establishment, Gwalior, Madhya Pradesh, India

M. K. Meghvansi, A. Varma (eds.), *Biology of Composts*, Soil Biology 58,
https://doi.org/10.1007/978-3-030-39173-7_9

9.1 Introduction

Composting is a procedure of decomposition of an organic material called compost. Compost is rich in vital nutrients and is used as a fertilizer for growing crops and for conditioning soil. A composite interaction among the waste and microbes found in the waste is called composting. The microbes that process composting is divided into three categories: fungi, actinomycetes, and bacteria. Fungi-like bacteria called actinomycetes decompose organic material. There are many possible uses of compost as a soil conditioner. In order to be sold in the market as a potential fertilizer, it should be of good quality and also it should be safe for the animal and plant life as well as the environment. It should have a minimum level of foreign matter, trace elements, and organic pollutants, should be free from pathogens and stable as well as mature in composition (Canadian Council of Ministers of the Environment (CCME) 2005; Baffi et al. 2007; Wichuk and McCartney 2010).

The words maturity and stability are used exchangeably but both these state various properties of compost. In the present review chapter, the expression stability is used for a progressive degree of biological material decay with defiance to additional decay. When used, a compost that has no detrimental effects on a plant-growing medium such as organic material and phytotoxic elements is called a mature compost. Plant bioassays usually regulate the maturity of compost bioassays (Delgado et al. 2002; Ge et al. 2006). A number of complications are caused by unstable and immature compost. Self-heating is caused by a moderately high degree of microbial action in an unstable matter (Mathur et al. 1993). Self-heating might be unsafe if huge amount of material gets heated. Due to the trapping of flammable gases, there is a possible threat of fires (Brinton 2000). Production of disease vector attraction and the smell is caused by continuous disintegration (Ge et al. 2006). The significance of consistently assessing compost maturity and stability is underlined by the presence of various possible adversative effects. In order to contribute to public and governing reception of compost, there should be a guarantee for the efficiency and safety of compost goods. There is an enhancement in the composting of organically degradable garbage by the schematization of European legislation. Compost is an important material for agricultural purposes but its use is very limited. In the current scenario, there is a substantial worldwide activity in several nations to progress ethics for boosting the composting business. Nowadays, there is an immense requirement to look into the regulation for the quality of composts and to develop and promote easy and economical maturity tests that can be applied by the operators of composting. Denmark, Finland, Norway, and Sweden are working together to find elucidation for problems related to the quality of compost (Itiivaara et al. 1998). Many approaches have been assessed by scholars and the last stage is to ring test the numerous preferred tests. Researchers are confronting difficulties to find appropriate tests that can consistently evaluate the compost maturity since compost undergoes consistent organic decay.

During maturation of compost, various chemical changes take place, for example, enhancement in the nitrate portion and a decline in the ammonium content of the compost (Standards Australia 1997), a decline in the liquefied biological carbon and variations in exchange capability of cations (Inbar et al. 1991). The action of various enzymes, for example, dehydrogenase (Saharinen and Vuorinen 1997), lipase, phosphatase, and glucosidase (Herrmann and Shann 1993), coenzymes (Deborah et al. 2017), ammonification of arginine (Forster et al. 1993), activities of various microbes (Tseng et al. 1996), and the carbon/nitrogen proportion (Hue and Liu 1995) indicate evident deviations throughout the maturity of compost. In this review chapter, we critically examine the various methods for measuring compost stability and maturity. Special attention is made to find out the most reliable and profitable method for determining compost stability and maturity with specific focus on an appropriate method for governing reasons and precise attention is made to check whether an adequate individual test for stability and maturity is possible. Additionally, we have categorized efforts to find out crucial areas where genuine research efforts are required to develop techniques for mature and stable compost that is essential for soil fertility.

9.2 Parameters Used for Evaluation Compost Maturity and Stability

Several methods of maturity and stability for compost have been recommended. These tests can be classified into physical, chemical, and biological parameters (Tiquia 2005). These methods can also be categorized as "maturity" or "stability" methods. The most normally mentioned methodologies to maturity and stability assessment, as well as some possibly favorable novel methods are presented here. In the succeeding text, the collective methods for maturity and stability of compost are described.

9.2.1 Physical and Sensory Parameters

In this section, physical parameters include temperature and sensory constraints include changes in color and odor. Signs of physical stability, for example, homogeneity, self-heating loss, overall appearance, murky color, and earthly odors are most consistent when correlated with other constraints.

9.2.1.1 Temperature

All the composting procedures do have a profile related to temperature, which is generally categorized as follows:

1. An initial sudden rise to thermophilic temperatures

2. A continuous high-temperature phase
3. A consequent decrease to near ambient temperatures

The initial sudden increase in temperature levels is due to exothermic reactions related to microbial decay of the easily decomposable biological material present in the compost raw material. Trends in temperature have been recommended as an index of the maturity of compost (Henry and Harrison 1996). During composting, three phases of temperature were detected. These are:

1. Mesophilic phase: This is the initial stage and during its starting 2 days temperature increases till 45 °C.
2. Thermophilic phase: In this phase temperature increased to 72 °C and more rise in temperature levels is evaded by adding compost to it, once in a week and this phase lasts for about 42 weeks.
3. Temperature levels decline and become equal to that of ambient, after 2 months of the composting process. The compost at this level is presumed to be of steady nature (Chefetz et al. 1996).

Consequent decrease in temperature levels happens due to a decline in microbe activity as the readily decomposable matter is used up and further bioresistant composites, for example, lignin and cellulose remain (Iglesias and Perez 1991). The compost is advancing toward a mature and steady form when temperature decreases to near ambient levels as recommended by Lasaridi et al. (2000). Since pile temperatures are a regular procedure for most of the programs related to the examining of compost so monitoring of temperature is an appealing procedure for assessing maturity and stability of compost as it is comfortable, swift and cost-effective (Boulter-Bitzer et al. 2006).

Tiquia (2005) and other researchers (Tiquia and Tam 2002) reported that a subsequent decrease in temperature of pile associated well with various indicators normally applied for the assessment of the maturity and stability levels of compost. They recommended that for assessing the steadiness and maturity of compost, monitoring of temperature is an economical, easy, and swift process. Though temperatures of the pile could also be disturbed by other circumstances, for example, very high temperature throughout the thermophilic stage, size of pile, lack of oxygen levels, absence of free air space, drying, ambient levels (especially in cold weathers), and very high moisture levels (Khan et al. 2009).

This may be a confusing concept that large pile which is still warm might be stable and small pile which has cooled might not be mature (Lasaridi et al. 2000). Similar reports are also published by other researchers when they noticed that enclosed piles take more time for cooling than exposed piles because of better exposure to ecological surroundings (Deportes et al. 1998). In addition, useful microflora found in compost gets destroyed, if the temperature reaches around 70–80 °C and thus the activity of microbe decreases and the pile gets cooled down although swiftly decomposable composites remain (Lasaridi et al. 2000).

Similarly, due to lack of moisture and oxygen, a decrease in the activity of microbes is reported and hence the pile gets cooled down quickly (Brewer and

Sullivan 2001a). Another similar reason for cooling the pile is elevated levels of water as it causes a lack of oxygen, reduction in the activity of microbes and restraining the circulation of gas (Tiquia et al. 1996). The above information covers some examples concerning the temperature of piles giving confusing evidence about the compost maturity and stability. As per the California Integrated Waste Management Board (2007), if the compost is having a temperature greater than 8 °C directly above ambient, it can be well-thought-out to be unstable and the contrary view that the temperature of pile within 8 °C of ambient designates maturity and stability of pile is not correct. Hence, most of the researchers consider that monitoring of the temperature of compost to be an invalid process for testing compost maturity or stability due to greater prospects of incorrect categorization of maturity or stability levels by this test.

9.2.1.2 Sensory Parameters: Change in Color and Odor

During composting procedure, changes in color were reported by Sughara and Inoko (1981). On higher maturity levels, composting matter color changes to grayish black or it becomes dark. By means of the normal colorimetric system CIE 1931, changes in the color of matter while composting is calculated. As the compost starts maturing and stabilizing, changes in its color and odor start appearing. Compost on maturing normally becomes dark in color and its smell becomes less unpleasant, undergoing a change from ammonia-like and rotten to plush and earthly. In addition, these sensible parameters are not specifically delicate procedures for defining compost stability and maturity, but these two parameters could be applied to find composts that are immature and unstable (Brewer and Sullivan 2001a). These two parameters when combined together may deliver a very poor assessment of maturity and stability (TMECC 2002d).

For evaluating four color tendency of compost, a moderately novel process CIELAB color space also known as “Lab” color space, has shown promising results as a parameter of stability and compost stability is designated by steadying of the color variables (Khan et al. 2009). Though additional study is required for higher application of CIELAB and both parameters (color and odor) should not be applied as a solitary parameter of maturity (ASCP 2001; Khan et al. 2009).

9.2.2 Chemical Parameters

Various chemical parameters considered for testing stability and maturity of compost. However, here we are going to discuss following parameters.

9.2.2.1 Carbon to Nitrogen (C:N) Proportion

For testing the maturity of compost, C:N proportion is applied as a potential parameter. The California Integrated Waste Management Board (2007) has reported that for the proportion of no greater than 25:1 to be applied in the assessment of compost maturity, it is essential to relate preliminary and last values. Charpentier and Vassout (1985) reported that the proportion of carbon to nitrogen of a municipal solid waste (MSW) compost was 30:1 for feedstock and it gets reduced to 13:1 with composting. Reduced tendency of C:N ratio was reported as composting proceeds because of carbon dioxide emission as biological matter are decayed and consequentially there is a decline in the carbon content as reported by Bio-Logic (2001) and Khan et al. (2009).

In the previous works, different values for C:N ratio of compost has been recommended varying from less then 20:1 to 10:1 (Goyal et al. 2005). As reported by Sullivan and Miller (2001), the range for C:N ratio should be dependent upon steady soil biological material having a range of 10:1 and 15:1. After more than 34 or 40 weeks of composting, Cayuela et al. (2008) observed a decrease in carbon-to-nitrogen proportion from 25:1 to 15:1. Carbon-to-nitrogen ratio is a better parameter for maturity of compost as described by Goyal et al. (2005). Carbon-to-nitrogen ratio would deliver better results if it is used in combination with other constraints, for example, release of carbon dioxide, water dissolving carbon, humic materials, or temperature of pile (Goyal et al. 2005). Nevertheless, the outcome of many research works advocated that carbon-to-nitrogen proportion is not a better constraint of maturity (Hutchinson and Griffin 2008) and therefore should not be applied as a single test for the assessment of maturity of compost (Boulter-Bitzer et al. 2006).

9.2.2.2 pH

Chefetz et al. (1998) observed that for composting procedure, pH displayed a distinctive design. During thermophilic phase, pH falls for the time being because the buildup of organic acids reveals the elevated proportion of organic matter decay. These acids are used subsequently as substrate by microbes. pH falls to 7 throughout maturation and cool down process. In order to evaluate the maturity of compost, pH is not a good constraint as its whole pattern is not defined by monotonic purpose (De Nobili and Petussi 1988). Few scientists have reported that on the maturity of compost, the value of pH reaches neutral values (Ko et al. 2008) whereas others have specified that a steady pH that is either neutral or not may indicate compost maturity (Cayuela et al. 2008). For the stability and maturity of compost, this constraint is considered as a possible parameter. In mature yard garbage composts, pH rises from an initial value of 5 to 7.0 as reported by Brewer and Sullivan (2003). They determined that in their research of yard garbage compost, it might be applied as a single constraint for the measurement of maturity of compost.

9.2.2.3 Humification Parameters

In the composting procedure, the existing biological material present in raw materials is changed into steady humic composites. The formation of non-phytotoxic and insoluble humic matter from the humic acid, as a transitional step, this procedure results in the generation of fulvic acids (Association of Swiss Compost Plants (ASCP 2001)), Test Methods for the Examination of Composting and Compost (TMECC 2002c). Various humification constraints have been considered. When manure compost matured, increase in humic acid and decline in the oxygen utilization percentage was reported by Tiquia (2005) whereas rise in humic acid and decline in fulvic acid, ratio the of humic acid: fulvic acid being over 1.6:1 for entirely mature composts in their research study was reported by Ko et al. (2008). Excellent association of humic matter with the evolution of CO_2 is reported by Goyal et al. (2005). Humic acid was associated with biomass of plant in plant growth experiment as concluded by Chefetz et al. (1996). Humification index was a better constraint of stability as stated by Namkoong et al. (1999).

Humification degree and index of humification are the two humification constraints reported by Mondini et al. (2003). At the final stage of the composting process, these two indicators reached steady levels representing the stable nature of the compost material. Many researchers have a view that humification constraint is not useful for the assessment of maturity and stability of compost. In addition, Wu et al. (2000) calculated that humic acid: fulvic acid proportion does not exactly define maturity and stability values.

9.2.2.4 Spectrophotometric Test

Inbar et al. (1990) reported that it is believed that phytotoxic impacts could be promptly expectable applying direct spectrophotometric procedures which can assess both stability and maturity of compost (Rynk 2003). Wang et al. (2004) stated that spectrophotometric procedures that have been applied composed of nuclear magnetic resonance (NMR) spectroscopy and Fourier transform infrared (FTIR).

Smidt and Meissl (2007) stated that spectrophotometric procedures enable the classification of several biological composites, granting a precise assessment of the state of decay of compost matter, comparative stages of chemical compounds differ as compost gets matured and this is revealed in the spectral designs formed by a sample. Khan et al. (2009) concluded that a decline in the intensity of the signal at wave numbers nearby 3300, 2930, 2852, and 1065 cm^{-1} has been detected to decline with increase in composting time. Some scientists have recommended that inclinations in the absorbance intensity at further wave numbers are also significant as investigated by Said-Pullicino and Gigliotti (2007). They observed that FTIR spectra permitted them to recognize the fluctuations in the biological material throughout composting and following incubation of sample, which represented a sign of stability of compost. Nevertheless, these processes are not competent of envisaging plant growth constantly in all kinds of compost material. High price and troublesome, as these procedures cannot be done on-site and necessitate a higher level of proficiency, accuracy, and knowledge of composting are the chief disadvantages of spectrophotometric investigations. Hence, spectrophotometric procedures are not endorsed for monitoring determinations (Wang et al. 2004).

9.2.3 *Biological Parameters*

Several biological parameters also play a very essential role for testing of stability and maturity of compost. These are discussed below.

9.2.3.1 Respiration

Barrena Gómez et al. (2016) reported that for assessment of stability of compost, respiration parameters are generally used. The generation of carbon dioxide and utilization of oxygen determined the amount of respiration because microorganisms consume oxygen and exhale carbon dioxide during aerobic decay. In regulated environments, reheating ability is also deemed as a respiration constraint since a huge quantity of swiftly decomposable matter enables a huge quantity of biodecay and generation of heat. On the contrary, when instantly decomposable matter is consumed, temperatures decrease and activity of microbes reduces (Mathur et al. 1993; Said-Pullicino and Gigliotti 2007). The utmost consistent test for assessing the maturity and stability of compost is the respiration frequency and reheating ability (Adani et al. 2006; Alexandros and Dimitrios 2018). While assessing the utility of other constraints, an evaluation is frequently made with respiration frequency (Baffi et al. 2007). Respiration percentage is reasonably constant for steady composts irrespective of the operating circumstances, raw material and preliminary state of the resources, and it reveals the present stage of decay of the garbage matter (Richard and Zimmerman 1995; Diana et al. 2017). Mostly lab respiration processes are done in consistent temperature and moisture environments, revealing the optimal atmosphere for respiration of microbes (Brewer and Sullivan 2003).

Tiquia (2005) reported that the quantity of oxygen consumed by microorganisms is used to calculate an oxygen uptake rate in a quantified mass of solids in a specified period of time. This constraint could be correlated to the stability of compost due to the activity of microbes since intake of oxygen declines, as decomposable biological material is consumed thereby indicating that compost reaches toward a steady phase. Lasaridi et al. (2000) assessed dry specific oxygen uptake rate (DSOUR) and specific oxygen uptake rate (SOUR) as index for stability. Specific oxygen uptake rate is assessed in an aqueous suspension of compost (Lasaridi et al. 2000), whereas DSOUR is assessed applying the solid sample of compost (Ianotti et al. 1993).

They reported that DSOUR was less subtle than SOUR and suggested a SOUR lesser than 2.5 mg $O_2.g^{-1}$ volatile solids (VS).h^{-1} as a level for stability. Highly steady compost is represented by SOUR $\leq$ 1.5 mg $O_2.g^{-1}$ $VS.h^{-1}$. A highest SOUR of 400 mg $O_2.kg^{-1}$ organic matter $(OM).h^{-1}$ is proposed by Bio-Logic (2001). Uptake rate of oxygen is not contemplated as a satisfactory method for estimating compost stability. Butler et al. (2001) reported that uptake rate of oxygen was not considered good to be applied as an independent parameter of stability, as it gets steady just after 29 days of composting whereas others parameters indicated that the compost was not mature. Sesay et al. (1997) stated that compost have a contrary outcome on the growth of plants even though specific oxygen uptake rate declined to lower levels. Since, not all microbe activity leads to the alteration of biological carbon to carbon dioxide, oxygen uptake rates are considered to be a better parameters for biological activity than carbon dioxide progression rates (TMECC 2002b). Bio-Logic (2001) stated that oxygen uptake rate is desirable for assessing aerobic respiration only, as carbon dioxide can be generated by both anaerobic and aerobic action.

Many researchers are of the view that carbon dioxide evolution is a better parameter for testing compost stability. Aslam and Vander-Gheynst (2008) reported that carbon dioxide evolution rate constraint can precisely envisage wheat growth and cress test outcomes. Hence, they form a suitable maturity-stability index. Carbon dioxide evolution is one among the most consistent indicators of testing the maturity of compost as concluded by Goyal et al. (2005). For most of the facilities for compost, Switzenbaum et al. (1997) suggested this process as the selected respiration test. As a pointer for maturity of compost, Goyal et al. (2005) applied a maximum limiting value of 500 mg $CO_2.100\ g^{-1}$ total biological carbon whereas Brewer and Sullivan (2001a) projected a value of lesser than or equal to 6 mg $CO_2.g^{-1}$ carbon $(C).day^{-1}$. For speedy, perfect, and economical method for assessing compost stability, Brewer and Sullivan (2001a) suggested the application of a colorimetric carbon dioxide detection tube technique. They recommended a carbon dioxide evolution rate of less than 2 mg $CO_2.g^{-1}$ $C.day^{-1}$ for representing stability.

However, there is no worldwide approval for these procedures. Respiration rate measurements by means of carbon dioxide evolution is not a suitable test for measuring compost maturity and stability as concluded by Forster et al. (1993). However, alkaline trap procedure delivers outcome only after 48 h (Brewer and Sullivan 2001b). In addition, it should also be taken into consideration that very less or greater amounts of moisture could influence outcomes of carbon dioxide respiration procedure (Brinton and Evans 2001).

9.2.3.2 Dewar Self-Heating Test

In 1982, the Dewar self-heating test was primarily presented in Europe and lately reassessed (Becker and Koter 1995). In 1984, German Department of Environment, accepted the Dewar self-heating method as an authorized standard for "ripeness" as a monitor up to the 1982 Sewage Sludge Order (LAGA 1984). To evaluate residual existing biological material and microbe respiration, compost self-heating procedures can be applied. The elementary concept behind the standardized procedure, which uses a distinct container known as vacuum flask named as Dewar flask or vessel is that within regulated settings (size of vessel as well as moisture present in compost) rise in temperature of compost within the container for some days is associated to activity of microbes and thus it leads to the stability of compost (Brinton et al. 1995; Bio-Logic 2001; Sullivan and Miller 2001; TMECC 2002b).

Some of the disadvantages of the Dewar test is that it is not very sensitive although it can differentiate between stable composts even from very young composts. Also, it is very hard to review the levels of stability between the two extreme limits (Brinton 2000; Bio-Logic 2001; Brinton and Evans 2001). Secondly, Switzenbaum et al. (1997) stated that the outcomes are not obtainable for a few days. Another drawback is that the correlation of microbe inhabitations in compost to nonoptimal situations, for example, salinity, pH, high temperature, and unequal contact of microbes to substrates may influence reheating (Brinton et al. 1995; Richard and Zimmerman 1995).

9.2.3.3 Solvita® Method

A color-coded test technique that regulates an index of maturity established on a two-tiered testing arrangement applying ammonia gas emission and respirometry is called as Solvita® test. A 4-h process for regulating the stability and maturity of compost is developed by Woods End Research Lab. The Solvita® test permits the difference amongst fresh, active, and finished composts and makes a rating on a scale of one to eight (Brinton 2000; Sullivan and Miller 2001). A compost that is graded as "finished" is regarded as mature as it is resilient to additional decay and it is without phytotoxic composites, for example, organic acids and ammonia (Woods End Research Laboratory 2000). Nevertheless, Solvita® test is quite similar to Dewar test and it is not very specific but it is exact and it is appropriate for practical compost procedures as contrary to research purposes (Brinton et al. 1995). Solvita® method in order to be precise and exact necessitates standardization to be correct. Moisture content centered on a "squeeze method" (Solvita 2009) to be moderate and this test is conducted in temperatures ranging amongst 20 and 25 °C (Seekins 1996; Brinton 2000; TMECC 2002b). Brinton and Evans (2001) reported that the Solvita® procedure was proficient in envisaging development in seedling tests. Changa et al. (2003) concluded that the Solvita® index was beneficial for on-site testing of maturity having a constant kind of raw material, as the maturity index associated finely with various maturity and stability index. Brewer and Sullivan (2001b) stated that Solvita® method was proficient in differentiating amongst new and mature compost. Additionally, Solvita® method was comfortable to apply in the field and the outcomes reveal a greater level of reproducibility within various operators (Seekins 1996; Hill et al. 2013).

9.2.3.4 Plant Bioassay (Phytotoxicity)

For maturity of compost, phytotoxicity test such as plant bioassay is assessed as a crucial test (Jagadabhi et al. 2018). Specially, when a compost material is to be cast-off in horticulture or agriculture. In compost extracts or mixtures, plant bioassays consists of the assessment of plant growth and seed germination. Plant bioassay tests were conducted under regulated environments and their outcomes are showed as a percentage of growth or germination as compared to control (Selim et al. 2012; Sullivan and Miller 2001). Which test is to be executed for the determination of the maturity of compost is a tough job. A suitable plant should be selected and a choice has to be made on how to make blending of the medium for the test. Chinese cabbage was the finest of the mostly applied plant bioassays as reported by Emino and Warman (2004).

Different countries are applying various types of plant bioassay test, for example, Germany is using barley test and the USA is applying cucumber test. Cucumber test is not considered as appropriate because they grow fine in fresh compost and tomato test has been rejected because tomatoes grow well in any media (Brinton and Evan 2001). Generally applied cress was more delicate than radish, Chinese cabbage and lettuce as reported by Aslam and VanderGheynst (2008) whereas Emino and Warman (2004) reported that cress was less delicate than carrot, Chinese cabbage, *Amaranthus tricolor*, and lettuce. Index for germination (root growth and number of seeds germinated) varies depending upon kind of compost and type of seeds as reported by Komilis and Tziouvaras (2009).

9.2.3.5 Seed Germination

Seed germination tests are proficient of representing whether or not there are substantial amount of phytotoxins in a compost as reported by TMECC (2002e). The development of seed is usually not disturbed unless there is a presence of substantial quantities of organic composites or ammonia. Khan et al. (2009) reported an anticipated enhancement in germination rate with compost time. Since these procedures are not principally subtle and thus are not considered as very good parameters of compost maturity (Brinton 2000). This has been proved in numerous findings.

Warman (1999) assessed three kinds of seed germination procedures:

1. A test in extracts of compost
2. A straight seed test in compost
3. The CCME germination test as stated in the CAN/BNQ 1996 article "Organic Soil Conditioners"—Composts. Seeds of Chinese cabbage, cress, and radish were tested. All these three tests were unable to reliably differentiate between mature and immature composts. Correspondingly as composting proceeded, there was no well-defined patterns in the germination of cress seeds as reported by Benito et al. (2005). Index of germination is not a consistent parameter for assessing the

maturity of compost since it is impossible to ascertain a maximum value that is appropriate for all kinds of compost as reported by Komilis and Tziouvaras (2009).

9.2.3.6 Plant Growth

Several compost materials are available for application for growth of plants, so it is sensible to apply a plant growth test as a method of assessing maturity of compost (Endale et al. 2018). Rynk (2003) reported that root elongation is one of the plant growth tests that can be applied as an easy and reliable method for measuring phytotoxicity. Nevertheless, outcomes are reliant on the formulation of medium such as soil blends and compost (Brinton 2000). Plant produce and growth of root keeps on varying with the degree of compost maturity as reported by Brinton and Evans (2001). They reported that root growth was repressed in semi-cured and immature composts although in well-cured composts roots of plants prolonged well to the bottom of the pots. Growth of plants is repressed in immature compost whereas it is not repressed in mature composts as reported by Chefetz et al. (1996). Neverthless, plant growth tests for example CCME germination and direct seed test were not a substantial parameter for assessing the maturity for Chinese cabbage, radish, and cress as reported by Warman (1999). Similarly, rye grass biomass was also not a good constraint for measuring maturity as described by Cooperband et al. (2003), as it varied amongst composts made from various raw materials.

9.2.3.7 Enzyme Activity

In the respiratory chain of all microbes, the role of catalyst is performed by enzymes. Different types of enzymes such as peroxidases, proteases, cellulases, and dehydrogenases are involved in various functions such as peroxidases perform a role in catalyzing the disintegration of benzyl alcohols and lignins, the hydrolysis of polysaccharides, for example, cellulose is done by cellulases, disintegration of proteins is performed by proteases and fermentation of glucose is aided by dehydrogenases (TMECC 2002a). Microbial action in a compost test sample could be assessed by enzymatic action. The pattern of enzymatic action rely upon which enzyme is taken. Pretend that, elevated amount of dehydrogenases action is evocative of substantial quantities of residual easily decomposable matter, for example, glucose and its action is projected to diminish as this substance is utilized and compost stabilizes. For investigation of compost maturity and stability, enzymatic action assessment is comparatively quick, easy, and economical (Gomez-Brandon et al. 2008; Deborah et al. 2017). Benito et al. (2003) reported that dehydrogenase activity has been completely associated with various other ways of compost maturity and stability, incorporating uptake of O_2, evolution of CO_2, and humification factors.

Dehydrogenase activity has been assessed by Benito et al. (2005). He reported that dehydrogenase activity can be applied as a parameter of reducing maturity of waste having a figure of 0.60 μg triphenyl formazan (TPF).g^{-1}dry weight.d^{-1} or a lesser amount representing a stable product. As composting proceeded, Tiquia (2005) examined that there is a decline in dehydrogenase activity representing a decrease in active decay having a familiar association amongst humification indicator and dehydrogenase activity. On the basis of above said findings, they advocated that a amount of lesser than 35 μg TPF.g^{-1} compost designates a mature pig manure compost. Since, Forster et al. (1993) detected better associations with the humic acid:fulvic acid ratio, he recommended that dehydrogenase activity might be utilized in stability testing. He suggested that stability testing can be applied in aggregation with ammonification tests. From the above discussion, it is very clear that throughout composting neither all results displayed the predictable trends in enzyme action nor all enzymes seem to be efficient as parameters of stability of compost.

Throughout the procedure of city waste composting, Saviozzi et al. (2004) observed relatively persistent enzymatic activity readings for catalase and dehydrogenase. As a matter of fact, initial readings were marginally lower than final readings. They determined that enzymatic activity was not a noble parameter for measuring of compost stability. Enzymatic activities, for example, arylsulfatase, fluorescein diacetate, and alkaline phosphatase associated well with other parameters of stability and could offer a method for the stability measurement Cayuela et al. (2008). Various raw materials may yield steady composts having a series of enzymatic activities and thus it is difficult to create a solitary threshold value (Gomez-Brandon et al. 2008).

From the above discussion, it is clear that for assessing stability and maturity of compost, no reliable tests are present. Many researchers are of the view that for envisaging the phase of maturity or stability of composts generated in a diversity of procedure circumstances from raw material (Khan et al. 2009; Monika 2013). Since, it delivers a more thorough knowledge of the stage of the compost, the application of two or more than two methods in aggregation is frequently anticipated as suitable (Inbar et al. 1990; CAN/BNQ/CCME/AAFC 1996; Brinton 2000; Mondini et al. 2003; Baffi et al. 2007). Sensor for activity for on-site testing established by Chikae et al. (2007) depend on the assessment of an amalgamation of constraints. A better methodology is to isolate the accessible methods into groups of maturity and stability of compost and to accomplish a single investigation from each group (Komilis and Tziouvaras 2009; Teshome and Amza 2017).

Since steady compost is not essentially mature so maturity and stability procedures are done in isolation and also compost which is mature is not essentially steady (Bio-Logic 2001; Brewer and Sullivan 2001b). High respiration frequencies are found in mature composts in few cases where as steady composts may need further curing to disintegrate residual phytotoxic composites. In a study conducted by Gomez-Brandon et al. (2008), respiration frequency exhibited stability after 80 days although greater than 6 months were required by the compost to have adequately lower phytotoxicity levels. Stability can be assessed by applying physical parameters such as temperature of pile, chemical constraints such as C:N ratio and pH or respirometric evaluation. This evaluation is contemplated as the most consistent and reliable method.

9.3 Guidelines and Regulations for Compost Stability and Maturity

Throughout the world, the subject of maturity and stability testing has been discussed in several instructions and rulings. A synopsis of international testing requisites is mentioned in Table 9.1. There is a great disparity among the nations

Table 9.1 Worldwide rules and guidelines for maturity test

Kind of test (Maturity/Stability)	Country	Details of test
Respiration—CO_2 evolution, as per test method ORG0020; AND Germination and growth test, as per BSI PAS 100:2005	UK (Russell and Best 2006)	In modified compost, decrease in germination of plants should be lesser than 20% when compared to peat control Decrease in plant mass that is above-surface should be less than 20% when compared to peat control No visible abnormalities are found
Self-heating	Germany	Unknown
Solvita1 or Self-heating	Sweden	Rottegrad[a] (amount of decay)
Conductivity, ammonium, pH, and nitrate	New Zealand	Unidentified
Solvita1 and oxygen demand	Denmark	Four stability groups are selected: notready, fresh, steady, or highly stable The smallest testing that should be done is total oxygen demand (96 h) and Solvita1
Nitrate:ammonium ratio Self-heating	Belgium (Hogg et al. 2002a, b)	NO_3-N:NH_4-N>1:1 for biowaste compost
Self-heating	Luxembourg	Rottegrad[a] (amount of decay)
Self-heating	Netherlands	Rottegrad (amount of decay)
Plant germination and growth	Austria	A cress (*Lepidium sativum*) method is applied. Growth (as biomass, germination rate, and germination delay) over a 9 day period should confirm to a lowest functioning
Self-heating	Australia	Recommended self-heating
1. Oxygen demand 2. Carbon dioxide evolution 3. Self-heating	Canada—CCME (CCME 2005)	Compost should be cured for at least 3 weeks and also meet one of the standards (a), (b), or (c). Test procedures to be applied are as per CAN/BNQ 0413-200-2005—"Biological Soil Conditioners—Composts" (BNQ 2005)
Must meet one of (1), (2), or (3) as listed for "Canada—CCME"	Canada—BNQ (BNQ 2005)	(a) Rate of respiration (demand for oxygen) regulated as in CAN/BNQ 0413-220; (b) Carbon dioxide evolution regulated as in TMECC 05.08-B; (c) self-heating regulated as in TMECC 05-08-D

[a]Rottegrad method is mainly an modification of a compost self-heating test

Table 9.2 Synopsis of maturity of compost indices in the Association of Swiss Compost plants (ASCP-2001) guiding principles

Maximum values for composts products projected for several applications			
Constraint	Application in agriculture	Application in horticulture	Application in sheltered cultures and private nursery
Decomposition	Raw material cannot be recognized except for wood		
Organic matter	S.B.S	<50%	<40%
pH	S.B.S	<8.2	<7.5
Extract color (humus number)	N.A.	Suggested test; an easy standard process has not been established	
Total N (Kjeldahl)	S.B.S	>10 $g.kg^{-1}$ DW	>12 $mg.kg^{-1}$ FW
C:N proportion	N.A.	S.B.S.	S.B.S.
NO_3-N	S.B.S	>40 $mg.kg^{-1}$ FW	>50 $mg.kg^{-1}$ FW
NH_4-N	<300 $mg.kg^{-1}$ FW	<300$mg.kg^{-1}$ FW	<300 $mg.kg^{-1}$ FW
NO_2-N	N.A.	<5$mg.kg^{-1}$ FW	<2.5 $mg.kg^{-1}$ FW
NO_3-N:NH_4-N ratio	N.A.	>2	>20
Plant compatibility test			
Ray grass	N.A.	N.A.	>70% of ref.
Bean	N.A.	N.A.	>70% of ref.
Salad	N.A.	>50% of ref.	>70% of ref.
Cress (closed test)	N.A.	>25% of ref.	>50% of ref.
Cress (open test)	N.A.	>70% of ref.	>90% of ref.

S.B.S Should be specified, *N.A.* Not applicable
Taken from ASCP (2001)

with varying requisites with regard to the numeral of tests done and test procedure. Generally, used procedures composed of Solvita® test, self-heating of compost, plant germination and growth processes, and respirometric methods (OUR or CO_2 emission). With isolated necessities for diverse end uses (ASCP 2001), the compost industry in Switzerland has developed an intricate methodology. The necessities of the Association of Swiss Compost Plants (ASCP) are mentioned in Table 9.2. ASCP (2001) has not precisely selected all the constraints mentioned in Table 9.2 as maturity procedures for compost but in the ASCP manuscript, all the parameters have been taken into account and in numerous studies to be probable constraints of the maturity of compost. Aqueous extract color, NO_3:NH_4^+ proportion, and amalgamation of biological material content is stipulated for assessing maturity. Plant compatibility procedures are essential for the assessment of the quality of compost. The requirements for compost to be used in covered culture and private gardens are particularly stringent, especially in terms of phytotoxicity testing but it was felt that this was necessary for such a high-end consumer market.

California Compost Quality Council (CCQC 2001) established a maturity index table to categorize compost into three classes: immature, mature, and very mature (Table 9.3). Based on a number of tests, a maturity rating is used for classifying

Table 9.3 Assessment groups of the California Compost Quality Council (CCQC'S) for Compost Maturity Index

Compost Maturity Index according to CCQC	Immature	Mature	Highly mature
Probable toxicity	High probable toxicity	Limited probable toxicity	Nil
Curing state	Uncured or raw compost	Cured compost	Highly
Influence on soil nitrogen	Substantial influence on plant accessible soil nitrogen	Marginal influences on plant accessible soil nitrogen	Nil
Odor features	Odors likely	Odor generation not likely	Nil

Taken from CCQC (2001) and TMECC (2002g)

compost into mature and very mature. It should have a carbon:nitrogen proportion of not greater than 25:1 and should clear stability (Group A) and phytotoxicity/maturity (Group B) procedures (Table 9.4). CCQC made the index because there was no single procedure that could constantly and precisely envisage the maturity of compost (CCQC 2001). A description of the CCQC's Index for the maturity of compost is encompassed in Test Methods for the Evaluation of Composting and Compost (TMECC 2002g). Buchanan (2002) reported that for assessing the superiority of a different compost matter, as it permitted recognition of immature composts which may be assumed to be mature by a single procedure, CCQC's index for maturity was a valuable means.

9.4 Conclusion and Future Perspectives

Industry for compost, scientists and researchers are searching for a unique, economical, and dependable test that can be applied for assessing maturity and stability of compost for composts produced from different raw materials applying a range of techniques for composting. Till now, not a single test exist that is economical, dependable and can be applied for assessing compost stability and maturity for different kinds of compost. Suppose, a unique, single, and economical procedure had to be selected for testing of maturity and stability of compost, the most preferred test under suitable conditions for stability testing is a respirometric test as indices for respiration are reliable and independent of raw materials. In order to find a more promising and suitable test, further research work is required. Few presently existing tests, for example, the mass-specific absorbance (MSA) of dissolved organic carbon (DOC) and few of the spectrophotometric procedures are appropriate methods, which require additional study. Few tests are more suitable for research work as compared to operational testing such as a suitable procedure for a field test should be quick and upfront.

Table 9.4 CCQC Compost Maturity Index constraint values as revealed in Test Procedures for the Investigation of Composting and Compost (TMECC)

Group A (stability procedures) perform one or more of the following				
		Rating		
Test (TMECC procedure)	Units	Highly stable	Stable	Unstable
SOUR (TMECC 05.08-A)	mg $O_2.g^{-1}$ $OM.d^{-1}$	<3	3–10	>10
CO2 evolution rate (TMECC 05.08-B)	mg $CO_2.g^{-1}$ $OM.d^{-1}$	<2	2–4	>4
Dewar (TMECC 05-08-D)	Dewar Index	V	V	<V
	Increase in Temperature (° C)	<10	10–20	>20
Solvita® CO_2 (TMECC 05.08-E)	Solvita color-code for CO_2	7–8	5–6	1–4
Biologically accessible carbon (TMECC 05-08-F)	mg $CO_2.g^{-1}$ $OC.d^{-1}$	<2	2–4	>4
Group B (maturity procedures) perform one or more of the following				
		Rating		
Test (TMECC procedure)	Units	Highly mature	Mature	Immature
Ammonium (TMECC 04.02-C)	mg $NH_4.kg^{-1}$ DW	<75	75–500	>500
NH_4:NO_3 ratio[a] (TMECC 05.02-C)	Unitless	<0.5	0.5–3.0	>3.0
Seedling emergence	% of control	>90	80–90	<80
Seedling vigor (TMECC 05.05)	% of control	>95	85–90	<85
In-vitro germination and root elongation (TMECC 05.05)	% of control	>90	80–90	<80
Earthworm bioassay (TMECC 05.05)	% weight gain	<20	20–40	>40
Solvita® ammonia (TMECC 05.08-E)	Solvita color-code for NH_4	5	4	3–1
Volatile fatty acids (TMECC 05.10-A)	mmol $VFA.g^{-1}$ DW	<200	200–1000	>1000

C:N ratio test; C:N should be ≤25:1, if not, compost is measured immature
Parameter for establishing a Compost Maturity Index rating
DW Dry weight, *VFA* Volatile fatty acids, *OC* Organic carbon, *OM* Organic matter
[a]If the sum of NH_4 and NO_3 (including NO_2-N) is below 75 to 100 mg/kg dry weight, the NH_4:NO_3 ratio should not be used as a maturity
Taken from CCQC (2001) and TMECC (2002g)

Acknowledgments The authors would like to express their gratitude to King Khalid University, Saudi Arabia for providing administrative and technical support. We are also thankful to Deanship of Scientific Research, King Khalid University. We are also grateful to Head of the Department of Biology, College of Science, King Khalid University, Abha, Saudi Arabia for providing facilities to carry out current work.

References

Adani F, Ubbiali C, Generini P (2006) The determination of biological stability of composts using the Dynamic Respiration Index: the results of experience after two years. Waste Manag 26:41–48

Alexandros E, Dimitrios K (2018) Liquid-phase respiration activity assays to assess organic waste stability: a comparison of two tests. Sustainability 10:1441

ASCP (2001) Quality criteria for composts and digestates from biodegradable waste management. Association of Swiss Compost Plants (ASCP) in collaboration with the Swiss Biogas Forum. Available from www.vks-asic.ch/acrobatreader/vks_richtlinie_english.pdf

Aslam DN, VanderGheynst JS (2008) Predicting phytotoxicity of compost-amended soil from compost stability measurements. Environ Eng Sci 25:72–81

Baffi C, Dell Abate MT, Nassisi A, Silva S, Benedetti A, Genevini PL, Adani F (2007) Determination of biological stability in compost: a comparison of methodologies. Soil Bio Biochem 39:1284–1293

Barrena Gomez R, Vázquez Lima F, Sánchez Ferrer A (2016) The use of respiration indices in the composting process: a review. Waste Manag Res 24:37–47

Becker G, Köter (1995) A standard measurement for compost maturity. In: Conference proceedings "Biological waste management: a wasted chance?", Bochum, 4–6 Apr 1995

Benito M, Masuguer A, Moliner A, Arrigo N, Palma RM (2003) Chemical and microbiological parameters for the characterization of the stability and maturity of pruning waste compost. Biol Fertil Soils 37:184–189

Benito M, Masaguer A, Moliner A, Arrigo N, Palma RM, Effron D (2005) Evaluation of maturity and stability of pruning waste compost. Soil Sci 170:360–370

Bio-Logic (2001) Report on assessing compost maturity. A final report for the Nova Scotia Department of Environment and Labour, Bio-Logic Environmental Systems, Dartmouth, NS

BNQ (2005) Organic soil conditioners – composts. CAN/BNQ 0413-200/2005, Bureau de normalisation du Quebec, SainteFoy, Que

Boulter-Bitzer JI, Trevors JT, Bol GJ (2006) A polyphasic approach for assessing maturity and stability in compost intended for suppression of plant pathogens. Appl Soil Ecol 34:65–81

Brewer LJ, Sullivan DM (2001a) Maturity and stability evaluation of composted yard debris. Master of Science Thesis, Oregon State University, Corvallis, OR

Brewer LJ, Sullivan DM (2001b) A quick look at quick compost stability tests. Biocycle 42:53–55

Brewer LJ, Sullivan DM (2003) Maturity and stability evaluation of composted yard trimmings. Compost Sci Util 11:96–112

Brinton WF (2000) Compost quality standards and guidelines. Woods End Research Laboratory, NY State Association of Recyclers

Brinton WF, Evans E (2001) How compost maturity affects container grown plants. Biocycle 42:56–60

Brinton WFJ, Evans E, Droffner ML, Brinton RB (1995) Standardized test for evaluation of compost self-heating. Biocycle 36:64–68

Buchanan M (2002) Compost maturity and nitrogen release characteristics in central coast vegetable production. Publication #44202-015, California Environmental Protection Agency, Integrated Waste Management Board, Sacramento, CA

Butler TA, Sikora LJ, Steinhilber PM, Douglass LW (2001) Compost age and sample storage effects on maturity indicators of biosolids compost. J Environ Qual 30:2141–2148

California Integrated Waste Management Board (2007) Compost quality: performance requirements—matching performance needs with product characteristics [online]. Available from http://www.ciwmb.ca.gov/Organics/Products/Quality/Needs

CAN/BNQ/CCME/AAFC (1996) Support document for compost quality criteria—National Standard of Canada (CAN/BNQ 0413-200), the Canadian Council of Ministers of the Environment (CCME) Guidelines, and Agriculture and Agri-Food Canada (AAFC) Criteria. Bureau de normalisation du Quebec, Environment

Cayuela ML, Mondini C, Sanchez-Monedero MA, Roig A (2008) Chemical properties and hydrolytic enzyme activities for the characterisation of two-phase olive mill wastes composting. Bioresour Technol 99:4255–4262
CCME (2005) Guidelines for compost quality. PN1340, Canadian Council of Ministers of the Environment, Winnipeg, MB
CCQC (2001) Compost maturity index. California Compost Quality Council, Nevada City, CA
Changa CM, Wang P, Watson ME, Hoitink HAJ, Michel FCJ (2003) Assessment of the reliability of a commercial maturity test kit for composted manures. Compost Sci Util 11:125–143
Charpentier S, Vassout FC (1985) Soluble salt concentrations and chemical equilibria in water extracts from town refuse compost during composting period. Acta Hortic 172:87–93
Chefetz B, Hatcher PG, Hadar Y, Chen Y (1996) Chemical and biological characterization of organic matter during composting of municipal solid waste. J Environ Qual 25:776–785
Chefetz B, Chen Y, Hadar Y (1998) Water extractable component released during composting of municipal solid waste. Acta Hortic 469:11–119
Chikae M, Kerman K, Nagatani N, TakamuraY TE (2007) An electrochemical on-field sensor system for the detection of compost maturity. Anal Chim Acta 581:364–369
Cooperband LR, Stone AG, Fryda MR, Ravet JL (2003) Relating compost measures of stability and maturity to plant growth. Compost Sci Util 11:113–124
De Nobili M, Petussi F (1988) Humification index as evaluation of the stabilization degree during compost. J Ferment Technol 66:557–583
Deborah A, Nehera L, Fanga B, Thomas RW (2017) Ecoenzymes as indicators of compost to suppress *Rhizoctonia solani*. Compost Sci Util 25:1–11
Delgado A, Garcıa-Morales JL, Solera del Rio R, Sales D (2002) Stability and maturity indexes of compost. Waste management and the environment (First international conference). WIT Press, Cadiz, pp 263–268
Deportes I, Benoit-Guyod JL, Zmirou D, Bouvier MC (1998) Microbial disinfection capacity of municipal solid waste (MSW) composting. J Appl Microbiol 85:238–246
Diana GF, Graham H, Thomas JA (2017) Improved precision and efficiency of a modified dynamic respiration test setup for compost stability assessment. Sustainability 9:2358–2370
Emino ER, Warman PR (2004) Biological assay for compost quality. Compost Sci Util 12:342–348
Endale W, Meron A, Debela T (2018) Maturity/phytotoxicity test of organic-fertilizer produced from coffee husk and pulp with other organic wastes on maize plant. Inter J Cre Res Thou 6:2320–2882
Forster JC, Zech W, Wiirdinger E (1993) Comparison of chemical and microbiological methods for the characterization of the maturity of composts from contrasting sources. Biol Fertil Soils 16:93–99
Ge B, McCartney D, Zeb J (2006) Compost environmental protection standards in Canada. J Environ Eng Sci 5:221–234
Gomez-Brandon M, Lazcano C, Domınguez J (2008) The evaluation of stability and maturity during the composting of cattle manure. Chemosphere 70:436–444
Goyal S, Dhull SK, Kapoor KK (2005) Chemical and biological changes during composting of different organic wastes and assessment of compost maturity. Bioresour Technol 96:1584–1591
Henry CL, Harrison RB (1996) Carbon fraction in compost and compost maturity tests. In: Magdoff FR, Tabatabai MA, Harlon EA (eds) Soil organic matter: analysis and interpretation, vol 46. SSA Publication, USA, pp 51–67
Herrmann RF, Shann JR (1993) Enzyme activities as indicators of municipal solid waste compost maturity. Compost Sci Util 1:54–63
Hill GB, Baldwin S, Vinneras B (2013) Evaluation of Solvita compost stability and maturity tests for assessment of quality of end-products from mixed latrine style compost toilets. Waste Manag 33:1602–1606

Hogg D, Barth J, Favoino E, Centremoero M, Caimi V, Amlinfer F, Devliegher W, Brinton W, Antler S (2002a) Comparison of compost standards within the EU, North America and Australia. In: Review of compost standards in Ireland. WRAP, Banbury, Oxon

Hogg D, Barth J, Favoino E, Centemero M, Caimi V, Amlinger F et al (2002b) Comparison of compost standards within the EU, North America and Australasia. Main Report. The Waste and Resources Action Programme, Banbury, Oxon

Hue NV, Liu J (1995) Predicting compost stability. Compost Sci Util 3:8–15

Hutchinson M, Griffin TS (2008) Evaluation of fiber content relative to other measures of compost stability. Compost Sci Util 16:6–11

Ianotti DA, Pang T, Toth BL, Elwell DL, Keener HM, Hoitink HAJ (1993) A quantitative respirometric method for monitoring compost stability. Compost Sci Util 1:52–65

Iglesias JE, Perez GV (1991) Composting of domestic refuse and sewage sludge. Evolution of temperature, pH, C/N ratio, and cation exchange capacity. Resour Conserv Recycl 6:45–60

Inbar Y, Chen Y, Hadar Y, Hoitink HAJ (1990) New approaches to compost maturity. Biocycle 31:64–69

Inbar Y, Chen Y, Hadar Y, Hoitink HAL (1991) Approaches to determining compost maturity In: Biocycle, the art and science of composting. JG Press, Emmaus, PA, pp 183–186

Itiivaara M, Venelampi O, Samsoe-Petersen L, Lystad H, Bjarnadottir H, Oberg L (1998) Assessment of compost maturity and ecotoxicity. Nordtest, Espoo, NT Techn Report 404, 84 p. NT Project No 1363-97

Jagadabhi PS, Wani SP, Kaushal M, Patil M, Vemula AK, Rathore A (2018) Physico-chemical, microbial and phytotoxicity evaluation of composts from sorghum, finger millet and soybean straws. Int J Recycl Org Waste Agric 8:1–15. https://doi.org/10.1007/s40093-018-0240-8

Khan MAI, Ueno K, Horimoto S, Komai F, Tanaka K, Ono Y (2009) Physiochemical, including spectroscopic, and biological analyses during composting of green tea waste and rice bran. Biol Fertil Soils 45:305–313

Ko HJ, Kim KY, Kim HT, Kim CN, Umeda M (2008) Evaluation of maturity parameters and heavy metal contents in composts made from animal manure. Waste Manag 28:813–820

Komilis DP, Tziouvaras IS (2009) A statistical analysis to assess the maturity and stability of six composts. Waste Manag 29:1504–1513

LAGA (1984) Quality criteria and application recommendations for compost obtained from refuse and refuse/sludge, Pamphlet M 10. German Federal Department of the Environment, Bonn

Lasaridi KE, Stentiford EI, Evans T (2000) Windrow composting of wastewater biosolids: process performance and product stability assessment. Water Sci Technol 42:217–226

Mathur SP, Owen G, Dinel H, Schnitzer M (1993) Determination of compost biomaturity. I. Literature review. Biol Agric Hortic 10:65–85

Mondini C, Dell Abate MT, Leita L, Benedetti A (2003) An integrated chemical, thermal, and microbiological approach to compost stability evaluation. J Environ Qual 32:2379–2386

Monika J (2013) Evaluation of maturity and stability parameters of composts prepared with sewage sludge. Fresenius Environ Bull 22:3398–3414

Namkoong W, Hwang EY, Cheong JG, Choi JY (1999) A comparative evaluation of maturity parameters for food waste composting. Compost Sci Util 7:55–62

Richard D, Zimmerman R (1995) Respiration rate–reheating potential: a comparison of measures of compost stability. Compost Sci Util 3:74–79

Russell S, Best L (2006) Setting the standards for compost. Biocycle 47:53–58

Rynk R (2003) The art in the science of compost maturity. Compost Sci Util 11:94–95

Saharinen MH, Vuorinen AH (1997) Evolution of microbiological and chemical parameters during manure and straw co-composting in a drum composting system. Agric Ecosyst Environ 66:19–29

Said-Pullicino D, Gigliotti G (2007) Oxidative biodegradation of dissolved organic matter during composting. Chemosphere 68:1030–1040

Saviozzi A, Cardelli R, Levi-Minzi R, Riffaldi R (2004) Evolution of biochemical parameters during composting of urban wastes. Compost Sci Util 12:153–160

Seekins B (1996) Field test for compost maturity. Biocycle 37:72–75

Selim SM, Mona SZ, Houssam MA (2012) Evaluation of phytotoxicity of compost during composting process. Nat Sci 10:69–77

Sesay AA, Lasaridi K, Stentiford E, Budd T (1997) Controlled composting of paper pulp sludge using the aerated static pile method. Compost Sci Util 5:82–96

Smidt E, Meissl K (2007) The applicability of Fourier transform infrared (FT-IR) spectroscopy in waste management. Waste Manag 27:268–276

Solvita (2009) Official Solvita guideline compost emissions test—V. 6.0. Woods End Laboratories

Standards Australia (1997) Composts, soil conditioners and mulches. Homebush, NSW (AS: 4454-2003)

Sughara K, Inoko A (1981) Composition analysis of humus and characterization of humic obtained from city refuse compost. Soil Sci Plant Nutr 27:213–224

Sullivan DM, Miller RO (2001) Compost quality attributes, measurements, and variability. In: Stofella PJ, Kahn BA (eds) Compost utilization in horticultural cropping systems. Lewis, Boca Raton, pp 95–120

Switzenbaum MS, Moss LH, Epstein E, Pincince AB, Donovan JF (1997) Defining bio solids stability. J Environ Eng 123:1178–1184

Teshome B, Amza J (2017) Evaluating methods of composting on date of maturity and quality of compost in Assosa, Western Ethiopia. Int J Waste Resour 7:4

Tiquia SM (2005) Microbiological parameters as indicators of compost maturity. J Appl Microbiol 99:816–828

Tiquia SM, Tam NFY (2002) Characterization and composting of poultry litter in forced-aeration piles. Process Biochem 37:869–880

Tiquia SM, Tam NFY, Hodgkiss IJ (1996) Microbial activities during composting of spent pig-manure sawdust litter at different moisture contents. Bioresour Technol 55:201–206

TMECC (2002a) Organic and biological properties—05.04 enzyme activity and analysis. In: Thompson WH, Leege PB, Millner PD, Wilson ME (eds) Test methods for the examination of composting and compost. United States Department of Agriculture, and Composting Council Research and Education Foundation, Holbrook, NY, pp 05.04-1–05.04-18

TMECC (2002b) Organic and biological properties—05.08 respirometry. In: Thompson WH, Leege PB, Millner PD, Wilson ME (eds) Test methods for the examination of composting and compost. United States Department of Agriculture and Composting Council Research and Education Foundation, Holbrook, NY, pp 05.08-1–05.07-24

TMECC (2002c) Organic and biological properties—05.07 organic matter. In: Thompson WH, Leege PB, Millner PD, Wilson ME (eds) Test methods for the examination of composting and compost. United States Department of Agriculture, and Composting Council Research and Education Foundation, Holbrook, NY, pp 05.07-1–05.07-17

TMECC (2002d) Organic and biological properties—05.06 odor. In: Thompson WH, Leege PB, Millner PD, Wilson ME (eds) Test methods for the examination of composting and compost. United States Department of Agriculture, and Composting Council Research and Education Foundation, Holbrook, NY, pp 05.06-1–05.06-13

TMECC (2002e) Organic and biological properties—05.05 biological assays. In: Thompson WH, Leege PB, Millner PD, Wilson ME (eds) Test methods for the examination of composting and compost. United States Department of Agriculture, and Composting Council Research and Education Foundation, Holbrook, NY, pp 05.05-1–05.05-11

TMECC (2002g) Organic and biological properties—05.02 indicator ratios. In: Thompson WH, Leege PB, Millner PD, Wilson ME (eds) Test methods for the examination of composting and compost. United States Department of Agriculture and Composting Council Research and Education Foundation, Holbrook, NY, pp 05.02-1–05.02-16

Tseng DY, Chalmers JJ, Tuovinen OH (1996) ATP measurement in compost. Compost Sci Util 4:6–17

Wang P, Changa CM, Watson ME, Dick WA, Chen Y, Hoitink HAJ (2004) Maturity indices for composted dairy and pig manures. Soil Biol Biochem 36:767–776

Warman PR (1999) Evaluation of seed germination and growth tests for assessing compost maturity. Compost Sci Util 7:33–37

Wichuk KM, McCartney D (2010) Compost stability and maturity evaluation—a literature review. Can J Civ Eng 37:1505–1523

Woods End Research Laboratory (2000) Guide to Solvita testing for compost maturity index. Mt Vernon ME. Manual provided with Solvita ® compost test master kit purchased from Woods End Research Laboratory Inc.

Wu L, Ma LQ, Martinez GA (2000) Comparison of methods for evaluating stability and maturity of bio solids compost. J Environ Qual 29:424–429

Chapter 10
Application of Nanotechnology to Research on the Microbiology of Composting

Chaney C. G. St. Martin, Judy Rouse-Miller, Piterson Vilpigue, and Richard Rampersaud

Abstract The inconsistent quality of compost across production cycles is often attributed to a lack of precise information on the complexity and heterogeneity of microecological systems and processes. Bulk measurements (physical, chemical and microbial) used to characterize compost further limit our ability to predict the efficacy of compost products since they obscure crucial information about variability or changes that occur at a deeper scale level. Advances in nanotechnology offer the opportunity to monitor, measure and manipulate microbial ecological systems and processes beyond the micro level. Exploring this opportunity will result in a more precise understanding of compost microbiology and strengthen predictive models for the efficacy of compost products. The aim of this review is to examine and highlight the prospective applications of nanotechnology to research work on compost microbiology. The focus is on the application of nanostructures (nanoparticles, nanodevices and nanosystems) to monitor and optimize composting processes, detect potential human and environmental health risks and enhance the effectiveness of compost products. To this end, the need for more studies using nanotechnology in compost research and application is recognized, within an economical framework that will better inform commercialization and on-farm adoption of the technology.

Keywords Nanoparticles · Quantum dots · Biological systems · Precision agriculture · Molecular barcoding · Biomolecules

C. C. G. St. Martin (✉) · R. Rampersaud
Inter-American Institute for Cooperation on Agriculture, Couva, Trinidad and Tobago
e-mail: chaney.stmartin@iica.int; richard.rampersaud@iica.int

J. Rouse-Miller
Department of Life Sciences, The University of the West Indies, St. Augustine, Trinidad and Tobago
e-mail: judy.rouse-miller@sta.uwi.edu

P. Vilpigue
Department of Food Production, The University of the West Indies, St. Augustine, Trinidad and Tobago

M. K. Meghvansi, A. Varma (eds.), *Biology of Composts*, Soil Biology 58,
https://doi.org/10.1007/978-3-030-39173-7_10

10.1 Introduction

A nanometre is a unit of spatial measurement that equates to one billionth of a metre. As such, nanoscience refers to the study of materials and structures on the scale of nanometres. Whereas, nanotechnology refers to manipulation of atoms and molecules to create particles/materials, devices and systems ranging from 1 to 100 nanometre (nm) (US Environmental Protection Agency 2007). Recognized as a "Key Enabling Technology" for sectoral growth by The European Commission (2012), nanotechnology is increasingly being used in various fields including medicine (Wang et al. 2011), engineering (Varadan et al. 2010) and agriculture (Duhan et al. 2017; Prasad et al. 2017) to gain unprecedented insights into various phenomena, at a scale, which affords the development of more precise information, analysis and application. By far, nanoparticles (naturally occurring and anthropogenic: engineered and incidental) are the most studied and advanced products of nanotechnology that have been used in microbiology-based research and product development. Engineered nanoparticles, which is the main focus of this chapter, refers to particles (e.g. iron oxide, PbS, Ca_2SiO_4 and quantum dots) that are intentionally manufactured with nanoscale dimensions and shapes (Baalousha et al. 2016). The size and shapes of these particles are often more controlled than incidental nanoparticles that are unintentionally produced as byproducts of man-made activities (e.g. industrial sources and vehicle exhaust) (Baalousha et al. 2016). In contrast, naturally occurring nanoparticles, e.g. certain clays, iron and manganese oxides and viruses, are already present or formed in the environment without human intervention (Tarafdar and Adhikari 2015).

Notwithstanding source or origin differences, the interest in nanomaterials has been fueled by their size-dependent, novel properties and functions that differ drastically from their bulk counterparts (Tarafdar and Adhikari 2015). These include changes in the properties of matter, such as energy and surface-area-to-volume ratio of nanomaterials, which influence their catalytic, electrochemical, mechanical, magnetic and optical properties. Harnessed effectively, these unique functional properties of nanomaterials have been used to create novel engineering possibilities with useful applications in biology. This is particularly so, since nanoparticles fall within the dimensions of biological macromolecules (double-stranded DNA—2 nm diameter) (Luo 2003), organelles, cells and other parts of cells, which constitute the fundamental blocks of life. This simple size comparison is suggestive of the idea of using nanoparticles and structures as cellular probes, which will allow for more detailed monitoring and analysis of cellular activity without introducing too much interference (Taton 2002). True to this suggestion, the mechanical properties of nanostructured composites, electronic properties of low-dimensional semiconductors, magnetic properties of single-domain particles and solution properties of colloidal suspensions have found use in the monitoring/characterizing (biosensing and molecular imaging) (Khan et al. 2017) and manipulating/optimizing (biological separation, anticancer therapy) (Wong et al. 2013) biological processes and systems beyond the limits afforded by other technologies. In the case of natural products, nanosizing generally improves their biological activity by increasing the rate of dissolution, enhancing saturation solubility, decreasing diffusional distance and improving adhesiveness (Prasad et al. 2017).

In this sense, the interface between functional properties of nanomaterials and biological systems or processes such as composting, can be structured in a conceptual framework of two overarching themes (Wang et al. 2011). That is, nanotechnology enables new ways to (1) monitor, measure and detects biological phenomena and processes and (2) manipulate cells/biomolecules, biological processes and treat or confer disease suppression abilities. With the development of nanosystems and devices that can simultaneously detect and manipulate biology based on feedback (abiotic and biotic), a bridging theme, which speaks to integration, adaptation, self-organization, and amplification is possible (Tarafdar et al. 2013). These capabilities speak to a potential automated and transformative level of precision for confronting optimization and predictability challenges of living systems as influenced by microecological factors, and spatial and temporal dynamics. Such advancements are critical to improve the understanding of compost microbiology as it relates to addressing the inconsistent quality and predictability problematic of compost products (St. Martin 2014; St. Martin and Ramsubhag 2015). This is particularly important since the practical and diverse uses of compost are related to most of the areas (plant protection and breeding, plant and animal nutrition, soil and water quality improvement and remediation and diagnostics) identified as relevant for the application of nanotechnology in agriculture (Bhagat et al. 2015). As such, the application of nanotechnology in compost research work is deemed important in monitoring and optimizing composting processes, detecting potential human and environmental health risks and controlling or guaranteeing the effectiveness of compost products used in the field and in other production systems.

To this end, limited studies have been done using nanotechnology in compost microbiological studies. Owing to this dearth of studies and the parallels between compost and soil, literature on the application of nanotechnology in soil science and other fields have been used to guide and inform this mini-review on the prospective application of nanotechnology in research on compost microbiology. Limitations and concerns on nanotechnology are addressed in a specific subsection although reference to such is made throughout the chapter. The chapter does not focus on approaches and methods for synthesizing, characterizing and measuring nanoparticles as these topics have been covered in sufficient details by Tarafdar and Adhikari (2015). Instead, the review focuses on the use of nano-sensors, -particles, -magnets, -filters, -composites, -injectors, -barcodes and quantum dots in research on compost microbiology. The potential applications of nanotechnology in studying compost microbiology listed in this chapter are not exhaustive.

10.2 Composting Process

The composting process, a key determinant of end-product quality, is largely driven by the activity of biological systems. As such, the characterization and functional assessment of microbes or microbial communities are important in formulating predictive models for the effect and quality of compost products. Interestingly, at the molecular level, all biological systems are made up of nanoscale components (the RNA, DNA, proteins, carbohydrates and lipids). Therefore, microbiology, in this case compost microbiology, is and has always been, studied at the nanoscale. However, the increased focus on the nanoscale has arisen from the interest to use knowledge of nanotechnology (nanodevices, nanoparticles or unique nanoscale phenomena) to tackle biological problems (nanobiology). During the composting process, the physical, chemical and engineering knowledge of nanotechnologies can be used to better understand the microbiology, visualize, detect and manipulate microbiological processes. The following subsections summarize potential applications of nanomaterials in studies on the microbiology of composting.

10.2.1 Biological Labelling

Nanoparticles with physical properties produced or driven by electromagnetic fields or possessing certain optical properties are of special interest for imaging purposes. For example, nanoparticles/structures (e.g. quantum and carbon dots, gold and silver and novel fluorescent latex/silica nanobeads) can be used for fluorescent imaging of cells (plain or targeted), chemical species (pH, oxygen, metal ions and glucose) and temperature. In biological analysis, the use of nanoparticles in fluorescent imaging has been found to permit exponentially greater sensitivity of cell detection with higher stability levels compared to older techniques that use organic fluorophores such as fluorescein and cyanine dyes(Wang et al. 2005; Wang et al. 2006). Moreover, quantum dots-based probes are considered ideal for time-gated detection with enhanced selectivity and sensitivity (Dahan et al. 2001). This is because light emission from quantum dots is slow enough to reduce background autofluorescence yet sufficient to allow/maintain high photon turn over (Yuan et al. 2001). More so, the suitability for time-gated detection and ability to intracellularly sense temperature (Wolfbeis 2015) is particularly useful in detecting cellular, intra- and extracellular changes across different phases of the composting process. This presents an interesting depth and detail (single-cell level) to compost microbial studies, which to the best of the authors' knowledge has not been explored in full. While interesting and possible, there is still a need to rationalize and weigh the practical relevance of such studies, given the complexity (microbial, chemical and physical) and dynamic nature of compost and composting. Despite the potential and success of biological labelling with nanomaterials, there are limitations specific to the type of materials and methods used. For example, quantum dots, of themselves, are not biocompatible and have to be surface modified before they are used in live cell or animal experiments (Hoshino et al. 2004). This had led to suggestions that the main problem associated with quantum dots is its potential toxicity since they are made of heavy metals (Li et al. 2011). Optical blinking has also been identified as a typical problem of using quantum dots, which makes the application of quantum dots in quantitative assays difficult (Michalet et al. 2005). Further details on the limitations of quantum dots and other types of nanomaterials used for biological labelling particles have been discussed by Michalet et al. (2005) and Zhong (2009).

10.2.2 Detection of Pathogens and Pesticides

Composts handled or used without analysis of quality pose a risk to human and environmental health since they may contain pathogens and hazardous chemicals and physical materials (Sunar et al. 2009). Traditionally, pathogen detection would normally be detected by culture-based methods which are time consuming. However, biosensors engineered from nanomaterials or nanoparticles, nanoemulsions and nanostructures including nanocapsules, nanotubes, fullerenes, quantum dots and nanowires (Jahanban and Davari 2014) can be used to rapidly detect and quantify pathogens. Studies have shown that these nanobiosensors have the potential to detect microbial contaminants like *E. coli, Salmonella typhimurium, Bacillus cereus, Campylobacter jejuni and Clostridium botulinum* present into water sources (Srivastava et al. 2014). Moreover, nanobiosensors have the potential to measure the lowest levels of contamination using restricted multiplexing for pathogens with very high sensitivity and specificity (Srivastava et al. 2014). Li et al. (2005) reported on a method for the rapid detection of pathogens and microbes using nanobarcodes. These nanobarcodes designed with dendrimer-like DNA, a built-in code and a probe for molecular recognition, rapidly detected the DNA of several pathogens. The practical relevance of these nanobiosensors is increasingly evident in light of cases of legionellosis disease in Scotland, which has been linked to compost products (Currie et al. 2014; Pravinkumar et al. 2010). Currie et al. (2014) identified the bacteria, *Legionella longbeachae* and *L. pneumophila*, the causal agent of Legionnaire's disease in 63% of commercial compost tested in Europe. Thus, a recommendation was made to label compost packages with warning of a potential public health risk. In a like manner, nanobiosensors have the potential to detect plant pathogens in compost, such as *Xanthomonas axonopodis* (bacterial spot disease) (Yao et al. 2009), *Candidatus Phytoplasma aurantifolia* (witches' broom disease of lime) (Rad et al. 2012), and Beet Necrotic Yellow Vein Virus (*Rhizomania*) (Safarpour et al. 2012a, b).

In contrast, silver, zinc oxide and titanium dioxide nanoparticles have the potential to kill pathogens in compost (Shyla et al. 2014). However, due to their non-specificity (antimicrobial) may result in the destruction of beneficial microorganisms. Moreover, the additions of heavy metals to compost for pathogen control may present a human and environmental concern, which is related to the accumulation of these compounds in soils, water and foods. Notwithstanding this, nanoparticles of zinc oxide, titanium dioxide and silver have been tested against *Macrophomina phaseolina*, a major soil borne pathogen of pulse and oilseed crops (Kannan et al. 2014). Silver nanoparticles were reported to have the highest anti-fungal effect. This may be due to the inactivation of cell wall thiol groups of the fungal cell wall by silver ions, resulting in the disruption of transmembrane, energy metabolism and electron transport chain (Duhan et al. 2017). Similarly, Ali et al. (2015) reported that the exposure of snails and soil matrix to silver nanoparticles significantly reduced the viability and activity of the snails. Moreover, it also reduced the frequency of fungal populations in the surrounding soil. Nanobiosensors can also be used to detect other biological molecules such as proteins, glucose, DNA, RNA and enzymes. All of which may be of interest when executing studies on compost microbiology. Other high throughput/massively paralleled nanobiosensors (that detect many targets at the same time or in a rapid manner) and ultra-small sample volume nanobiosensors (that use very small volumes of chemicals or reagents) may have similar potential applications as previously described, however, may be more efficient in terms of resource-use efficiency.

As it relates to hazardous chemicals, nanobiosensors also have the potential to detect pesticide levels in composts. For example, Dasary et al. (2008) and Dong et al. (2013) successfully developed and/ tested nanobiosensors for the detection of organophosphate pesticides in food and water samples. Detection limit for these nanobiosensors was as low as 0.2 nmol/dm^3, which is sufficient to screen compost to inform possible precautionary measures (Dong et al. 2013).

10.2.3 Analytic Systems: Sensing and Manipulating Biological Processes

Until recently, the technology for miniaturizing analytical systems for use in biological systems was not available. Advances in nanotechnology and photolithography, a process used in microfabrication to pattern parts of a thin film or bulk substrates (Merkoçi and Kutter 2012; Voropaeva et al. 2012) have permitted the development of nanochips or nanopatterned devices. These nanochips or devices can be used to screen large numbers of biological targets or serve as stimuli-responsive interfaces for spatially and temporally controlling extracellular physical and biochemical cues (Voropaeva et al. 2012) during composting. Due to the small size of these systems, researchers can use smaller sample sizes, perform faster analyses, or use smaller amounts of expensive chemicals and reagents. Lab-on-a-chip (LOC), which refers to “low-power” devices and methods for controlling and manipulating fluid flows at microlevels, is an example of such a system. By leveraging microfluidics, LOC is able to function as an automated platform for performing one or a range of analyses steps including separations and detections biological molecules with high resolution and sensitivity (Timm et al. 2016).

Aleklett et al. (2018) identified microfluidics systems as ideal for studying soil systems as it mitigates challenges related to studying these systems. These include spatial and temporal heterogeneity, lack of technologies that realistically simulate soil, no visualization of natural soil systems and the complexity of studying the interaction among soil microbiome. Given the parallels between soils and compost, it is likely that microfluidics systems may prove useful in studying and manipulating the compost microbiome, single-cell transcriptome analysis, and plant–microbe interaction in rhizospheres amended with compost. The same potential use is expected with the advancement of nanofluidics systems, which will permit the level of precision and resource-use efficiency. Though similar to LOC, Biological Micro Electrical Mechanical Systems (BioMEMS) often integrate decision-making capability with sensing and a higher degree of control elements on a chip than LOC. That is, techniques and processes from BioMEMS often provide a means to control a wide variety of microenvironmental cues to better understand the nature of the cellular response and ways to control cell behaviour. As such, BioMEMS tends to offer greater scope for manipulating biological processes. Although potentially useful in investigating how this technology can be used to create consistent quality compost across production cycles, presently there is not sufficient interest to establish scientific clarity on how this can be achieved. What is certain is that such study would require the complementary use of other nano-based and advanced technologies such as Oxford Nanopore Technologies, a fourth-generation gene sequencer (Simpson et al. 2017) and IChip (Nichols et al. 2010; Balagadde et al. 2005), which can be used to culture previously unculturable bacteria during the composting process. The Oxford Nanopore Technologies will allow direct measurement and sequencing of transcriptionally active cells without requiring library preparation. This sequencer has great potential for the future as it is able to determine the non-consensus sequences (Woyke and Rubin 2014) and can scan a whole genome in 15 min (Ku and Roukos 2013). The main drawback of the sequencer is it is not readily available. Ongoing development of microfluidic devices for genomics, proteomics and metabolomics should prove useful in addressing previous availability, cost and convenience challenges.

10.3 Limitations/Concerns of Nanotechnology

Though useful, nanotechnology is not without limitations and concerns. Studies have shown that the insoluble and possible toxicological nature of some nanoparticles may pose harmful effects to human, animal, plant and environmental health (Orthen et al. 2007). Gaurab and Dattatrya (2015) reported that toxicological, ethical and social concerns about the use of nanotechnology in medicine (nanomedicine), particularly for diagnostics and treatment of disease is still evident. An overarching concern is that exposure of biological systems to nanomaterials may produce oxidative stress, inflammation, immunotoxicity, genotoxicity and carcinogenicity, particularly since most engineered nanoparticles are not biocompatible.

With the rapid expansion of nanotechnology, the accumulation of manufactured nanomaterials and their possible entry into the food chain (Priester et al. 2012) is also a concern.

To this extent, researchers reported that the engineered nanomaterials can negatively affect soils, plants/food and water since they are mainly made of heavy metals. According to Agrawal and Rathore (2014), the most commonly encountered engineered nanoparticles in the environment fall into one of the five categories such as carbonaceous nanoparticles, metal oxides, quantum dots, zero-valent metals and nanopolymers. Due to their minuscule features, these nanoparticles can adhere to plant roots via uptake and bioaccumulation and exert physical and chemical toxicity to plants. Humans are exposed to all those harms cited above since these nanoparticles have the potential to stay in the environment and enter the food chain and being consumed by humans and animals. Studies have shown the toxic nanoparticles may be able to suspend in the air and enter the human body easily through skin or inhalation.

10.4 Economic and Financial Viability of Nano-based Compost Products

Information on economic or financial viability of using nanotechnology on-farm level or in composting operations or application is very limited. However, several researchers have indicated that the cost of nanotechnology is extremely high and will translate into higher production or operations costs. To this extent, Agrawal and Rathore (2014) stated that the full potential of nanotechnology in the agricultural and food industry is yet to be realized and is gradually moving from theoretical knowledge towards practical, on-farm application. Most of the work on agro-nanotechnology is still in the laboratories and has not been translated into physical products or structures that can be implemented in composting systems. Parisi et al. (2015) indicated that agro-nanotech innovative products are experiencing difficulties in reaching the market and this continues to make agriculture a marginal sector in nanotechnology. They suggested that this is mainly due to the high production costs of nanotech products, which are required in high volumes in the agricultural sector. More so, the practical benefits remain relatively unproven. This was also confirmed by Faruqi et al. (2015) who reiterated that the high costs were high partly due to the novelty of the technology but also to the complexity of the equipment and specialized personnel used to produce the nanoparticles, nanomaterials and structures. In contrast to high production costs, the implications of nanotechnology in determining the farm-gate selling price of compost products must be seriously considered.

10.5 Conclusions

There is tremendous scope to gain a more precise understanding of compost microbiology by using nanotechnologies in compost studies. Moreover, the potential of this technology to detect and suppress pathogens also makes it a useful tool in ensuring high-quality compost with minimal risk to human, plant and environmental health. Further to this, the potential of nano-based LOC and BioMEMS to simultaneously detect, analyze and manipulate biology based on feedback (abiotic and biotic) mechanisms presents an exciting prospective to address the problem of inconsistent quality compost across production cycles. Despite these potential benefits, toxicological (human, animal, plant and environment), ethical and social concerns (Prasad et al. 2017) may limit the widespread use of nanotechnologies in compost studies and applications. However, there is a need to further substantiate some of these toxicological concerns through well-designed studies. Indeed, well-designed studies are needed in all areas related to the development and testing of compost amended with the use of nanotechnologies. This does not only pertain to research focused on studying the effects or usefulness of nanotechnology on compost microbiology but should extend to when compost is applied to the soil and put to other uses. When applicable, these studies should be done within a cost/benefit framework to better develop a stronger economic argument for the adoption of nanotechnologies by compost producers and users. To this end, accessibility to nanotechnology remains a major issue, mainly due to cost aversions, health and environmental concerns, and relatively modest pace of commercializing ideas and prototypes into viable on-farm applications (Mukhopadhyay 2014). Access to such technologies is in alignment with precision agriculture and will surely do much to increase resource-use efficiency and overall enterprise productivity.

References

Agrawal S, Rathore P (2014) Nanotechnology pros and cons to agriculture: a review. Int J Curr Microbiol App Sci 3:43–55

Aleklett K, Kiers ET, Ohlsson P, Shimizu TS, Caldas VEA, Hammer E (2018) Build your own soil: exploring microfluidics to create microbial habitat structures. ISME J 12(2):312–319. https://doi.org/10.1038/ismej.2017.184

Ali SM, Yousef NMH, Nafady NA (2015) Application of biosynthesized silver nanoparticles for the control of land snail *Eobania vermiculata* and some plant pathogenic fungi. J Nanomat 2015:218904., 10 pp. https://doi.org/10.1155/2015/218904

Baalousha M, Yang Y, Vance ME (2016) Outdoor urban nanomaterials: the emergence of a new, integrated, and critical field of study. Sci Total Environ 557–558:740–753. https://doi.org/10.1016/j.scitotenv.2016.03.132

Balagadde FK, You L, Hansen CL, Arnold FH, Quake SR (2005) Long-term monitoring of bacteria undergoing programmed population control in a microchemostat. Science 309(5731):137–140. https://doi.org/10.1126/science.1109173

Bhagat Y, Gangadhara K, Rabinal C, Chaudhari G, Ugale P (2015) Nanotechnology in agriculture a review. J Pure Appl Microbiol 9:737–747

Currie SL, Beattie TK, Knapp CW, Lindsay DSJ (2014) *Legionella* spp. in UK composts—a potential public health issue? Clin Microbiol Infect 20:O224–O229. https://doi.org/10.1111/1469-0691.12381

Dahan M, Laurence T, Pinaud F, Chemla DS (2001) Time-gated biological imaging by use of colloidal quantum dots. Opt Lett 26:825–827

Dasary SSR, Rai US, Yu H et al (2008) Gold nanoparticle based surface enhanced fluorescence for detection of organophosphorus agents. Chem Phys Lett 460:187–190. https://doi.org/10.1016/J.CPLETT.2008.05.082

Dong J, Zhao H, Qiao F et al (2013) Quantum dot immobilized acetylcholinesterase for the determination of organophosphate pesticides using graphene-chitosan nanocomposite modified electrode. Anal Methods 5:2866–2872. https://doi.org/10.1039/c3ay26599d

Duhan JS, Kumar R, Kumar N (2017) Nanotechnology: the new perspective in precision agriculture. Biotechnol Rep (Amst) 15:11–23. https://doi.org/10.1016/j.btre.2017.03.002

Faruqi M, Castillo L, Sai J (2015) State-of-the-art review of the applications of nanotechnology in pavement materials. J Civ Eng Res 5:21–27. https://doi.org/10.5923/j.jce.20150502.01

Gaurab R, Dattatrya S (2015) Nanomedicine: therapeutic applications and limitations. In: Shiyani S, Amandeep S, Mrutyunjay S (eds) Handbook of research on diverse applications of nanotechnology in biomedicine, chemistry and engineering. IGI Global, Hershey, p 820

Hoshino A, Fujioka K, Oku T, Suga M (2004) Physicochemical properties and cellular toxicity of nanocrystal quantum dots depend on their surface modification. Nano Lett 4(11):2163–2169. https://doi.org/10.1021/nl048715d

Jahanban L, Davari M (2014) Organic agriculture and nanotechnology. In Rahaman G, Aksoy U (eds) Proceedings of the 4th ISOFAR scientific conference 'Building organic bridges', at the Organic World Congress 2014, 13–15 Oct, Istanbul (eprint ID 23620)

Kannan N, Rajendran V, Yuvakkumar R et al (2014) Application of silica nanoparticles in maize to enhance fungal resistance. IET Nanobiotechnol 8:133–137. https://doi.org/10.1049/iet-nbt.2013.0004

Khan I, Saeed K, Khan I (2017) Nanoparticles: properties, applications and toxicities. Arab J Chem 12(7):908–931. https://doi.org/10.1016/j.arabjc.2017.05.011

Ku C-S, Roukos DH (2013) From next-generation sequencing to nanopore sequencing technology: paving the way to personalized genomic medicine. Expert Rev Med Devices 10(1):1–6. https://doi.org/10.1586/erd.12.63

Li Y, Cu YTH, Luo D (2005) Multiplexed detection of pathogen DNA with DNA-based fluorescence nanobarcodes. Nat Biotechnol 23:885–889. https://doi.org/10.1038/nbt1106

Li M, Cushing SK, Wang Q, Shi X (2011) Size-dependent energy transfer between CdSe/ZnS quantum dots and gold nanoparticles. Chem Lett 2(17):2125–2129. https://doi.org/10.1021/jz201002g

Luo D (2003) The road from biology to materials. Mater Today 6:38–43

Merkoçi A, Kutter JP (eds) (2012) Analytical miniaturization and nanotechnologies. Lab Chip 12 (11):1915–1916. https://doi.org/10.1039/c2lc90040h

Michalet X, Pinaud FF, Bentolila LA, Tsay JM, Doose S, Li JJ, Sundaresan G, Wu AM, Gambhir SS, Weiss S (2005) Quantum dots for live cells, in vivo imaging, and diagnostics. Science 307 (5709):538–544

Mukhopadhyay SS (2014) Nanotechnology in agriculture: prospects and constraints. Nanotechnol Sci Appl 7:63–71. https://doi.org/10.2147/NSA.S39409

Nichols D, Cahoon N, Trakhtenberg EM, Pham L (2010) Use of Ichip for high-throughput *in situ* cultivation of "uncultivable" microbial species. Appl Environ Microbiol 76(8):2445–2450. https://doi.org/10.1128/AEM.01754-09

Orthen B, Rappolder M, Zimmer R (2007) Nanotechnology: health and environmental risks of nanomaterials. https://www.bfr.bund.de/cm/349/nanotechnology_health_and_environmental_risks_of_nanomaterials_research_strategy_final_version.pdf

Parisi C, Vigani M, Rodriguez-Cerezo E (2015) Agricultural nanotechnologies: what are the current possibilities? Nano Today 10(2):124–127. https://doi.org/10.1016/j.nantod.2014.09.009

Prasad R, Bhattacharyya A, Nguyen QD (2017) Nanotechnology in sustainable agriculture: recent developments, challenges and perspectives. Front Microbiol 8:1014. https://doi.org/10.3389/fmicb.2017.01014

Pravinkumar SJ, Edwards G, Lindsay D, Redmond S (2010) A cluster of Legionnaires' disease caused by *Legionella longbeachae* linked to potting compost in Scotland, 2008–2009. Euro Surveill 15:19496

Priester JH, Ge Y, Mielke RE et al (2012) Soybean susceptibility to manufactured nanomaterials with evidence for food quality and soil fertility interruption. Proc Natl Acad Sci 109:E2451–E2456. https://doi.org/10.1073/pnas.1205431109

Rad F, Mohsenifar A, Tabatabaei M, Safarnejad MR, Shahryari F, Safarpour H, Foroutan A, Mardi M, Davoudi D, Fotokian M (2012) Detection of *Candidatus* Phytoplasma aurantifolia with a quantum dots fret-based biosensor. J Plant Pathol 94:525–535

Safarpour H, Hasanzadeh F, Safarnejad MR (2012a) Development of a specific serological kit for detection of *Polymyxa betae*, transmitting agent of sugar beet rhizomania disease. J Food Agric Environ 10:729–732

Safarpour H, Safarnejad MR, Tabatabaei M et al (2012b) Development of a quantum dots FRET-based biosensor for efficient detection of Polymyxa betae. Can J Plant Pathol 34:507–515. https://doi.org/10.1080/07060661.2012.709885

Shyla KK, Natarajan N, Nakkeeran S (2014) Antifungal activity of zinc oxide, silver and titanium dioxide nanoparticles against *Macrophomina phaseolina*. J Mycol Plant Pathol 44:268–273

Simpson JT, Workaman RE, Zuzarte PC, David M (2017) Detecting DNA cytosine methylation using The Oxford Nanopore Technologies MinION sequencer. Nat Methods 14(4):407–410. https://doi.org/10.1038/nmeth.4184

Srivastava P, Pandey S, Singh P, Singh KP (2014) Nanotechnology and its role in pathogen detection: a short review. Int J Curr Sci 13:E 9–E15

St. Martin CCG (2014) Potential of compost tea for suppressing plant diseases. CAB Rev 9:1–38. https://doi.org/10.1079/PAVSNNR20149032

St. Martin CCG, Ramsubhag A (2015) Potential of compost for suppressing plant diseases. In: Sangeetha G, Kurucheve V, Jayaraman J (eds) Sustainable crop disease management using natural products. CABI, Boston, p 424

Sunar NM, Stewart DI, Stentiford E, Fletcher LA (2009) The process and pathogen behavior in composting: a review. In: Proceeding UMT-MSD 2009 post graduate seminar 2009. Universiti Malaysia Terengganu, Malaysian Student Department UK & Institute for Transport Studies University of Leeds, pp 78–87 https://www.researchgate.net/publication/261761610_The_Process_and_Pathogen_Behavior_in_Composting_A_Review. Accessed 13 Apr 2019

Tarafdar JC, Adhikari T (2015) Nanotechnology in soil science. In: Rattan RJ, Katyal JC, Dwivedi BS, Sarkar AK, Bhattacharyya T et al (eds) Soil science: an introduction. Indian Society of Soil Science, New Delhi, pp 775–807

Tarafdar JC, Sharma S, Raliya R (2013) Nanotechnology: interdisciplinary science of applications. Afr J Biotechnol 12(3):219–226. https://doi.org/10.5897/AJB12.2481

Taton TA (2002) Nanostructures as tailored biological probes. Trends Biotechnol 20(7):277–279. https://doi.org/10.1016/S0167-7799(02)01973-X

The European Commission (2012) Communication from the Commission to the European Parliament, the Council, the European Economic and Social Committee and the Committee of the Regions. A European strategy for key enabling technologies—a bridge to growth and jobs. http://ec.europa.eu/transparency/regdoc/rep/1/2012/EN/1-2012-341-EN-F1-1.Pdf

Timm AC, Shankles PG, Foster CM et al (2016) Toward microfluidic reactors for cell-free protein synthesis at the point-of-care. Small 12:810–817. https://doi.org/10.1002/smll.201502764

US Environmental Protection Agency (2007) Nanotechnology white paper. Report EPA 100/B-07/001.Washington, DC. http://www.epa.gov/osainter/pdfs/nanotech/epa-nanotechnology-whitepaper-0207.pdf

Varadan VK, Pillai AS, Mukherji D, Dwivedi M et al (2010) Nanoscience and nanotechnology in engineering. World Scientific, New Jersey, p 324. https://doi.org/10.1142/7364

Voropaeva N, Figovsky O, Ibraliu A, Shehu I, Kadiasi N et al (2012) Innovative nanotechnology for agriculture. SITA J 14(1):98–105

Wang L, Yang C, Tan W (2005) Dual-luminophore-doped silica nanoparticles for multiplexed signaling. Nano Lett 5(1):37–43. https://doi.org/10.1021/nl048417g

Wang F, Tan WB, Zhang Y et al (2006) Luminescent nanomaterials for biological labelling. Nanotechnology 17:R1–R13. https://doi.org/10.1088/0957-4484/17/1/R01

Wang J, Byrne JD, Napier ME, DeSimone JM (2011) More effective nanomedicines through particle design. Small 7:1919–1931

Wolfbeis OS (2015) An overview of nanoparticles commonly used in fluorescent bioimaging. Chem Soc Rev 44:4743–4768. https://doi.org/10.1039/C4CS00392F

Wong IY, Bhatia SN, Toner M (2013) Nanotechnology: emerging tools for biology and medicine. Genes Dev 27:2397–2408. https://doi.org/10.1101/gad.226837.113

Woyke T, Rubin EM (2014) Evolution searching for new branches on the tree of life. Science 346 (6210):698–699. https://doi.org/10.1126/science.1258871

Yao KS, Li SJ, Tzeng KC et al (2009) Fluorescence silica nanoprobe as a biomarker for rapid detection of plant pathogens. Adv Mater Res 79–82:513–516. https://doi.org/10.4028/www.scientific.net/amr.79-82.513

Yuan J, Wang G, Majima K, Matsumoto K (2001) Synthesis of a terbium fluorescent chelate and its application to time-resolved fluoroimmunoassay. Anal Chem 73:1869–1876. https://doi.org/10.1021/ac0013305

Zhong W (2009) Nanomaterials in fluorescence-based biosensing. Anal Bioanal Chem 394 (1):47–56. https://doi.org/10.1007/s00216-009-2643-x

Part III
Composting Applications

Chapter 11
Bioremediation of Pesticides in Soil Through Composting: Potential and Challenges

Sunita Varjani, M. Chaithanya Sudha, N. Manoj Kumar, B. Basant Kumar Pillai, Vijay Kumar Srivastava, Mukesh Kumar Awasthi, Sanjeev Kumar Awasthi, and Zengqiang Zhang

Abstract Agriculture plays a major role in India's socioeconomic fabric and pesticide use has become standard practice. The pesticides are primarily utilized for increased food production by controlling agricultural pests. Each year, 50 million tonnes of pesticides are consumed across India. However, only less than 0.1% of employed pesticide will effectively reach the target. The remaining pesticides are found to be tenacious in the environment having a half-life of 15–30 years. This residual pesticide causes toxic effects in the nontarget organisms by contaminating the soil, water, and air through several pathways such as runoff, sorption, photolysis, and leaching. The soil contamination is highly crucial because it is the part of many biological cycles and serves as a habitat for humans, flora, fauna, and other living organisms. Hence, it is of prime importance to mitigate or possibly eliminate pesticide concentration from the soil. Several physical, chemical, and biological techniques are available for removing pesticides. The physical and chemical techniques can separate, immobilize, and transform pesticides into a less toxic form. The use of physical and chemical techniques in agriculture is limited as it modifies the properties of soil and makes it unfit for crop production. Thus, an environmentally

S. Varjani
Gujarat Pollution Control Board, Gandhinagar, Gujarat, India

M. C. Sudha (✉)
Department of Environmental Science, S.V. University, Tirupati, India

N. M. Kumar
Vellore Institute of Technology, Vellore, India

B. B. K. Pillai
NIT Warangal, Warangal, Telangana, India

V. K. Srivastava
Sankalchand Patel Vidyadham, Sankalchand Patel University, Visnagar, Gujarat, India

M. K. Awasthi · S. K. Awasthi · Z. Zhang
College of Natural Resources and Environment, Northwest A&F University, Yangling, Shaanxi Province, People's Republic of China

M. K. Meghvansi, A. Varma (eds.), *Biology of Composts*, Soil Biology 58,
https://doi.org/10.1007/978-3-030-39173-7_11

friendly and economically viable biological technique, viz. composting has gained importance for remediating pesticide-contaminated soils. The composting practice is one of the promising techniques, as it has high removal efficiency and yields useful end products. Although composting has several benefits the studies available in the Indian context are scarce. Hence to highlight the importance, this chapter reviews the present status of pesticide consumption in India, pesticide transport mechanism in the soil, its effect on human health, and case studies related to pesticide composting.

Keywords Agriculture · Pesticide · Contamination · Composting · Organophosphates

11.1 Introduction

Agriculture gets paramount importance in the Indian economy and contributes 18% to the country's GDP (Bhardwaj and Sharma 2013; Varjani et al. 2019). Ensuring food security for more than 1.27 billion Indian population with decreasing cultivable land resource is a phenomenal and challenging task. The UN FAO (Food and Agriculture Organization) estimated that weeds, pests, and diseases destroy approximately 40% of field crops and around 6–7% of the crops following the harvest. More than 50% of pre-harvest losses are estimated in Africa and Asia (FAO 1975; Oerke et al. 1995).

Agrochemicals also known as pesticides have a significant role in productivity enhancement and crop protection. Pesticides are any substance or mixture of substances envisioned to prevent, destroy, repel or mitigate any rodent, insect, nematodes, fungi or weed, or any other life forms declared to be pests (Azmi and Naqvi 2011). The application of these chemicals increased many folds after the green revolution of 1970s in India, and the usage of the chemical pesticides in agriculture turn into an essential part of crop production in most parts of the country (Atreya 2007; Shetty et al. 2008; Devi 2009; Varjani 2018). In recommended dosages, they are diluted and applied to seeds, cultivation soil, water used for irrigation, and crops to prevent pests and disease damage. Insecticides form the major sub-segment of agrochemicals having 60% of market share, while herbicides hold 16% of market share are the segments that are rising fast in India (FICCI Report 2015).

In India, pesticides use began in 1948 with the import of dichloro diphenyl trichloroethane (DDT) and benzene hexachloride (BHC) to eradicate malaria and locust, respectively (Gupta 2004). The usage of both the pesticides (DDT and BHC) is extended to the farming sector in 1949. India presently accounts for about 3% of the total intake of pesticides worldwide and is rising at a rate of 2–5% annually (Bhadbhade et al. 2002). Agriculture and horticulture consume 67% of the total Indian pesticide consumption (Puri et al. 1998). The consumption of pesticide has increased multi-folds, between 154 metric tonnes in 1954 and 41,822 metric tonnes in 2009–2010 (Chauhan and Singhal 2006). The most commonly used pesticides are shown in Table 11.1.

Table 11.1 Major pesticides used for various crops in India

Molecule	Type	Crop
Acephate	Insecticide	Chillies, vegetables, fruits and cereals, tobacco
Chlorpyrifos	Insecticide	Cotton, pulses, oilseeds, rice
Dinotefuran	Insecticide	Rice
Fipronil	Insecticide	Cole crops, sugarcane, chilli
Flonicamid	Insecticide	Apples, peaches, wheat, potato, vegetables
Imidacloprid	Insecticide	Cotton, rice, vegetable crops
Carbamates	Insecticide	Lawns and gardens
Organochlorines	Insecticide	Variety of crops
Glyphosate	Herbicide	Variety of crops
Quizalofop	Herbicide	Broad leaf crops
Dipyridil	Herbicide	Variety of crops
Triazines	Herbicide	Maize (corn) and sugarcane
Hexaconazole	Fungicide	Cereals, oil seeds, horticultural, plantation crops
Tricyclazole	Fungicide	Rice
Organotin	Fungicide	Potato, sugar beet, peanuts

Source: Subash et al. (2017)

The actual utilization of the applied pesticides to the target pest is estimated to be less than 0.1%, and the remaining pesticides are found to be persistent in the environment with a half-life period ranging between 15 and 30 years. The unused pesticides that entered into the environment contaminate the soil, water, air, and cause adverse effects on nontarget organisms (Pimentel and Levitan 1986; Kumar et al. 2018). Further, the ineffective chemical management and the residual pesticides had resulted in harmful health impacts, flora, fauna, and subsequently the entire environment. In 2004, 8.3% of global deaths were attributed to environmental exposure of selected toxic chemicals, and a higher percentage is accounted by pesticides (Pruss-Ustun et al. 2011). It is of prime importance to mitigate or possibly eliminate the pesticide concentration, especially from the soil. The current chapter primarily aims at understanding the present status of pesticide consumption in India, pesticide transport mechanisms in the soil, and its effect on human health. It also highlights various treatment technologies with a special focus on composting technology as an eco-friendly and cost-effective biological method for remediating the pesticide-contaminated soils.

11.2 Pesticide Production and Consumption Patterns in India

India is Asia's second largest pesticide producer after China and ranks fourth after the USA, Japan, and China (Agnihotri 1999; Kumar et al. 2012; Yadav et al. 2015). India has an annual production capacity of over 139,000 MT of pesticides, with over

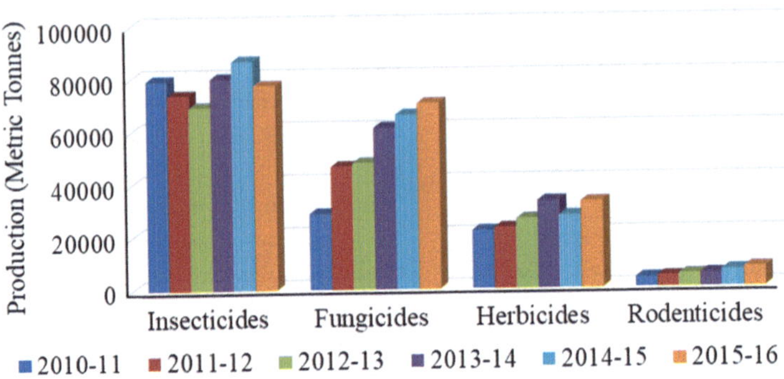

Fig. 11.1 Production volume of pesticides in India from 2010 to 2016 (Source: Ministry of Agriculture & Farmers Welfare statistical database, Government of India)

219 technical grades and manufacturing units and over 4000 formulation units (Yadav et al. 2015). The volume of pesticides produced during the fiscal years between 2013 and 2017 is shown in Fig. 11.1. Further, the state-wise consumption of pesticides is shown in Fig. 11.2. India also exports and imports the pesticides for various applications (Table 11.2). The pesticide industries in India produce two groups of pesticides viz., the technical grade and the formulated. Pesticides of technical grade are incredibly venomous and composed of harmful substances; the formulations are made using emulsifiers to process pesticides of a technical grade.

The consumption patterns of pesticides are significantly unreliable and differ in different states (SEEP 2010). Andhra Pradesh, a southern state of India has the maximum consumption of the pesticides (9289 MT), followed by Uttar Pradesh in northern part of India (8839 MT), and Maharashtra in western part of India (6723 MT) (Yadav et al. 2015). The state of Punjab located in western part of India consumes 5770 MT of the pesticides yearly for an area of 50,000 km^2. The usage of insecticide-resistant management (IRM) and non-IRM zones are 5.6 and 8 kg ha^{-1}, respectively. This average usage of pesticides is very high compared to the average national usage (0.5^{-1} kg ha^{-1}) (Gupta 2004; Peshin 2005). Annually, the state of Tamil Nadu and Maharashtra, uses 12,500 metric tonnes and 6000 metric tonnes of these pesticides, which have an agricultural area of 130,000 km^2 and 307,000 km^2 respectively (SEEP 2010).

11.3 Classification of Pesticides

The pesticide is a common name and may be generally classified based on organic chemical complexes and pests (Varjani et al. 2019). Pesticides can be grouped majorly into five groups based on synthetic organic chemical composites viz., Organo Chlorins (OC), chlorophenoxy acids, organophosphates (OP), carbamates (CB), and synthetic pyrethroids (SP). The chlorinated hydrocarbons such as DDT,

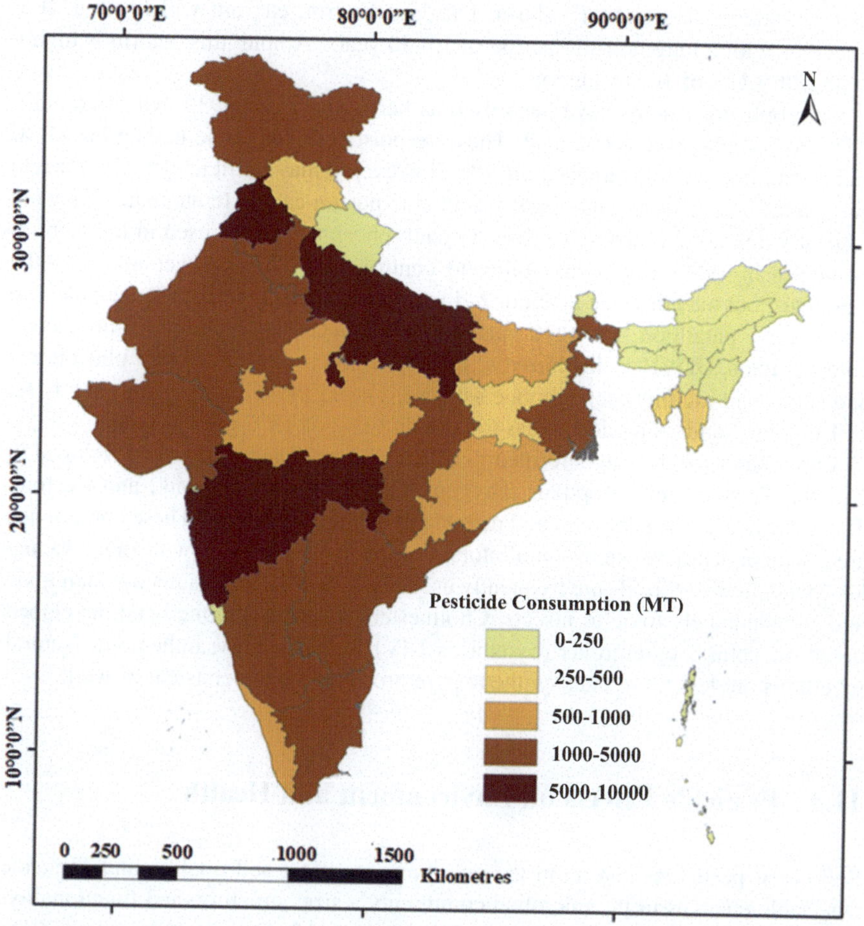

Fig. 11.2 State-wise consumption of pesticides in India (Source: Ministry of Agriculture & Farmers Welfare statistical database, Government of India)

Table 11.2 Export and import of pesticides in India (in Tonnes)

Year	Insecticide		Fungicide		Herbicide		Plant growth regulator	
	Export	Import	Export	Import	Export	Import	Export	Import
2010–2011	57,742	18,679	36,081	3961	23,072	4048	984	361
2011–2012	59,591	19,598	57,511	3983	30,692	4691	1720	217
2012–2013	55,443	18,140	66,321	121	34,142	7138	1953	307
2013–2014	65,823	20,450	66,820	4705	32,200	10,056	1721	731
2014–2015	63,550	19,941	105,613	6868	32,411	12,329	891	6009
2015–2016	43,286	15,732	133,708	7021	53,108	10,243	1175	6986

Source: Ministry of Agriculture & Farmers Welfare statistical database, Government of India

BHC (Lindane), Heptachlor, Aldrin, Dieldrin, Endrin, etc. are widely used. It is tenacious with a long half-life period of 15–30 years. Among this, endrin is highly toxic with LD_{50} of 10–15 mg kg^{-1}.

Chlorophenoxy acids have been used as herbicides since 1930, which contains 2, 4, dichlorophenoxy acetic acid. They are poisonous for florae having broadleaf and weeds, but less toxic toward animals. However, it may result in eye irritation and gastrointestinal disorder, and regular animal exposure can be teratogenic. They are generally degraded in about 15 days. Organo-phosphates were used in the form of pesticides in 1950s due to its ability to control pest. They are venomous to the mammals, but may degrade in about 2–4 weeks. The most prevalent insecticides are malathion, parathion, methyl parathion, dimecron, diazinon, dimethoate, chlorpyriphos, DDVP or dichlorovos, TTEP, etc. These are made up of phosphorus and sulfur and inhibits acetylcholine esterase. The TEPP is highly lethal with LD_{50} of 1 mg kg^{-1} and malathion is less toxic with LD_{50} of 1500 mg kg^{-1}.

Carbamates are the esters formed from carbonic acid and primarily developed in Nigeria. They contain propoxur (baygon), carbaryl (sevin), Temik, and Zectran. These esters are less persistent and degrade in about 25–28 days. These compounds are also acetylcholine esterase inhibitors. Propoxur is more toxic with LD_{50} 30 mg/kg. Synthetic pyrethroids are frequently used because they are less mammalian toxic and are adequately toxic to insects. Cholinesterase is also inhibited and developed based on natural pyrethroids (pyrethrins I-IV) found in chrysanthemum. Natural pyrethrins are less toxic than synthetic pyrethroids and less persistent as well.

11.4 Pesticide Effects on Environment and Health

Soil-based pesticides can harm the environment, affect soil quality, and persuade detectable changes in the microbial community's size, structure, and functionality, thereby changes in soil life functions, dynamics, and biodiversity (Chen et al. 2015; Varjani 2018). The widespread use of pesticides and improper use, handling, or storage of pesticides poses threats to human health. Aldrin, chlordane, DDT, dieldrin, endrin, heptachlor, hexachlorobenzene, mirex, and toxaphene were recognized under the Stockholm Convention to cause fatal effects on mankind and also on the ecosystem. Environmental contamination of pesticides is directly related to its persistent nature, which determines the degree of persistence of the pesticide compound itself as chemical characteristics (Fumio 1982; Varjani 2018). Firstly, whether or not they are intended for crops, most pesticides are known to end up in soil; secondly, they are spread in the aquatic environment, thirdly, many of the pesticides have volatility thus pesticides contaminates the air.

The fate of the pesticide employed to soil mainly depends on the solubility and persistence of pesticides (Varjani et al. 2019). The "lasting-power" of a pesticide is described by its persistence. Majority of the pesticides decimate by sunlight or may undergo degradation with time owing to various chemical and microbiological reactions in the soils. Usually, pesticides degrade partially through chemical pathways, while microorganisms of soil can convert many pesticides into carbon dioxide, water, and inorganic elements. However, the pesticides that leached underneath the root zone would be less probable to degrade due to the less microbial activity (Rao et al. 1992; Kerle et al. 1994). Organic pesticide residues with persistent nature were largely dispersed in Indian soils (Yadav et al. 2015; Kumar et al. 2018).

The solubility of the pesticide is the most significant property affecting its movement within the water. When the pesticide is applied to the soil, some of the organic matter will remain in the soil, by the adsorption process and some of them will dissolve in soil-water. The pesticides adsorbed in the soil may get detached by the process of desorption. Solubility of pesticide and sorption on soil are inversely related. The coefficient of partitioning is determined by the chemical properties of a pesticide such as solubility and melting point (Rao et al. 1992). The Partition coefficient (PC) is a useful index to quantify the pesticide adsorption on soils. The PC value can be defined as the ratio between pesticide concentration existing in the adsorbed state and the solution phase. Thus, the lower the value of PC, the greater the pesticide concentration in solution. Pesticides having lesser PC values may leach more probably when compared to those having higher PC values. The extensive application of these pesticides to ecosystems has caused many environmental problems globally. Once the pesticide is employed to the cropland, the following things may happen:

- Plants may uptake.
- Animals, worms, insects, or micro-organisms may be ingested from the soil.
- May enter in the soil and either dissolve or adhere to the particles.
- It can go into the atmosphere and vaporize.
- It may break down into other, less poisonous compounds through microbial and chemical processes.
- Rain or irrigation water may leach from the root zone or wash off the surface of the soil.

Pesticides have serious impacts on nontarget species and impact the biodiversity of animals and plants, the webs and ecosystems of aquatic and terrestrial food. Insecticides are considered to be the most toxic among all types of pesticides, followed by fungicides and herbicides. Pesticides that are soluble in water get dissolved and enter soil, streams, rivers, and lakes, causing damage to species that are not targeted. Through the bio-amplification process, fat-soluble pesticides enter animal bodies. They are absorbed into animal fatty tissues, leading to long-term persistence of pesticides in food chains (Warsi 2018).

The health consequences connected with exposure to pesticides include neurological, reproductive, developmental, ophthalmic, genotoxic, and carcinogenic impacts. Most of the insecticides are made of organophosphates or carbamates and these are powerful inhibitors of cholinesterase (Rao et al. 1992). The poisoning signs relate to the pesticide's level, identity, and solubility of fat. The symptoms include such as salivation, lacrimation, urination, defecation, gastric emesis, bronchorrhea, bronchospasm, bradycardia, respiratory depression, cardiac arrhythmias, miosis, neuropathy, and altered mental status (Wolff et al. 1992; Mahmood et al. 2016; Kumar et al. 2018). A report by UNEP and WHO says, around 2,00,000 people face death, and about 3 million people are poisoned every year through the exposure of pesticides, globally. The major (95%) instances come mainly from the developing countries (WHO/UNEP 1990; FAO/WHO 2000; Pope et al. 1994). It is assessed that globally 3 million instances of serious acute pesticide poisoning may occur every year, accounting for more than 2,20,000 deaths yearly, and in the developing countries, 99% of these fatalities may happen (Bhardwaj and Sharma 2013). Severe pesticide poisoning is therefore a major health problem in developing nations.

The first report in India on the pesticide poisoning was from Kerala in 1958, about 100 individuals died following the consumption of parathion-contaminated wheat meal (Karunakaran 1958). A survey of food contaminants in India during 1993 discovered that DDT residues were present in about 82% of the 2205 bovine milk specimens gathered in the 12 states of India. DDT residue samples were higher than the tolerance limit in Maharashtra (74%), followed by Gujarat (70%), Andhra Pradesh (57%), Himachal Pradesh (56%), and Punjab (51%) (Wasim et al. 2009). The Centre for Environment and Science reported the high pesticide levels in blood samples of villagers in Punjab (Bhardwaj and Sharma 2013).

Pesticides also posed a severe risk to the aquatic and terrestrial ecosystems and to its biological integrity. A higher percentage of pesticides are used in standing crops in India. Soil contamination thus becomes a serious issue, particularly with the persistent kinds of pesticides, primarily when used for soil drainage for several years (van Werf 1996). They decrease soil fertility, soil acidification, nitrate leaching, amplified floral and faunal resistance, groundwater as well as surface water pollution, and contamination of agricultural soils signifies the importance of remediation of soil and water-contaminated with pesticides (Kumar et al. 2018).

11.5 Legal Framework for Indian Pesticides

The Insecticides Act, 1968 and the Insecticides Regulations, 1971 govern the process of importing, registering, manufacturing, selling, transporting, distributing, and using insecticides (pesticides) throughout India to avoid danger to humans or livestock and all related matters. India's Central Insecticides Boards and Registration Committee (CIBRC) is the Indian government's legislative body responsible for registering pesticides before they can be made accessible for use or sale. India's Food Safety and Standard Authority of India (FSSAI) establishes the licensed pesticide maximum residual limits (MRLs). Indian standards permit considerably higher concentrations of persistent organic pesticides compared to EU (European Union) standards. Lindane's regulatory threshold is 100 times greater than the EU and acceptable aldrin and dieldrin concentrations in Indian foods are ten times greater than in the Czech Republic (IPEN 2006). The residual limit for lindane in Indian drinking water is almost 20 times greater than in the Czech Republic and

30 times greater for heptachlor than in the EU, which is indicating the alarming deviation of MRLs for the pesticides in India.

11.6 Mechanism of Pesticide Transport in Soil

In terms of energy input, the soil ecosystem is predominantly allochthonous and receives energy chiefly in the form of organic carbon residues from herbal and animal debris. Pesticides used in the soil become entangled in the procedures of transportation and degradation affecting all the organics that added to the soil system (Sawhney and Brown 1989; Kenneth 1990). Many processes are engaged in transporting residues of pesticides inside or outside the soil ecosystem.

First class of transport procedures includes aerial transport into the atmosphere of pesticides from the soil. Volatilization is an important method whereby pesticides can migrate from the soil to the atmosphere during the vapor phase (Biggar and Seiber 1987). The second class of transport process of pesticide is the movement of dissolved or particulate-sorbed pesticides in water. The soil's chemical and physical characteristics affect the behaviour of pesticides being transported to aquifers. In specific, clay and organic soil contents play a significant role (McEwen and Stephenson 1979). Pesticides are connected by dipole–dipole attraction, hydrogen bonding, or ionic binding to negatively loaded clay and organic matter locations. Ionizable compounds readily carry deeper layers of soil than hydrophobic chemicals. A pesticide's octanol–water partition ratio (Kow) and the organic carbon content of soil or aquifer sediments predict the velocity of travel relative to water (Schwarzenbach and Westall 1981; Schwarzenbach et al. 1983). The greater the content of Kow and organic carbon, the slower the flow of pesticides through a matrix of soil or sediment. The quantity and sequence of rainfall occurrences and the depth of the vadose zone are some of the variables that affect pesticides' transport conduct through soils to aquifers. A pesticide poses a potential danger to groundwater if the chemical has water solubility higher than 30 mg L^{-1}, a partition coefficient less than 5, a half-life of hydrolysis longer than 25 weeks, and a half-life of soil longer than 2–3 weeks (Creeger 1986).

Pesticide leaching was perilous and trace amounts of pesticide can led to groundwater contamination (Biggar and Seiber 1987). The transport of pesticides flowing over the surface of the soil in water is also a significant method that can affect the quality of the surface water. Convection and diffusion are concurrent procedures involving the transport of pesticides in gravitationally moving water or with a gradient of concentration. Pesticides with water solubility exceeding 10 μg g^{-1} were assumed to move mainly in the solution stage (Wauchope 1978).

Pesticides can also subject to processes of abiotic and microbiological transformation. The abiotic processes in soil partially degrade the pesticides. Significant abiotic degrading responses happen within the soil profile between pesticides and inorganic and organic reactive compounds. The pesticide residues that get exposed to ultraviolet radiation undergo significant photodegradation (Biggar and Seiber 1987). Hydrolytic reactions are one of the transformation mechanisms for most of the pesticides in the soil such as organophosphorus and carbamate insecticides and herbicide esters (Tinsley 1979; Drossman et al. 1988). These hydrolytic reactions are triggered by catalyzed base or acidic relationships in clay mineral surfaces with reactive chemical organizations (Yaron 1978), organic reactive products (e.g., enzymes) immobilized on or in aggregates of soil, or inorganic elements (e.g., Cu^{+}) in the soil solution. Microbial degradation often requires hydrolysis of selected functional groups (Kenneth 1990).

The significance of the microbial community of soil in facilitating the pesticide's degradation has well understood by many studies. The environmental conditions such as temperature, moisture influence microbial activity which impacts microbial soil degradation of pesticides (Chapman et al. 1986). The microbial metabolism of pesticides occurs in two processes namely "catabolism" and "co-metabolism." The catabolism process is due to the mineralization of certain parts of the organic compound by enzymatic routes to simpler products. In few cases, one part of the molecule can be mineralized and another part can accumulate in the soil, microbial degradation of carbofuran is an excellent example of this process (Lal 1984; Kenneth 1990). Hence, catabolism does not involve incomplete mineralization or pesticide destruction. Catabolism is principally a process driven by the microbial quest for energy. Therefore, catabolism process helps in the use of a pesticide as a source of energy and therefore a growth substrate for the microbial population in soil (Kenneth 1990). Pure crops of soil isolated pesticide-degrading microorganisms have been able to exploit pesticides as only sources of phosphorus, nitrogen, or sulfur. The cometabolism is associated with the operations of microbial enzymes. The cometabolism phenomenon is described as the degradation of materials without the use of their oxidation energy to promote growth (Kenneth 1990).

11.7 Remediation Techniques

The pesticide-contaminated soils can be remediated by physical, chemical, and biological treatments or a combination of these approaches, performed either in situ or ex situ (Pascal-Lorber and Laurent 2011; Varjani 2017b). Table 11.3 shows various techniques adopted for pesticide removal and its respective removal efficiency. The physical treatment involves the separation of pesticides from soil whereas, in chemical and biological treatment, the contaminants are entirely destructed or transformed into a less toxic substance (Varjani 2017a). Conventionally, pesticide contamination has been remediated using physicochemical technologies where soils are excavated and consequently transported to secured landfills; incineration or stabilization of the material may also be carried out on-site (McGuinness and Dowling 2009). The various physical, chemical, and biological treatments used for remediation of pesticide-contaminated soils are shown in Fig. 11.3.

Table 11.3 Pesticide removal using physical, chemical, and biological methods

Pesticide	Type	Scale	Technique	Key points	References
Metolachlor, alachlor, atrazine, pendimethalin, and chlorpyrifos	Herbicides and insecticide (chlorpyrifos)	Field scale	Zero-valent iron and $Al_2(SO_4)_3$	Within 90 days – Zero-valent iron reduces >60% of all pesticides – Zero-valent iron and $(Al_2SO_4)_3$ combination reduces >90%	Shea et al. (2004)
Atrazine	Herbicide	Lab scale	Supercritical extraction	96–99% of atrazine were extracted	Castelo-Grande et al. (2005)
Dinoseb	Herbicide	Lab scale	Zero-valent iron, $FeSO_4$, and $Al_2(SO_4)_3$	100% removal within 7 days	Satapanajaru et al. (2009)
DDT	Insecticide	Lab scale	Cattle manure compost	The thermophilic stage exhibits a high degradation potential of 46 and 88%	Purnomo et al. 2010a
DDT	Insecticide	Lab scale	Bioremediation by spent mushroom waste	During 28 days incubation – Degraded DDT by 40% in sterilized soil – Degraded DDT by 80% in unsterilized soil	Purnomo et al. (2010b)
DDT and DDE	Insecticide	Lab scale	Soil washing (TX-100 solution) and photo-Fenton oxidation (solar)	After 3 sequential washings – 66% DDT and 80% of DDE were removed – After 6 h irradiation of wastewater – 99% DDT and 95% of DDE were removed	Villa et al. (2010)
DDT	Insecticide	Lab scale	Bioremediation by brown rot fungi (*Gloeophyllum trabeum*, *Fomitopsis pinicola*, and *Daedalea dickinsii*)	Removal in sterilized soil – 41% by *Gloeophyllum trabeum* – 9% by *Fomitopsis pinicola* – 15% by *Daedalea dickinsii* – % removal in unsterilized soil – 43% by *Gloeophyllum trabeum*	Purnomo et al. (2011)

(continued)

Table 11.3 (continued)

Pesticide	Type	Scale	Technique	Key points	References
				– 29% by *Fomitopsis pinicola* – 32% by *Daedalea dickinsii*	
Oxyfluorfen	Herbicide	Lab scale	Bioremediation	After 45 days – Biodegradation is high in soil incubated at 40 °C than 28 °C – Highest degradation achieved by *Bacillus* spp. (95.6%), *Pseudomonas* sp. and *Arthrobacter* spp. (82.2%).	Mohamed et al. (2011)
Chlorpyrifos	Insecticide	Field	Bioremediation (biobeds)	Degradation greater than 50%	Tortella et al. (2012)
DDT	Insecticide	Lab scale—column experiment	Nano size zero valent iron	45% of DDT is degraded	El-Temsah et al. (2013)
DDT	Insecticide	Lab scale	Nano size zero valent iron	– Degraded DDT by 50% in spiked soil – Degraded DDT by 24% in ages soil	El-Temsah and Joner (2013)
Thiocarbamate molinate	Herbicide	Lab scale	Bioaugmentation	After 42 days – 41% by *Gloeophyllum trabeum* – 9% by *Fomitopsis pinicola*	Lopes et al. (2013)
Lindane	Insecticide	Lab scale	Phytoremediation by *Jatropha curcas* L.	After 300 days – Residual concentration reduced to 72–89% of initial lindane amounts	Abhilash et al. (2013)
Chlordanes, endosulfans, and mirex	Insecticide	Lab scale	Soil washing with maize oil and carboxylmethyl-β-cyclodextrin	Removal after first wash – Chlordanes-81.2% – Endosulfan-95.6% – Mirex-69.2% Removal after second wash – Chlordanes-92.3% – Endosulfan-98.5%	Ye et al. (2014)

				– Mirex-87.2% Removal after third wash – Chlordanes-94.2% – Endosulfan-99.2% – Mirex-94.3%	
Triazophos	Insecticide	Lab scale	Normal solar radiation and enhanced solar radiation	After 50 days the degradation efficiency is – 71% in normal solar radiation – 92% in enhanced solar radiation	Rani and Sud (2015)
2,4-Dichlorophenoxyacetic acid	Herbicides	Bench scale—Pilot plant	Electrokinetic soil flushing	After 40 days – Removed 50% 2,4-D from the soil – 25% remained in the soil – 25% volatilized	Risco et al. (2016)
2,4-Dichlorophenoxyacetic acid	Herbicides	Bench scale—Pilot plant	Electrokinetic soil flushing powered by solar panels and DC current	After 15 days – 73.6% removal by solar panel powered electrokinetic soil flushing – 90.2% removed by DC-powered electrokinetic soil flushing	Souza et al. (2016)
2,4-Dichlorophenoxyacetic acid, atrazine, chlorsulfuron, and oxyfluorfen	Herbicides	Bench scale	Electrokinetic soil flushing	After 15 days – 95% of 2,4-D is removed – 80% of atrazine and oxyfluorfen are removed – Chlorsulfuron highly remained in the soil	Vieira dos Santos et al. (2016)
Endosulfan α, endosulfan β, and aldrin	Insecticide	Pilot scale	Rotary drum and windrow composting Muntjeer	Removal % in the rotary drum – Endosulfan α—83.3 – Endosulfan β—85.3 – Aldrin—86.8 Removal % in windrow composting – Endosulfan α—77.7	Ali et al. (2016)

(continued)

Table 11.3 (continued)

Pesticide	Type	Scale	Technique	Key points	References
				– Endosulfan β—67.2 – Aldrin—66.6	
Triazine (atrazine, terbuthylazine, terbutryn) and chlorpyrifos	Herbicide and insecticide	Lab scale	Bioaugmented and non-bioaugmented biomixture (biomixture volumetric composition: 13% compost, 45% coconut fiber, and 42% soil)	After 60 days Bioaugmented removal – Atrazine-45.4% – Terbuthylazine-36.5% – Terbutryn—No significant reduction – Chlorpyrifos-72.1% Non-bioaugmented removal – Atrazine-97.9% – Terbuthylazine-90.3% – Terbutryn-35.5 – Chlorpyrifos-94.4%	Lizano-Fallas et al. (2017)
Oxyfluorfen, diuron, tebuconazole, imidacloprid, and dimethoate	Herbicide	Field	Bioremediation (biobeds with vermicompost)	Pesticide removed – 50% of Oxyfluorfen – 73% of tebuconazole – 75% of diuron – 80% of imidacloprid – 100% dimethoate	Delgado-Moreno et al. (2017)
Atrazine and oxyfluorfen	Herbicides	Bench scale	Reversible electrokinetic adsorption barrier	After 15 days – 10% of herbicides remained in soil – 45–50% evaporation loses	Rodrigo et al. (2018)
Beta-hexachlorocyclohexane and chlordecone	Chlorinated pesticides	Lab scale	Photolysis and Fenton oxidation	After 5 h – Photolysis removed 100% of both pesticide – Fenton oxidation removed 15% of beta-hexachloro cyclohexane and has less significant removal in chlordecone	Cruz-González et al. (2018)

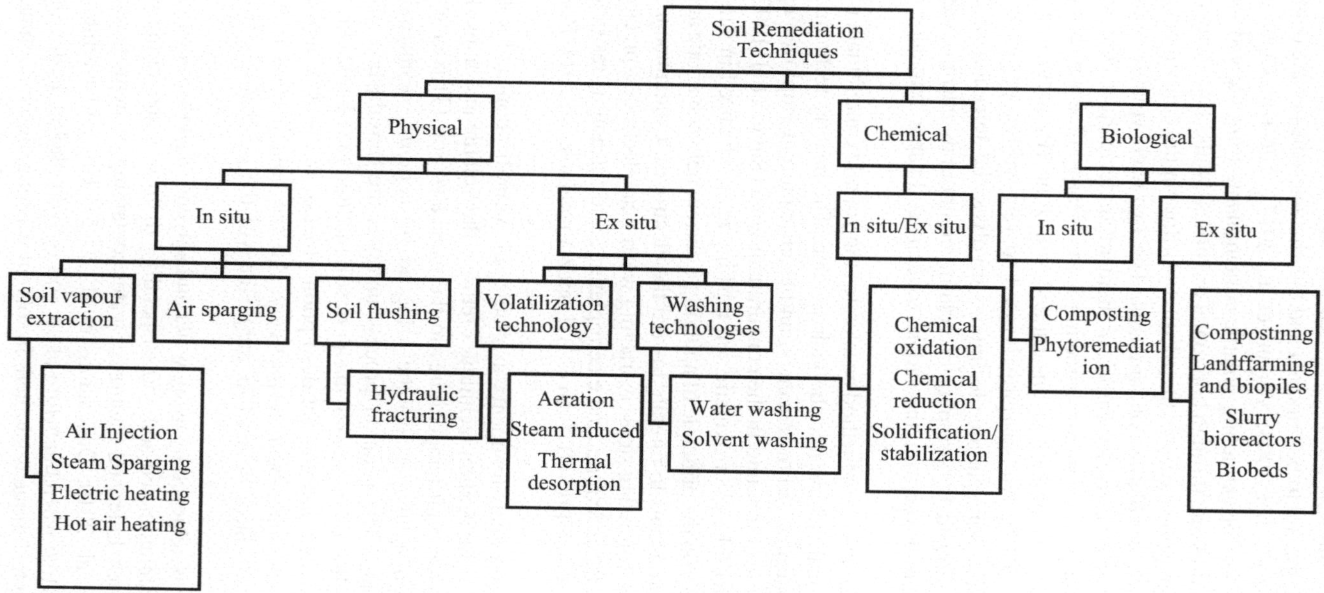

Fig. 11.3 Physical, chemical, and biological techniques for contaminated soil remediation (Source: Koustas and Fischer 1998; Morillo and Villaverde 2017)

In physical treatment, the widely adopted in situ treatments is soil vapor extraction, air sparging, and soil flushing. In general, the selection of any technique is based on the contaminant to be removed and the characteristics of the contaminated soil (Koustas and Fischer 1998; Varjani et al. 2017). The soil vapor extraction technique is adopted for extracting volatile contaminants from the vadose zone by applying vacuum pressure through horizontal or vertical wells. The vacuum pressure applied by vacuum pumps creates the adequate airflow in the contaminated zone to extract the subsurface volatile contaminants. The extracted volatile organic compounds are removed by activated carbon or incineration techniques (Varjani 2017b). The major drawback of soil vapor extraction is the cleanup time which may extend up to years, and it is limited to unsaturated soil zone (Rathfelder et al. 1995; Simpanen et al. 2017). Air sparging is used to remove the pesticide from the saturated zone. The air sparging technique involves the injection of air in the saturated zone followed by stripping of volatile compounds into the vadose zone and subsequently removed from the vadose zone, by soil vapor extraction (Brusseau and Maier 2004).

The ambient air injection method accelerates the volatile contaminant removal by supplying additional airflow into the vadose zone. The heating techniques such as hot air, steam sparging, radio frequency, and electric current employed to vaporize the volatile organic compound in the contaminant zone (Sittler et al. 1992; Koustas and Fischer 1998). The electrokinetic soil flushing is an advanced in situ treatment method that works on the principle of applying a current between anode and cathode, causing the contaminants to migrate toward the electrodes. Contaminant separation is achieved by the following steps: extracting water near electrodes, precipitating the contaminants on the electrode, and electroplating. Soil flushing is restricted to soils having a high permeability (10^{-2} cm s^{-1}) (López-Vizcaíno et al. 2014; Rodrigo et al. 2014).

The ex situ physical treatments available are volatilization and soil washing techniques. The volatilization techniques remove the volatile and nonvolatile compounds by applying high temperatures to the excavated soil (Troxler et al. 1993). The volatilization techniques can be achieved by aeration, steam induced, and thermal desorption. The aeration process involves the introduction of air to the excavated soil followed by heating the air or soil medium to volatilize the pesticides. In the steam induced and thermal desorption process, the heat is applied through steam and desorber chambers, respectively. All the volatilization techniques are followed by gas control devices to purge the volatile compounds (Koustas and Fischer 1998).

Soil washing is achieved by using water and solvent. Water washing involves the excavation of contaminated soil and suspending them in water in the presence of chemical additives. Later soil fraction or water containing pesticide is separated and treated (Villa et al. 2010; Dos Santos et al. 2015). This soil washing technique becomes ineffective when the nature of the soil is hydrophobic. In this situation, the solvent washing technique is adopted. In solvent washing, the contaminated soil and organic solvent are introduced in the extractors where the contaminants get readily dissolved and separated (Koustas and Fischer 1998; Varjani 2017b).

The transformation and stabilization of pesticides can be achieved by the chemical treatments viz., chemical oxidation, chemical reduction, and stabilization/solidification (Lin et al. 1996; Dolfing et al. 2007; Tsitonaki et al. 2010). All the chemical techniques can be adopted in situ and ex situ. The chemical oxidation employs strong chemical oxidants such as hydrogen peroxide to oxidize the organic contaminant (Cruz-González et al. 2018). The addition of ferrous ion to hydrogen peroxide (Fenton's reagent) is also widely used in the chemical oxidation technique. The transformation of organic contaminants into the less toxic compound is achieved through reduction reactions. The stabilization/solidification utilizes the cement or pozzolanic material to solidify the given soil matrix. By solidification, the contaminant leaching potential is significantly reduced (Castelo-Grande et al. 2010). Although effective, these technologies have significant limitations such as high cost, labor-intensive process, and ecosystem disruption (Dijkgraaf and Vollebergh 2004). The viable alternatives to these traditional remediation technologies are essential for sustainable ecosystems.

The biological treatment is one such alternative (Varjani et al. 2018a) and various methods are in use to degrade the pesticides in soil and water (Varjani and Upasani 2017). Among various biological methods, composting is an eco-friendly and promising technology for soil bioremediation due to its advantages over physical and chemical technologies (Varjani et al. 2015). Composting has been successfully applied to the bioremediation of contaminated soils with polyaromatic hydrocarbons and pesticides (Chen et al. 2015; Varjani and Upasani 2016; Varjani et al. 2018b).

11.7.1 Composting

Composting is a process where the organic matter gets degraded microbially under aerobic conditions to attain a stable material that can be used as organic fertilizer (Bustamante et al. 2010; Awasthi et al. 2018). Composting can help to stabilize and/or degrade pesticides in contaminated soils. In the soil composting, a high catabolic activity by microbial populations composed of a wide variety of mesophilic, thermotolerant, and thermophilic aerobic microorganisms are involved (López-González et al. 2015). The notion actinomycetes bacteria (NAB), actinomycetes, and fungi are the chief pollutant-degrading microbes in soils and composts and considered to be the crucial governing factors in the remediation of contaminated soils (Insam et al. 2015; Singh 2008). The understanding of microbial interactions and their roles during the composting process of pesticide-contaminated soils are very crucial (Bandala et al. 2006; Paul et al. 2006). A deeper understanding of the dynamics of microbial communities found in soil/compost mixtures is necessary, in order to assess the effect of composting in the remediation of contaminated soils (Cruz-Ruíz et al. 2015; Nadia et al. 2015).

Compost is well-suited for pesticide degradation due to the thermophilic temperatures achieved during composting. This permits faster biochemical reactions than under ambient temperatures, which accelerates the pesticide degradation. The high temperatures may increase the bioavailability of pesticides thereby increasing the chance of microbial degradation. Few microorganisms may co-metabolize the pesticides, where the microbes depend on the feedstock for food and energy while breaking down an adjacent pesticide. Different organic matter structures in compost help to endorse co-metabolism of numerous offensive compounds, even recalcitrant xenobiotics such as DDT, PCB, and TCE (Buyuksonmez et al. 2000).

11.7.2 Mechanism of Pesticide Degradation in Composting

The behavior of pesticides in the soil is governed by a variety of physical, chemical, and biological properties (Fig. 11.4). The main useful properties are water solubility, vapor pressure (VP), Henry's law constant (K_H), octanol/water partition coefficient (K_{OW}), soil organic carbon/water partition coefficient (K_{OC}), acid dissociation constant (K_a), and half-life ($t_{1/2}$). The nature and position of functional groups, substituent, and unsaturated bonds present in the pesticide can be used to estimate its sorption behaviors and the risks of environmental pollutions (Espinoza-Navarro and Bustos-Obregón 2005).

The persistence of a pesticide in the environment largely depends on its chemical structure and on the presence of unusual functional groups. The water solubility and bioavailability of pesticides get influenced by its chemical structure. The pesticides can degrade rapidly if its functional groups are made of weak or labile bonds. Today, many pesticides have been designed with weak bonds to avoid problems of extended persistence into the environment. Malathion, an insecticide is one such example containing many labile bonds that may be broken through hydrolysis with the help of hydrolytic enzymes (Bezdicek et al. 2001; Buyuksonmez et al. 2000). The hydrolysis is the first step in the composting process. The microbial enzymes then react with the simpler compounds that degrade to lesser toxicity. The mono and dioxygenases are the enzymes that are commonly associated with pesticide degradation. These enzymes help in the oxidation of pesticide, which in turn increases its water solubility, thereby increasing its bioavailability.

The microbes produce enzymes either extracellularly or intracellularly. The enzymes produced extracellularly will have very low specificity. Hence, it reacts with many different chemical compounds with different structures. If pesticides are reacted with these enzymes before the intended substrates such as cellulose, hemicellulose, and lignin, the enzymes may react with the pesticide and convert into a possibly less toxic and less hazardous form. This type of co-metabolism plays a significant role in degrading pesticides found in compost and soil (Gustavo et al. 2016).

Fungi are the vital source of most extracellular enzymes, and they grow as the long strings of the cell (Hyphae), which extend throughout the composting process. *Trichoderma*, *Gliocladium*, *Penicillium*, and *Phanerochaete* are the genera often associated with compost and soil organic matter. The extracellular enzymes released from hyphae, break down the pesticide and allow it to pass into the cells. This

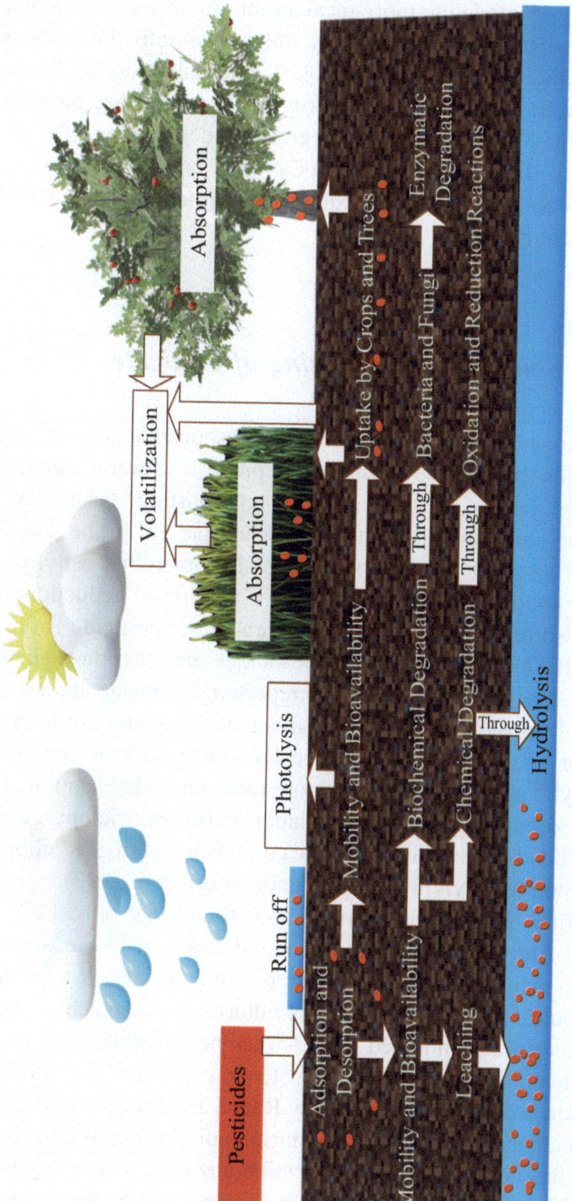

Fig. 11.4 Pesticide degradation mechanisms in soil (Source: Gustavo et al. 2016)

process allows the production of additional hyphae and energy. Although fungi are present in compost feedstock, they contribute more to composting in its later stages. At first, the bacteria consume the easily degraded organic matter from the feedstock; fungi then begin to degrade the more recalcitrant polymeric organic matter.

The pesticide dissolved in water can only enter into the cells of microbes. Pesticides containing more oxygen, nitrogen, and sulfur are inclined to be more water soluble due to the presence of hydrogen bonding. Once the pesticide enters inside a cell, it may undergo varying degrees of degradation. The pesticide gets reduced to carbon dioxide, water, and other inorganic components through the mineralization process. However, it accounts for only a small portion of the pesticide loss through the composting.

11.7.3 Case Studies on Composting of Pesticide

The soil and exogenous microorganisms are responsible for the degradation of pesticides in soils during the composting process. Several studies have been conducted on the fate of pesticides during composting. Pesticides will behave differently during composting than they do when present in the soil. This may be due to increased temperature, microbial diversity, microbial activity, and organic matter during composting. Studies showed a significant reduction in detectable pesticides and also reported that a small percentage of pesticide is typically lost to mineralization. Besides mineralization, pesticides also lost through volatilization, adsorption, and leaching. The findings reported by many studies showed that concentrations of organophosphate and carbamate pesticides are found to be lower after composting. Also, it is found that recalcitrant organochlorine insecticides and pyridine carboxylic acid herbicides are more resistant to degradation. The chlorpyrifos, diazinon, isofenphos, and pendimethalin, isoxaben, triclopyr, clopyralid, and fluprimidol are found to be degraded by the composting with grass clippings in about 128 days (Lemmon and Pylypiw 1992; Vandervoort et al. 1997).

On the contrary, Muller and Korte (1976) found that only 12% of the initial concentration of Aldrin, 3% of the dieldrin, and less than 15% of monolinuron and imugan added to municipal solid waste and biosolids feedstock were degraded after composting. About 55% of the herbicides buturon and heptachlor were degraded. However, the study was conducted for a 3-week period which may be very short to evaluate the potential for removal of persistence pesticides. The study on mineralization of pesticides during composting by Racke and Frink (1989) showed 97% of the insecticide carbaryl was transformed during composting of municipal biosolids, but only 5% of this could be attributed to mineralization. This study was conducted for a period of 20 days, which is found to be a very short period for the composting process. A study by Petruska et al. (1985) showed the importance of volatilization in the remediation of pesticide-contaminated compost. Losses due to volatilization reached 22% for diazinon and 50% for chlordane after 3 weeks of cow manure and sawdust composting. Chlordane was not substantially mineralized, and diazinon was considerably transformed, but with a low rate of mineralization.

The interaction between pollutants and microbes also influences the effect of composting on soil remediation. The undesirable properties of pollutants are adverse for the growth of microorganisms, interfering with their normal metabolic function. A study carried out by Scelza et al. (2008) showed that the soil microflora was difficult to recover from its inhibitory state induced by the toxicity of pentachlorophenol (PCP). Composting exhibited a strong effect on the removal of chlorophenol with a degradation rate of about 85.85% (from 212 to 30 mg kg^{-1}). Bacterial populations in compost- manure- and cornstalk-amended soils were increased at a rate of 5% compared to non-amended soils. These amendments showed no effect on fungal or Actinomycete populations (Moorman et al. 2001).

The addition of pollutant-degrading microorganisms can improve the degrading progress during composting. The addition of *Rhodococcus chlorophenolicus* a microorganism that can degrade several types of chlorophenols as an amendment, the decomposition of chlorophenols was faster than in the absence of *R. chlorophenolicus* (Valo and Salkinoja-Salonen 1986). *P. chrysosporium*, a basidiomycete, exhibited high degradation ability to lignin and PCP (Chen et al. 2011; Martinez et al. 2005; Yu et al. 2011). So, *P. chrysosporium* inoculants could increase lignocellulose biodegradability and eventually improve the quality of compost products. Cattle manure compost (CMC) is studied for its degradability of 1,1,1-trichloro-2,2-bis (4-chlorophenyl) ethane (DDT). DDT was degraded during composting, and 1,1-dichloro-2,2-bis (4-chlorophenyl) ethane (DDD) was detected as a metabolic product. Degradation of DDT at 60 °C was the most effective of all the stages of composting (Purnomo et al. 2010a).

The non-actinomycetes bacteria (NAB) are heterotrophic bacteria, which gets their energy from pesticides as an only carbon source. A large group of Gram-negative and positive NAB genera isolated from soil have been reported to degrade organophosphate compounds. *Serratia* sp. SPL- 2 can degrade methidathion, *Pseudomonas aeruginosa* Is-6 can degrade acephate, methamidophos, methyl parathion, dimethoate, and malathion. Soil bacterial communities containing isolates of *Agrobacterium* sp., *Bacillus cereus*, *Bacillus subtilis*, *Brucellamelitensis*, *Klebsiella* species, *Pseudomonas aeruginosa*, *P. fluorescens,* and *Serratia marcescens* are capable of degrading chlorpyriphos as a sole carbon source after an incubation of 20 days. Diflubenzuron (100–500 μg/g) had a stimulatory effect on *Azotobacter vinelandii* in soil (Deng et al. 2015; Gustavo et al. 2016). Actinomycetes are organic matter decomposers that have potential use as agents for composting and biodegradation of pesticides such as organochlorines, s-triazines, carbamates, acetanilides, organophosphorus, and sulfonylurea in polluted soils (Fuentes et al. 2010).

Soil fungi produce extracellular ligninolytic enzymes like manganese peroxidase, lignin peroxidase, and laccase. Laccase and peroxidase, degrade bentazon (a very recalcitrant herbicide), by the copresence of a variety of humic materials. Fungal enzymes have a greater ability to resist the application of pesticides, except for fungicides (Jilani 2013). White-rot fungi have been reported to have a high capacity of pesticide removal such as monocrotophos, methamidophos, dimethoate, fenpropathrin, acetamide, profenophos, chlorpyriphos, and carbosulfan during composting (Yu et al. 2011). A consortium of *Phanerochaete chrysosporium, Trametes versicolor, Bjerkandera adusta,* and *B. fumosa* were assessed for pentachlorophenol removal (Zeng et al. 2011). Fly larvae composting is gaining greater attention as it is found to be a novel and efficient organic waste management strategy. High resource recovery efficiency can be achieved in this closed-looped system, but pharmaceuticals and pesticides in waste could potentially accumulate in every loop of the treatment system and spread to the environment. The study by Lalander et al. (2016) evaluated the fate of three pharmaceuticals viz., carbamazepine, roxithromycin, trimethoprim, and two pesticides, namely azoxystrobin and propiconazole in a fly larvae composting system and in a control treatment with no larvae. The study found that the half-life of all five substances was shorter in the fly larvae compost (<10% of control) and no bioaccumulation was detected in the larvae. Fly larvae composting could thus hinder the spread of pharmaceuticals and pesticides into the environment.

All the studies are not positively related to the degradation of pesticides during composting. Some studies reported that the concentration of pesticides remains unchanged or increased after composting. This is due to a particular group of organic pesticide which is recalcitrant in nature. Also, few reports indicated that pesticides experience only partial degradation and lead to the formation of secondary compounds that are more toxic than the parent compound. Hence for the efficient removal of considered pesticides, it is essential to select the right compost material and composting method (Buyuksonmez et al. 1999).

11.8 Conclusions and Perspectives

The consumption of pesticides is increasing each year due to the massive demand for agricultural products. These pesticides are persistent in nature and have negative impacts on the environment and human health. Although several techniques are available for removing the residues of pesticide from the soil, the bioremediation techniques gained importance as they are eco-friendly and do not modify properties of soil. Composting is one such remediation technique in which degradation of the pesticide is achieved by microbial degradation. Despite having several advantages, composting received less attention as it takes a long time, requires optimum conditions for the microbial activity, and protection from rodents and pests. It is therefore essential to understand the properties of microbes and genetics in developing microbes that can grow in wider environmental conditions. Future research on the genetically modified microbes in enhancing the composting process for wider pesticide degradation and in situ studies on pesticide-contaminated soils will be the need of the hour.

References

Abhilash PC, Singh B, Srivastava P, Schaeffer A, Singh N (2013) Remediation of lindane by *Jatropha curcas* L: utilization of multipurpose species for rhizoremediation. Biomass Bioenergy 51:189–193

Agnihotri NP (1999) Pesticide: safety evaluation and monitoring. All India Coordinated Research Project on Pesticide Residues, Division of Agricultural Chemicals, Indian Agricultural Research Institute, New Delhi

Ali M, Gani KM, Kazmi AA, Ahmed N (2016) Degradation of aldrin and endosulfan in rotary drum and windrow composting. J Environ Sci Health B 51:278–286

Atreya K (2007) Pesticide use in Nepal: understanding health costs from short-term exposure, South Asian Network for Development and Environmental Economics (SANDEE), Working Paper No 28-07. Available from http://citeseerx.ist.psu.edu/viewdoc/download?doi=10.1.1.924.8990&rep=rep1&type=pdf. Accessed 12 July 2018

Awasthi SK, Joshi R, Dhar H, Verma S, Awasthi MK, Varjani S, Sarsaiya S, Zhang Z, Kumar S (2018) Improving methane yield and quality via co-digestion of cow dung mixed with food waste. Bioresour Technol 251:259–263

Azmi MA, Naqvi SNH (2011) Pesticide pollution, resistance and health hazards. In: Stoytcheva M (ed) Pesticides - the impacts of pesticide exposure. InTech, Rijeka, pp 1–24

Bandala ER, Andres-Octaviano J, Pastrana P, Torres LG (2006) Removal of aldrin, dieldrin, heptachlor, and heptachlor epoxide using activated carbon and/or *Pseudomonas fluorescens* free cell cultures. J Environ Sci Health B 41:553–569

Bezdicek D, Fauci M, Caldwell D, Finch R, Lang J (2001) Persistent herbicides in compost. Biocycle 42:25–30

Bhadbhade BJ, Sarnaik SS, Kanekar PP (2002) Bioremediation of an industrial effluent containing monocrotophos. Curr Microbiol 45:346–349

Bhardwaj T, Sharma JP (2013) Impact of pesticides application in agricultural industry: an Indian scenario. Int J Agric Food Sci Technol 4:817–822

Biggar JW, Seiber JN (1987) Fate of pesticides in the environment. Publication 3320 of the Agricultural Experiment Station, Division of Agriculture and Natural Resources. University of California, Oakland, CA

Brusseau ML, Maier RM (2004) Soil and groundwater remediation. In: Environmental monitoring and characterization. Academic Press, Burlington, pp 335–356

Bustamante MA, Suárez-Estrella F, Torrecillas C, Paredes C, Moral R, Moreno J (2010) Use of chemometrics in the chemical and microbiological characterization of composts from agro-industrial wastes. Bioresour Technol 101:4068–4074

Buyuksonmez F, Rynk R, Hess TF, Bechinski E (1999) Occurrence, degradation and fate of pesticides during composting. Part 1: Composting, pesticides, and pesticide degradation. Compost Sci Util 7:66–82

Buyuksonmez F, Rynk R, Hess TF, Bechinski E (2000) Occurrence, degradation and fate of pesticides during composting. Part II: Occurrence and fate of pesticides in compost and composting systems. Compost Sci Util 8:61–81

Castelo-Grande T, Augusto PA, Barbosa D (2005) Removal of pesticides from soil by supercritical extraction - a preliminary study. Chem Eng J 111:167–171

Castelo-Grande T, Augusto PA, Monteiro P, Estevez AM, Barbosa D (2010) Remediation of soils contaminated with pesticides: a review. Int J Environ Anal Chem 90:438–467

Chapman RA, Harris CR, Harris C (1986) Observations on the effect of soil type, treatment intensity, insecticide formulation, temperature and moisture on the adaptation and subsequent activity of biological agents associated with carbofuran degradation in soil. J Environ Sci Health B 21:125–141

Chauhan RS, Singhal L (2006) Harmful effects of pesticides and their control through cowpathy. Int J Cow Sci 2:61–70

Chen F, Ying GG, Kong LX, Wang L, Zhao JL, Zhou LJ (2011) Distribution and accumulation of endocrine-disrupting chemicals and pharmaceuticals in wastewater irrigated soils in Hebei, China. Environ Pollut 159:1490–1498

Chen M, Xu P, Zeng G, Yang C, Huang D, Zhang J (2015) Bioremediation of soils contaminated with polycyclic aromatic hydrocarbons, petroleum, pesticides, chlorophenols and heavy metals by composting: applications, microbes and future research needs. Biotechnol Adv 33:745–755

Creeger SM (1986) Considering pesticide potential for reaching ground water in the registration of pesticides. In: ACS Symposium Series, American Chemical Society, Washington, DC

Cruz-González G, Julcour C, Chaumat H, Bourdon V, Ramon-Portugal F, Gaspard S, Jáuregui-Haza UJ, Delmas H (2018) Degradation of chlordecone and beta-hexachlorocyclohexane by photolysis, (photo-)fenton oxidation and ozonation. J Environ Sci Health B 53:121–125

Cruz-Ruíz A, Cruz-Ruíz E, Vaca R, Del Águila P, Lugo J (2015) Effects of pumice mining on soil quality. Solid Earth Discuss 7:1375–1398

Delgado-Moreno L, Nogales R, Romero E (2017) Biodegradation of high doses of commercial pesticide products in pilot-scale biobeds using olive-oil agroindustry wastes. J Environ Manag 204:160–169

Deng S, Chen Y, Wang D, Shi T, Wu X, Ma X, Li X, Hua R, Tang X, Li QX (2015) Rapid biodegradation of organophosphorus pesticides by *Stenotrophomonas* sp. G1. J Hazard Mater 297:17–24

Devi IP (2009) Pesticide use and crop productivity in food crops of Kerala. Research Project Report, Kerala State Council for Science, Technology and Environment, Government of Kerala

Dijkgraaf E, Vollebergh HRJ (2004) Burn or bury? A social cost comparison of final waste disposal methods. Ecol Econ 50:233–247

Dolfing J, Eekert M, Seech A, Vogan J, Mueller J (2007) In situ chemical reduction (ISCR) technologies: significance of low Eh reactions. Soil Sediment Contam Int J 17:63–74

Dos Santos EV, Sáez C, Martínez-Huitle CA, Cañizares P, Rodrigo MA (2015) Combined soil washing and CDEO for the removal of atrazine from soils. J Hazard Mater 300:129–134

Drossman H, Johnson H, Mill T (1988) Structure activity relationships for environmental processes. 1: Hydrolysis of esters and carbamates. Chemosphere 17:1509–1530

El-Temsah YS, Joner EJ (2013) Effects of nano-sized zero-valent iron (nZVI) on DDT degradation in soil and its toxicity to *Collembola* and ostracods. Chemosphere 92:131–137

El-Temsah YS, Oughton DH, Joner EJ (2013) Effects of nano-sized zero-valent iron on DDT degradation and residual toxicity in soil: a column experiment. Plant Soil 368:189–200

Espinoza-Navarro O, Bustos-Obregón E (2005) Effect of malathion on the male reproductive organs of earthworms, *Eisenia foetida*. Asian J Androl 7:97–101

FAO (Food and Agricultural Organization) (1975) Report of the first training course in crop Pest control with special reference to desert locust control and research, Dakar, Senegal, 17 Feb–21 Mar (progress report). Rome, Italy

FAO/WHO (2000) Pesticide residues in food - 1999 evaluations. Part II—toxicological. Joint FAO/WHO Meeting on Pesticide Residues. World Health Organization, Geneva. Available from http://www.fao.org/agriculture/crops/thematic-sitemap/theme/pests/jmpr/jmpr-rep/en/. Accessed 15 July 2018

FICCI Report (2015) Ushering in the 2nd Green Revolution – role of crop protection chemicals. Available from http://ficci.in/spdocument/20662/Agrochemicals-Knowledge-report.pdf. Accessed 23 July 2018

Fuentes MS, Benimeli CS, Cuozzo SA, Amoroso MJ (2010) Isolation of pesticide-degrading actinomycetes from a contaminated site: bacterial growth, removal and dechlorination of organochlorine pesticides. Int Biodeterior Biodegrad 64:434–441

Fumio M (1982) Degradation of pesticides in the environment by microorganisms and sunlight. In: Biodegradation of pesticides. Springer, New York, pp 67–87

Gupta PK (2004) Pesticide exposure—Indian scene. Toxicology 198:83–90

Gustavo YO, Wong-Villarreal A, Del P, Águila-Juárez JL, Fuente RVP (2016) Composting of soils polluted with pesticides: a microbial approach and methods for monitoring. JSM Environ Sci Ecol 4:1–9

Insam H, Gómez-Brandón M, Ascher J (2015) Manure-based biogas fermentation residues–friend or foe of soil fertility? Soil Biol Biochem 84:1–14

IPEN (2006) Establishing the prevalence of POPs pesticide residues in water, soil and vegetable samples and creating awareness about their ill-effects. International POPs Elimination Network. Janhit Foundation, India. Available from http://www.ipen.org. Accessed 25 July 2018

Jilani S (2013) Comparative assessment of growth and biodegradation potential of soil isolate in the presence of pesticides. Saudi J Biol Sci 20:257–264

Karunakaran CO (1958) The Kerala food poisoning. J Ind Med Assoc 31:204–205

Kenneth DR (1990) Pesticides in the soil microbial ecosystem. In: ACS Symposium Series, American Chemical Society, Washington, DC

Kerle EA, Jenkins JJ, Vogue PA (1994) Understanding pesticide persistence and mobility for groundwater and surface water protection. Available from https://passel.unl.edu/Image/Robles%20VazquezWilfredo1129928587/EFATE_general.pdf. Accessed 20 Aug 2018

Koustas RN, Fischer D (1998) Review of separation technologies for treating pesticide-contaminated soil. J Air Waste Manag Assoc 48:434–440

Kumar B, Kumar S, Mishra M, Prakash D, Singh SK, Sharma CS (2012) Persistent chlorinated pesticide residues in selected market vegetables of root and leaf origin. Asian J Plant Sci Res 2:232–236

Kumar S, Kaushik G, Dar MA, Nimesh S, Lopez-Chuken UJ, Villarreal-Chiu JF (2018) Microbial degradation of organophosphate pesticides: a review. Pedosphere 28:190–208

Lal R (1984) Insecticide microbiology. Springer, Berlin

Lalander C, Senecal J, Gros Calvo M, Ahrens L, Josefsson S, Wiberg K, Vinnerås B (2016) Fate of pharmaceuticals and pesticides in fly larvae composting. Sci Total Environ 565:279–286

Lemmon CR, Pylypiw HM (1992) Degradation of diazinon, chlorpyrifos, isofenphos, and pendimethalin in grass and compost. Bull Environ Contam Toxicol 48:409–415

Lin SL, Cross WH, Chian ESK, Lai JS, Giabbai M, Hung CH (1996) Stabilization and solidification of lead in contaminated soils. J Hazard Mater 48:95–110

Lizano-Fallas V, Masís-Mora M, Espinoza-Villalobos D, Lizano-Brenes M, Rodríguez-Rodríguez CE (2017) Removal of pesticides and ecotoxicological changes during the simultaneous treatment of triazines and chlorpyrifos in biomixtures. Chemosphere 182:106–113

Lopes AR, Danko AS, Manaia CM, Nunes OC (2013) Molinate biodegradation in soils: natural attenuation versus bioaugmentation. Appl Microbiol Biotechnol 97:2691–2700

López-González JA, Suárez-Estrella F, Vargas-García MC, López MJ, Jurado MM, Moreno J (2015) Dynamics of bacterial microbiota during lignocellulosic waste composting: studies upon its structure, functionality and biodiversity. Bioresour Technol 175:406–416

López-Vizcaíno R, Alonso J, Cañizare P, León MJ, Navarr V, Rodrigo MA, Sáez C (2014) Removal of phenanthrene from synthetic kaolin soils by eletrokinetic soil flushing. Sep Purif Technol 132:33–40

Mahmood I, Imadi SR, Shazadi K, Gul A, Hakeem KR (2016) Effects of pesticides on environment. In: Plant, soil and microbes. Springer, New York, pp 253–269

Martinez ÁT, Speranza M, Ruiz-Dueñas FJ, Ferreira P, Camarero S, Guillén F, Martinez MJ, Gutiérrez Suárez A, Río Andrade JCD (2005) Biodegradation of lignocellulosics: microbial, chemical, and enzymatic aspects of the fungal attack of lignin. Int Microbiol 8:195–204

McEwen FL, Stephenson GR (1979) The use and significance of pesticides in the environment. Wiley, New York

McGuinness M, Dowling D (2009) Plant-associated bacterial degradation of toxic organic compounds in soil. Int J Environ Res Public Health 6:2226–2247

Ministry of Agriculture & Farmers Welfare Statistical Database, Government of India. Available from http://ppqs.gov.in/statistical-database. Accessed 9 Oct 2018

Mohamed AT, Hussein AA, El Siddig MA, Osman AG (2011) Degradation of oxyfluorfen herbicide by soil microorganisms biodegradation of herbicides. Biotechnology 109:274–279
Moorman TB, Cowan JK, Arthur EL, Coats JR (2001) Organic amendments to enhance herbicide biodegradation in contaminated soils. Biol Fertil Soils 33:541–545
Morillo E, Villaverde J (2017) Advanced technologies for the remediation of pesticide-contaminated soils. Sci Total Environ 586:576–597
Muller WP, Korte F (1976) Ecological chemical evaluation of waste treatment procedures. In: Environmental quality and safety: global aspects of chemistry, toxicology and technology as applied to the environment. Academic Press, New York, pp 215–236
Nadia OF, Xiang LY, Lie LY, Anuar DC, Afandi MPM, Baharuddin SA (2015) Investigation of physico-chemical properties and microbial community during poultry manure co-composting process. J Environ Sci 28:81–94
Oerke EC, Dehne H, Schohnbeck F, Weber A (1995) Crop production and crop protection: estimated losses in major food and cash crops. Elsevier, Amsterdam
Pascal-Lorber S, Laurent F (2011) Phytoremediation techniques for pesticide contaminations. In: Alternative farming systems, biotechnology, drought stress and ecological fertilisation. Springer, New York, pp 77–105
Paul D, Pandey G, Meier C, Roelof van der Meer J, Jain RK (2006) Bacterial community structure of a pesticide-contaminated site and assessment of changes induced in community structure during bioremediation. Microbiol Ecol 57:116–127
Peshin R (2005) Evaluation of dissemination of insecticide resistance management technology in cotton crop in Punjab (PhD Dissertation). Punjab Agricultural University, Ludhiana, India
Petruska JA, Mullins DE, Young RW, Collins ER Jr (1985) A benchtop system for evaluation of pesticide disposal by composting. Nucl Chem Waste Manag 5:177–182
Pimentel D, Levitan L (1986) Pesticides: amounts applied and amounts reaching pests. Bioscience 36:86–91
Pope JV, Skurky-Thomas M, Rosen CL (1994) Toxicity. Organochlor Pestic (New York Medscape) 259–278
Pruss-Ustun A, Vickers C, Haefliger P, Bertollini R (2011) Knowns and unknowns on burden of disease due to chemicals: a systematic review. Environ Health 10(1):9
Puri SN, Murthy KS, Sharma OP (1998) Affordable basis and compatible tactics. Seminar on integrated pest management. Indian Crop Protection Association, India International Centre, New Delhi, 3rd April, 1998. Pestology (Special issue) 22:34–46
Purnomo AS, Koyama F, Mori T, Kondo R (2010a) DDT degradation potential of cattle manure compost. Chemosphere 80:619–624
Purnomo AS, Mori T, Kamei I, Nishii T, Kondo R (2010b) Application of mushroom waste medium from *Pleurotus ostreatus* for bioremediation of DDT-contaminated soil. Int Biodeterior Biodegrad 64:397–402
Purnomo AS, Mori T, Takagi K, Kondo R (2011) Bioremediation of DDT contaminated soil using brown-rot fungi. Int Biodeterior Biodegrad 65:691–695
Racke KD, Frink CR (1989) Fate of organic contaminants during sewage sludge composting. Bull Environ Contam Toxicol 42:526–533
Rani S, Sud D (2015) Role of enhanced solar radiation for degradation of triazophos pesticide in soil matrix. Sol Energy 120:494–504
Rao PSC, Mansell RS, Baldwin LB, Laurent MF (1992) Pesticides and their behaviour in soil and water. Long Island Horticulture News
Rathfelder K, Lang JR, Abriola LM (1995) Soil vapor extraction and bioventing: applications, limitations, and future research directions. Rev Geophys 33:1067–1081
Risco C, López-Vizcaíno R, Sáez C, Yustres A, Cañizares P, Navarro V, Rodrigo MA (2016) Remediation of soils polluted with 2,4-D by electrokinetic soil flushing with facing rows of electrodes: a case study in a pilot plant. Chem Eng J 285:128–136
Rodrigo MA, Oturan MA, Oturan N (2014) Electrochemically assisted remediation of pesticides in soils and water: a review. Chem Rev 114:8720–8745

Rodrigo S, Saez C, Cañizares P, Rodrigo MA (2018) Reversible electrokinetic adsorption barriers for the removal of organochlorine herbicide from spiked soils. Sci Total Environ 640–641:629–636

Satapanajaru T, Onanong S, Comfort SD, Snow DD, Cassada DA, Harris C (2009) Remediating dinoseb-contaminated soil with zerovalent iron. J Hazard Mater 168:930–937

Sawhney BL, Brown K (1989) Reactions and movements of organic chemicals in soils. Soil Science Society of America, Madison, WI

Scelza R, Rao MA, Gianfreda L (2008) Response of an agricultural soil to pentachlorophenol (PCP) contamination and the addition of compost or dissolved organic matter. Soil Biol Biochem 40:2162–2169

Schwarzenbach RP, Westall J (1981) Transport of nonpolar organic compounds from surface water to groundwater. Laboratory sorption studies. Environ Sci Technol 15:1360–1367

Schwarzenbach RP, Giger W, Hoehn E, Schneider JK (1983) Behavior of organic compounds during infiltration of river water to groundwater. Field studies. Environ Sci Technol 17:472–479

Subash SP, Chand P, Pavithra S, Balaji SJ, Pal S (2017) Pesticide use in Indian agriculture: trends, market structure and policy issues (Policy brief, ICAR-National institute of agricultural economics and policy research). Indian Council of Agricultural Research, New Delhi

SEEP (2010) Factors influencing the use of pesticides in cotton in India. Expert Panel on Social, Environmental and Economic Performance of Cotton Production, International Cotton Advisory Committee, Washington, DC

Shea PJ, Machacek TA, Comfort SD (2004) Accelerated remediation of pesticide-contaminated soil with zerovalent iron. Environ Pollut 132:183–188

Shetty PK, Murugan M, Sreeja KG (2008) Crop protection stewardship in India: wanted or unwanted. Curr Sci:457–464

Simpanen S, Yu D, Mäkelä R, Talvenmäki H, Sinkkonen A, Silvennoinen H, Romantschuk M (2017) Soil vapor extraction of wet gasoline-contaminated soil made possible by electroosmotic dewatering lab simulations applied at a field site. J Soils Sediments 1:1–7

Singh DK (2008) Biodegradation and bioremediation of pesticide in soil: concept, method and recent developments. Indian J Microbiol 48:35–40

Sittler SP, Swinford GL, Gardner DG (1992) Use of thermal-enhanced soil vapor extraction to accelerate remediation of diesel-affected soils. In: Proceedings of Conference on Petroleum Hydrocarbons and Organic Chemicals in Ground Water: Prevention, detection, and Restoration, Houston, TX

Souza FL, Saez C, Llanos J, Lanza MRV, Cañizares P, Rodrigo MA (2016) Solar-powered electrokinetic remediation for the treatment of soil polluted with the herbicide 2,4-D. Electrochim Acta 190:371–377

Tinsley IJ (1979) Chemical concepts in pollutant behaviour. Wiley, New York

Tortella GR, Rubilar O, Castillo MDP, Cea M, Mella-Herrera R, Diez MC (2012) Chlorpyrifos degradation in a biomixture of biobed at different maturity stages. Chemosphere 88:224–228

Troxler WL, Goh SK, Dicks LWR (1993) Treatment of pesticide-contaminated soils with thermal desorption technologies. Air Waste 43:1610–1617

Tsitonaki A, Petri B, Crimi M, Mosbaek H, Siegrist RL, Bjerg PL (2010) In situ chemical oxidation of contaminated soil and groundwater using persulfate: a review. Crit Rev Environ Sci Technol 40:55–91

Valo R, Salkinoja-Salonen M (1986) Bioreclamation of chlorophenol-contaminated soil by composting. Appl Microbiol Biotechnol 25:68–75

van der Werf HMG (1996) Assessing the impact of pesticides on the environment. Agric Ecosyst Environ 60:81–96

Vandervoort C, Zabik MJ, Branham B, Lickfeldt DW (1997) Fate of selected pesticides applied to turfgrass: effect of composting on residues. Bull Environ Contam Toxicol 58:38–45

Varjani SJ (2017a) Microbial degradation of petroleum hydrocarbons. Bioresour Technol 223:277–286

Varjani SJ (2017b) Remediation processes for petroleum oil polluted soil. Indian J Biotechnol 16:157–163

Varjani SJ (2018) Bioremediation: a sustainable solution for complex environmental issues. In: Patel BN, Nagar R (eds) Sustainable development and India: convergence of law, economics, science, and politics. Oxford University Press, New Delhi, India, pp 104–115

Varjani SJ, Upasani VN (2016) Biodegradation of petroleum hydrocarbons by oleophilic strain of *Pseudomonas aeruginosa* NCIM 5514. Bioresour Technol 222:195–201

Varjani SJ, Upasani VN (2017) A new look on factors affecting microbial degradation of petroleum hydrocarbon pollutants. Int Biodeterior Biodegrad 120:71–83

Varjani SJ, Rana DP, Jain AK, Bateja S, Upasani VN (2015) Synergistic ex-situ biodegradation of crude oil by halotolerant bacterial consortium of indigenous strains isolated from on shore sites of Gujarat, India. Int Biodeterior Biodegrad 103:116–124

Varjani SJ, Gnansounou E, Pandey A (2017) Comprehensive review on toxicity of persistent organic pollutants from petroleum refinery waste and their degradation by microorganisms. Chemosphere 188:280–291

Varjani SJ, Gnansounou E, Baskar G, Pant D, Zakaria ZA (2018a) Introduction to waste bioremediation. In: Varjani S, Gnansounou E, Gurunathan B, Pant D, Zakaria Z (eds) Waste bioremediation. Energy, environment, and sustainability. Springer, Singapore

Varjani SJ, Agarwal AK, Gnansounou E, Gurunathan B (2018b) Introduction to environmental protection and management. In: Bioremediation: applications for environmental protection and management. Springer, New York, pp 1–6

Varjani S, Kumar G, Rene ER (2019) Developments in biochar application for pesticide remediation: current knowledge and future research directions. J Environ Manag 232:505–513

Vieira dos Santos E, Souza F, Saez C, Canizares P, Lanza MRV, Martinez-Huitle CA, Rodrigo MA (2016) Application of electrokinetic soil flushing to four herbicides: a comparison. Chemosphere 153:205–211

Villa RD, Trovó AG, Nogueira R (2010) Soil remediation using a coupled process: soil washing with surfactant followed by photo-fenton oxidation. J Hazard Mater 174:770–775

Warsi F (2018) How do pesticides affect ecosystems. In: Pesticides. Available from http://farhanwarsi.tripod.com/id9.html. Accessed 14 Aug 2018

Wasim A, Sengupta D, Chowdhury A (2009) Impact of pesticides use in agriculture: their benefits and hazards. Interdiscip Toxicol 2:1–12

Wauchope RD (1978) The pesticide content of surface water draining from agricultural fields—a review. J Environ Qual 7:459–472

WHO/UNEP (1990) Public health impact of pesticides used in agriculture. World Health Organization, Geneva. Available from http://apps.who.int/iris/handle/10665/39772. Accessed 28 Aug 2018

Wolff MS, McConnell R, Cedillo L, Rivera M (1992) Dermal levels of methyl-parathion, organochlorine pesticides, and acetylcholinesterase among formulators. Bull Environ Contam Toxicol 48:671–678

Yadav IC, Devi NL, Syed JH, Cheng Z, Li J, Zhang G, Jones KC (2015) Current status of persistent organic pesticides residues in air, water, and soil, and their possible effect on neighbouring countries: a comprehensive review of India. Sci Total Environ 511:123–137

Yaron B (1978) Some aspects of surface interactions of clays with organophosphorus pesticides. Soil Sci 125:210–216

Ye M, Sun M, Liu Z, Ni N, Chen Y, Gu C, Kengara FO, Li H, Jiang X (2014) Evaluation of enhanced soil washing process and phytoremediation with maize oil, carboxymethyl-β-cyclodextrin, and vetiver grass for the recovery of organochlorine pesticides and heavy metals from a pesticide factory site. J Environ Manag 141:161–168

Yu Z, Zeng GM, Chen YN, Zhang JC, Yu Y, Li H, Liu ZF, Tang L (2011) Effects of inoculation with Phanerochaete chrysosporium on remediation of pentachlorophenol-contaminated soil waste by composting. Process Biochem 46:1285–1291

Zeng G, Yu Z, Chen Y, Zhang J, Li H, Yu M, Zhao M (2011) Response of compost maturity and microbial community composition to pentachlorophenol (PCP)-contaminated soil during composting. Bioresour Technol 102:5905–5911

Chapter 12
Current Trends and Insights on Compost Utilization Studies: Crop Residue Composting to Improve Soil Organic Matter in Sugarcane Cultivation, Tamil Nadu, India

J. Siva Rani

Abstract Vermicomposting is very popular worldwide due to its ability to convert the organic waste to manure as a substitute for chemical fertilizers. Organic agriculture aims to reduce the dependence of chemical fertilizers and attempts to incorporate crop residues or other forms of organic materials to provide nutrients, which improves soil structure and maintains the soil fertility. During crop cultivation, crop residues are produced in large quantities. The sugar industry is a major agro-industry in India and it is water-intensive crop. The sugarcane crop produces large quantities of crop residues which have high nutrient value. These crop residues can be a great help in improving soil fertility. Thus, the recycling of these crop residues gain worthwhile importance. This chapter discusses the findings of a case study wherein the decomposed bagasse was used as organic compost in cane fields as 3 Mt per acre. The compost was recommended to use continuously for 3 years to increase the organic carbon content in soil. In, drought prone areas, 5 Mt of compost per acre of land was recommended for water conversation. Bagasse has high water holding capacity and it has a water holding capacity of 1:5 ratio by its weight. Bagasse when applied to soil it slowly release the moisture to the soil and holds the soil water capacity. Hence, the frequency of irrigation in cane in red soil is reduced from alternate days to once in 4 days of irrigation. The sugarcane crop was applied with sugarcane bagasse compost which was low in cost followed by cane trash in situ composting. It improved the organic carbon, water absorption, and texture of the soil, which resulted in higher productivity of the crop. The bagasse compost also gave vigorous growth in young seedlings. The results described the advantages of bagasse composting which has high nutrient value and improved the organic carbon content of the soil. The high growth of cane seedlings will enhance the cane production and its various economic benefits can be realized in future.

J. Siva Rani (✉)
EID Parry, Chennai, India
e-mail: sivaranij@parry.murugappa.com

M. K. Meghvansi, A. Varma (eds.), *Biology of Composts*, Soil Biology 58,
https://doi.org/10.1007/978-3-030-39173-7_12

Keywords Organic agriculture · Microbes · Waste · Biomass · Earthworm's

12.1 Introduction

Organic agriculture aims to reduce the dependence of chemical fertilizers and attempts to incorporate crop residues or other forms of organic materials to provide nutrients, which improves soil structure and maintains the soil fertility. Vermicomposting is very popular worldwide due to its ability to convert the organic waste to manure as a substitute for chemical fertilizers. Earthworm's intestine consists of a variety of microbes, enzymes, and hormones that support the fast breakdown of semi-digested material that transforms into vermicompost in a tiny time (around 4–8 weeks) (Nagavallemma et al. 2004). Convectional process of composting gains the advantage of microorganisms alone which extend the time period of compost production to nearly 20 weeks (Sánchez-Monedero et al. 2001).

A wide range of wastes are produced through various agricultural and crop deposits such as straw, husk, stover, biomass of weeds, other plants and forest, and wastes from animals and human habitation wastes, (Table 12.1). During crop cultivation, large quantities of crop residues are produced. The plant's economic part (s) are harvested and the remaining parts of the crops are considered as wastes and dumped toward the field side. These wastes cause disagreeable odors and create

Table 12.1 Recyclable organic wastes from various waste sources

Group	Type of wastes	Source of wastes
Plant wastes	Crop residues	Field crop residues and biomass
	Kitchen wastes	Daily kitchen wastes
	Green market wastes	Fruits and vegetable market wastes
	Coconut-areca nut/perennials wastes	By-products of these crops
	Forest biomass	Natural forest biomass and by-products
	Roadside vegetation	Weeds and invasive plants biomass
	Aquatic plant biomass	Biomass of aquatic plants
Animal wastes	Animal dung and urine	Feces and urine of domestic animals and dairies
	Poultry excreta	Poultry droppings of boiler and layering farm
	Fish meal and fish wastes	Fish wastes arise from freshwater fish and sea fish industries
Other wastes	City garbage	City garbage and municipal solid wastes
	Biogas slurry	By-product of biogas plant
	Sewage and sludge	Industrial/municipal wastewater-treatment plants
	Sugar industry and distillery wastes	Spent and effluent of the sugar industry
	Paper mill industrial wastes	Spent and effluent of the paper mill
	Fly ash	Fly ash generated from thermal power plants

Source: Chatterjee et al. (2017)

problems related to disposal. These waste falls into the environment and spread the foul odors, becomes breeding home for most of the pathogenic microorganisms and also to mosquito vectors. But, most of these wastes are rich in nutrients that are eco-friendly, having good organic carbon sources.

The sugar industry in India is an important agro-industry and a water-intensive crop. Farmers in subtropical areas irrigate the crop four to five times while their counterparts in the tropical regions irrigate it at least 20 times due to geographical conditions, although water is already a scarce commodity. Sugarcane is grown primarily throughout the state during the main season (December–May), it is also raised during the special season (June–September) (Sharma 2005).

Sugarcane productivity is cyclical in Tamil Nadu, India. In the past seven decades, the growth of sugarcane farming in the country has been consistent (Priyanka and Alka 2017). The cultivation farm of sugarcane had increased between 1.18 million ha and 5.0 million ha, from 1930–1931 to 2010–2011. The productivity also raised between 31 tonnes and 70 tonnes per ha. The total production of sugarcane increased from 37 million tonnes to 340 million tonnes. Currently, the production of sugar in the country is around 24.5 million tonnes (Co-operative sugar 2011). Among India's sugarcane-growing states, Tamil Nadu ranks third in the area (0.37 M.ha), production (3.5 million tons) and ranks first in productivity (105 t/ha). The productivity of sugarcane is 40% higher in Tamil Nadu compared to national productivity (69.5 t/ha). The sugarcane crop produces large quantities of crop residues, which have high nutrient value. These crop residues can be a great help in improving soil fertility. Thus, the recycling of these crop residues gains significant importance.

In the present study, the crop residue composting for the sugarcane residues is discussed and the role of composting is highlighted. The sugarcane bagasse contains a very high nutrient value and holds a good amount of water content. The bagasse as fuel to boiler with the specifications of 45–50% moisture as received basis from the mill with Kcal value of 2200 Kcal, ash content of 3–4%, sulfur concentration of 0.01% and Cl of 002%. This bagasse has great potential to improve the soil nutrient, especially soil carbon which in turn helps for the high yielding crops. Here, the EID Parry attempts to convert the waste bagasse to manure using composting as a sustainable option and use the same to improve the yield of sugarcane.

12.2 Current Trends in Vermicomposting

Vermicomposts are eco-friendly and valuable by-products from the organic wastes and their usage improves the soil environment. The role of food waste vermicompost (FWV), cow manure vermicompost (CMV), and paper sludge vermicompost (PSV) on various ginseng root production was studied by Jinu and Kee-Choon (2019). The results showed that the treatment with vermicompost from paper sludge increased the yield of roots by 40 t ha^{-1} in comparison to the untreated control. The root yield and concentration of nitrate are correlated negatively. The vermicompost treatments did not vary significantly in terms of root loss when compared to control. Vermicompost was used safely with regard to root rot disease, and the compost satisfactorily raised the pH levels of the fields that are converted from the rice paddies to the ginseng production. Vermicompost application reformed the allocation of resources and chemical properties of soil that led to the novel interactions among the root parameters and other components.

The changes that happened in the bakery industry sludge (BIS) during the vermicomposting were studied by Yadav and Garg (2019). Six different mixtures of cow dung (CD) and bakery industry sludge (BIS) comprising 10–50% of BIS were studied. Significant increase in the NPK content and electrical conductivity, whereas decreased pH, total organic carbon, and C: N ratio of BIS. The total Kjeldahl nitrogen, total available phosphorus, and total potassium contents increased 2.0–3.5, 1.2–1.9 and 1.2–1.4 times, respectively, when compared to the initial blends. The metal concentration was higher in the vermicomposts. The results indicated that the bakery industry sludge spiked with cow dung can be transformed into biologically valuable manure with the help of earthworms.

Effective removal of vermicompost over heavy metals depends primarily on the binding ability of humic substances to the metal ions. In simulated in situ conditions using column testing, Zhang et al. (2017) investigated the immobilization effect of vermicomposted sewage sludge for Pb, Cd, and Cr in the sediment. The average decrease in leachate of Pb, Cd, and Cr of 93, 97, and 75.5%, with the accumulated adsorbed amount of 11.80, 4.81, and 5.62 mg g^{-1}, respectively, was caused by placing 10% dw of vermicompost on the column's bottom layer. With Pb, Cr, and Cd, the humic substances formed the organo–metal complexes. The sorbent sequestered metal ions. Vermicompost can therefore be used to remediate heavy metal contaminated sediments as a promising in situ sorbent.

Vermicomposting was also used as an efficient distillery sludge waste (DSW) management technology combined with tea leaf residues (TLR) using an eco-biological tool, vermicomposting (Mahaly et al. 2018). Based on the earthworm, *Eisenia fetida*, DSW, and TLR were united in different ratios and subject to vermicomposting for a period of 45 days. The results indicated that the values of pH, total, and available N, K, P, Mg, and Ca were increased with a declining trend in electrical conductivity, total organic carbon, and C/N ratio from its initial value. The distillery solid wastes were converted into nutrient-rich vermicompost successfully.

The efficacy of the vermicomposting process on the three different sewage sludge (pre-composted with sawdust, grass clippings, and municipal solid waste) using three earthworm species was studied by Hanine et al. (2017). The immune parameters, such as biomarkers of stress and the metal body burdens, were used to biomonitor the process of vermicomposting and to assess the influence of contaminants on the physiology of earthworms. The process as well as the quality of the end product were also monitored using biotic and abiotic parameters. In all experimental conditions, *Dendrobaena veneta* showed much lower resistance as the body weight and the total number of circulating immune cells decreased under the most contaminated conditions. The accumulated heavy metals occurred in the following order Cd > Co > Cu > Zn > Ni > Pb > Cr. *Eisenia* sp. showed maximum ability to accumulate various heavy metals. The vermicompost achieved after a period of 45 days was suitable as per the agronomic parameters and to the compost quality norms of France and Poland.

Sustainable farming aims at managing plant and soil health while being less dependent on chemical inputs. Two tomato cultivars of dissimilar resistance to root-knot nematodes (RKNs, *Meloidogyne incognita*) were experimented with three modifications including inorganic fertilizer (IF), conventional compost (CC), and vermicompost (VC) (Xiao et al. 2016). Simulation of the root-knot nematode disease in field conditions was carried out with the treatments that were inoculated with the second-stage juveniles of *M. incognita*. Plant growth, root defense metabolites (phenolics) and their associated genes expression, and properties of soil viz., pH, EC, available nutrients, 3-indoleacetic acid (IAA), microbial biomass, and its activity were examined at 14 and 30 days of post inoculation (dpi). Significant reduction in the numbers of nematode-induced galls on susceptible (Sus) and resistant (Res) cultivar roots was observed; the reduction was seen by 77% for Sus and 42% for Res at 14 dpi, respectively, and by 59% and 46% at 30 dpi, respectively, when compared to inorganic fertilizer. Microbial activity of soil, pH, and IAA concentrations were positively allied with plant defense metabolites production and biomass for susceptible and resistant cultivars. Vermicompost could conquer root pests significantly, primarily for the susceptible plant, through moderating soil properties as well as plant defenses.

12.3 Case Study: Sugarcane Crop Residue

The soil organic matter is declining due to so many reasons and ultimately it resulted in low productivity of the crop (Zhenggao et al. 2016). To enhance the soil productivity, especially for sugarcane crop production the bagasse composting is considered and discussed here in this chapter. The main objectives of the study is to evaluate the effect of bagasse compost on crop yield, effect of factors influencing the yield, and assessment of productivity per unit area. A scientific study was conducted to analyze the status and root cause for 60 Mt from the existing achieved farmers. Phase manner implementation strategy is planned, based on the study results. Twenty-seven farmers of >60 Mt yield per acre of EID parry. The status of the soil properties in the study area is shown in Table 12.2.

12.3.1 Soil Conditions

12.3.1.1 Variety of the Crop

40.74% of farmers (11 farmers in 25.90 acres) planted HSV varieties like Co-86032, Co-94012, and P-00-1110. About 29.62% planted HSV + OV (8 farmers in

Table 12.2 Soil properties in Pudukottai District of Tamil Nadu

Parameters	Unit	Min.	Max.	Average	Status remarks	Requirement
Organic carbon	%	0.15	0.68	0.38	Low	>0.75
Available phosphorus	Kg/acre	4	15.6	9.5	Medium	10.56
Available potassium	Kg/acre	41	160	94	Medium	83
Iron	mg/kg	4.9	28.26	12.71	Sufficient	6.3
Maganese	mg/kg	2.02	14.28	6.56	Sufficient	2
Zinc	mg/kg	0.4	1.98	0.9	Deficient	1.2
Copper	mg/kg	0.38	1.74	0.93	Deficient	1.2

Source: EID parry (India) ltd

24.4 acres) and 29.62% planted OV varieties like MC-707, CoSi-94045, COA8001, and PI96-843 (8 farmers in 19 acres).

12.3.1.2 Rainfall in the Study Area

It is very important to analyze the pattern of rainfall in the study area, which helps in understanding crop productivity. In the past 10 years, the study area has received good rains. In 2007, the rainfall recorded as 1034 mm, while it is 1301 mm in 2008 followed by 1248 mm in 2010, and 980 mm in 2015.

12.3.1.3 Yield of the Crop

The average yield graph is declining from 40.42 Mt/acre from 2006 to 27.22 in 2009–2010, again the line is showing an upward trend to 32 Mt in 2010–2011 and declining to 29.02 Mt in 2013–2014 and again upward trend to 30.75 Mt in 2014–2015 to 29.18 in 2016–2017.

12.3.1.4 Land Preparation and Irrigation

All the farmers prepared the land to fine tilth condition. All the farmers adopted the furrow method of irrigation with alternate drying and wetting system. In wetlands, they followed once in 6–7 days irrigation, Garden land—once 3–4 days irrigation based on the climatic condition.

12.3.1.5 Soil and Soil Organic Matter

37.03% of selected farmers tested their soil and soil types classified as red soil (11%), sandy loam soil (63%), sandy soil (11%), and clay soil (15%). About 85% of lands fell under Upland, and 15% under low land. None of the farmers has adopted soil test-based nutrient application. The soil organic matter (SOM) in the study area is shown in Fig. 12.1. Approximately 66% of farmers applied SOM in the form of chicken waste, pressmud, and FYM. 33.33% farmers did not apply any organic matter. Among the farmer who applied the SOM, around 25.92% of farmers applied 5–8 Mt SOM (7.40%), 10 Mt/acre (7.4%), and 15 Mt/acre (7.40%). When the organic carbon (OC) percentage was 0.38–0.53%, the nutrient uptake of the crop was high. The two farmers of 0.36–0.38% OC in soil received 62.22 Mt, 67.28 Mt, while the other two farmers of 0.42%, soil received 61.49, 62.49 Mt, while with 0.51–0.59% of OC in soil resulted in 66.40, 71.59, 64.49%, although they have applied lesser dose of nitrogen and potassium.

12.3.1.6 NPK and Micronutrient

About 70% farmers applied more concentration of nitrogen than EID soil lab recommendation. Approximately 44% farmers were applied more concentration of phosphorus (52%), potassium (48%), and applied sulfur in addition to micronutrient. 60% of farmers out of 37.03% who tested soil properties have applied a lesser concentration of nitrogen than recommendation received the yield ranges from 62 to 67.28 Mt due to their organic carbon content in soil ranges from 0.38 to 0.53%. All the farmers applied fertilizer in a split manner up to seventh month of crop growth. The cost spent for the crop was as follows: 59% of the cost spent on harvest followed by 12%—fertilizer, 4% for SOM, land preparation 5%, and 0% on pest and disease.

12.3.2 Crop Residue Composting

Compost feeds the soil and fertilizer feeds the plants. The optimum level of organic compost application supported the crop to absorb the nutrients and enhanced the plant growth. Thus, the EID Parry decided to increase the organic matter through in situ composting (Fig. 12.2).

12.3.2.1 Composting

The awareness on composting to the farmers is very important in its successful implementation. The farmers who did not show interest in in situ composting were advised to do the composting nearer to their field (Yaxin et al. 2019). This indirectly reduces the cost of cultivation in the purchase of farmyard manure. As the sugarcane crop residue contains potash and silica, when it decomposes it becomes potash rich organic compost with silica and sulfur which is very much essential for the sugar-cane growth.

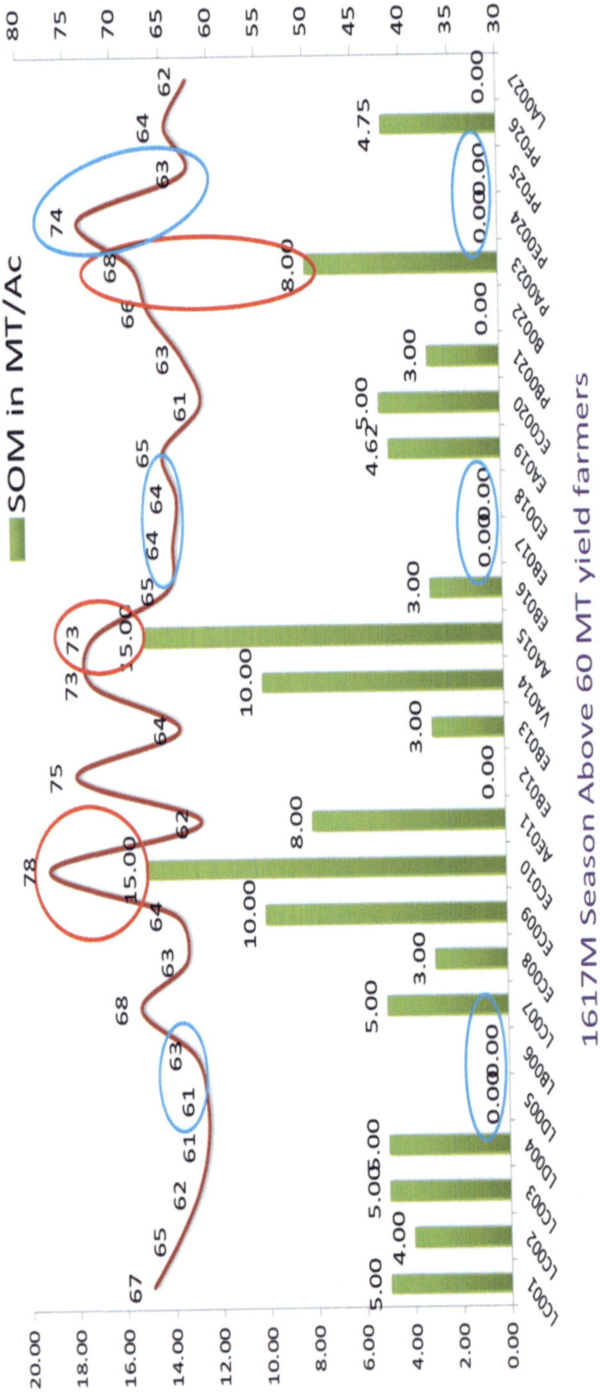

Fig. 12.1 Soil organic matter in the sugarcane production area (Source: EID Parry (India) Ltd.)

Fig. 12.2 Sugarcane cultivation using drip irrigation

12.3.2.2 Bagasse Compost

Bagasse was heaped in rectangle shape in layers up to 5 ft. Each feet was spread with one bottle of *Pleurotus sajarcaju* fungus and 5 kg of urea alternatively as decomposing agent (Figs. 12.3 and 12.4). The heap was tilted 2 times for aeration and decomposed in 45–50 days period. C:N ratio was 20:1. 30% as bagasse was produced per Mt of sugarcane during milling, which contains cellulose, hemicellulose, organic carbon, and lignin with low EC and neutral pH.

The decomposed bagasse was used as organic compost in cane fields as 3 Mt per acre. The compost was recommended to use continuously for 3 years to increase the organic carbon content in the soil. In, drought prone areas, 5 Mt of compost is recommended, as it helps for water retention in soil. Bagasse have 1:5 ratio water holding capacity to its weight. It also helps in maintaing soil microclimate by maintianing root atmospheric mositure thru slow release of water to crops. Hence, the frequency of irrigation in cane in red soil is reduced from alternate days to once in 4 days of irrigation.

Bagasse mixed with filter cake was decomposed and used in soil for improving organic matter for sustainable cane yield. Since the bagasse was having the good porosity, expansion, lightweight, low EC, and pH, their compost was highly suitable to use as grow media for chip bud seedling production in sugarcane. The shoot growth was observed to be very good within 15 days. Compare to coco pith compost

Fig. 12.3 Bed preparation for bagasse composting

Fig. 12.4 Bagasse compost (**a**) Enrichment with nutrients after 45 days (**b**) Chip bud seedling

cost, bagasse compost cost was very low and easily available all the time to cane growers to get quality seedlings.

12.4 Conclusions

Organic agriculture aims to reduce the dependence on chemical fertilizers and attempts to incorporate crop residues or other forms of organic materials to provide nutrients, which improves soil structure and maintains soil fertility. During crop cultivation, large quantities of crop residues are produced. Plant's economic part (s) are harvested and the remaining parts of the crops are considered as wastes and dumped on the field side. Sugar industry is an important agro-based industry in India and it is an water-intensive crop. The sugarcane crop produces large quantities of crop residues that have high nutrient value. These crop residues can be a great help in improving soil fertility. Thus, the recycling of these crop residues gains importance. Bagasse mixed with filter cake was decomposed and used in the soil for improving organic matter for sustainable cane yield. Since the bagasse was having the good porosity, expansion, lightweight, low EC, and pH, their compost was highly suitable to use as a growth media for chip bud seedling production in sugarcane. The shoot growth was observed to be very good within 15 days. Compare to coco pith compost cost, bagasse compost cost was very low and easily available all the time to cane growers to get quality seedlings. The decomposed bagasse was used as organic compost in cane fields as 3 Mt per acre. The compost was recommended to use continuously for 3 years to increase the organic carbon content in soil. Water conservation is essential in drought prone areas and it is important to provide optimum quantity of compost to the soil. 5 Mt of compost per acre is recommended to maintain water in the soil. Bagasse have 1:5 ratio water holding capacity to its weight. It also helps in maintaining soil microclimate by maintaining root atmospheric moisture through slow release of water to crops. Hence, the frequency of irrigation in cane in red soil is reduced from alternate days to once in 4 days of irrigation.

References

Chatterjee R, Gajjela S, Thirumdasu RK (2017) Recycling of organic wastes for sustainable soil health and crop growth. Int J Waste Resour 7:296. https://doi.org/10.4172/2252-5211.1000296

Hanine S, Agnieszka R, Anna G, Anna G, Marcin M, Barbara P, Małgorzata K, Franck V (2017) Determination of the performance of vermicomposting process applied to sewage sludge by monitoring of the compost quality and immune responses in three earthworm species: *Eisenia fetida*, *Eisenia andrei* and *Dendrobaena veneta*. Bioresour Technol 241:103–112

Jinu E, Kee-Choon P (2019) Effect of vermicompost application on root growth and ginsenoside content of *Panax ginseng*. J Environ Manag 234:458–463

Mahaly M, Abbiramy K, Senthilkumar A, Chitrapriya K, Nagarajan K (2018) Vermicomposting of distillery sludge waste with tea leaf residues. Sust Environ Res 28:223–227

Nagavallemma KP, Wani SP, Stephane L, Padmaja VV, Vineela C, Babu Rao M, Sahrawat KL (2004) Vermicomposting: recycling wastes into valuable organic fertilizer. Global Theme on Agroecosystems Report no. 8: International Crops Research Institute for the Semi-Arid Tropics, Patancheru, Andhra Pradesh, p 20

Priyanka U, Alka S (2017) An economic analysis of sugarcane cultivation and its productivity in major sugar producing states of Uttar Pradesh and Maharashtra. Econ Aff 62:711–718

Sánchez-Monedero MA, Roig A, Paredes C, Bernal MP (2001) Nitrogen transformation during organic waste composting by the Rutgers system and its effects on pH, EC and maturity of the composting mixtures. Bioresour Technol 78:301–308

Sharma HR (2005) Agricultural development and crop diversification in Himachal Pradesh: understanding the patterns, processes, determinants and lessons. Indian J Agric Econ 60:71–93

Xiao Z, Liu M, Jiang L, Chen X, Griffiths BS, Li H, Hu F (2016) Vermicompost increases defense against root-knot nematode (Meloidogyne incognita) in tomato plants. Appl Soil Ecol 105:177–186

Yadav A, Garg VK (2019) Biotransformation of bakery industry sludge into valuable product using vermicomposting. Bioresour Technol 274:512–517

Yaxin Z, Ye T, Duofei H, Jinshi F, Maocai S, Guangming Z (2019) Is vermicompost the possible *in situ* sorbent? Immobilization of Pb, Cd and Cr in sediment with sludge derived vermicompost, a column study. J Hazard Mater 367:83–90

Zhang X, Wang X, Wang D (2017) Immobilization of heavy metals in sewage sludge during land application process in China: a review. Sustainability 9:1–19

Zhenggao X, Manqiang L, Linhui J, Xiaoyun C, Bryan Griffiths S, Huixin L, Feng H (2016) Vermicompost increases defense against root-knot nematode (*Meloidogyne incognita*) in tomato plants. Appl Soil Ecol 105:177–186

Chapter 13
Applications of *Streptomyces* spp. Enhanced Compost in Sustainable Agriculture

Laura Buzón-Durán, Eduardo Pérez-Lebeña, Jesús Martín-Gil, Mercedes Sánchez-Báscones, and Pablo Martín-Ramos

Abstract *Streptomyces* is the most abundant genus among actinomycetes and holds great potential for sustainable agriculture, both now and in the future, given its role as a source of antibiotics, bioactive compounds, and enzymes. This mini-review discusses the role of these microorganisms in the degradation by enzymatic hydrolysis of lignocellulosic residues during the composting process. Examples of soil amendment with compost bioaugmentated with populations of *Streptomyces* are reviewed. The advantages derived from the combined use of organic compost and microorganisms of this genus (and other members of Actinobacteria phylum) as biofertilizers to increase plants growth and yield are also presented. Finally, strategies aimed at biocontrol or at the improvement of the capacity for suppression of diseases through an increase in organic matter and Actinobacteria levels are discussed.

Keywords Actinobacteria · Bioaugmentation · Biocontrol · Biofertilizer · Bioremediation · Plant growth promotion · Secondary metabolites · Soil amendment

L. Buzón-Durán · E. Pérez-Lebeña · J. Martín-Gil
Agriculture and Forestry Engineering Department, ETSIIAA, Universidad de Valladolid, Palencia, Spain
e-mail: laura.buzon@uva.es; eplebena@gmail.com; mgil@iaf.uva.es

M. Sánchez-Báscones
Department of Agroforestry Sciences, ETSIIAA, University of Valladolid, Palencia, Spain
e-mail: msanchez@agro.uva.es

P. Martín-Ramos (✉)
Department of Agricultural and Environmental Sciences, Instituto Universitario de Investigación en Ciencias Ambientales de Aragón (IUCA), EPS, Universidad de Zaragoza, Huesca, Spain
e-mail: pmr@unizar.es

M. K. Meghvansi, A. Varma (eds.), *Biology of Composts*, Soil Biology 58,
https://doi.org/10.1007/978-3-030-39173-7_13

13.1 Introduction

Actinobacteria are widely distributed in the water, soil, and surfaces of colonized plants and have an enormous importance in composting processes, provided that they can degrade a wide variety of biopolymers by hydrolytic enzymes (Diaz et al. 2007). The presence of organic matter makes them dominant in dry, humic, and calcareous soils, and the physical characteristics of those soils conform the size of the population (20–60%) and their composition (Araragi 1979).

The Actinomycetes phylum encompasses 6 classes, 19 orders, 50 families, and 221 genera, but new taxa are still being discovered. Actinomycetes have morphologies that range from cocci to mycelia and differentiated spores (Timková et al. 2018). The *Streptomyces* genus is the most abundant among soil bacteria and actinomycetes, and they are aerobic microorganisms that feature a high content of G+C (75%) in their DNA and large genomes in comparison with those of other microorganisms (Sanglier et al. 1993).

The *Streptomyces* genus has been exploited in the pharmaceutical industry and in commercial biocontrol products for agriculture, but their importance and the mechanisms that govern their complicated interactions with plants and other organisms are still a very active area of research (Yu et al. 2011).

Actinomycetes degrade organic matter and produce secondary metabolites, and have the ability to solubilize phosphate, to produce organic acids, siderophores, and phytohormones (Locatelli et al. 2016).

Actinobacteria, particularly those of the *Streptomyces*, *Pseudomonas*, *Agrobacterium*, and *Bacillus* genera, can be used for the biocontrol of phytopathogens in fertile soils (Wang et al. 2013b) and have shown their efficacy in controlling plant diseases both in vitro and in vivo (Virolle et al. 2015). The main mechanisms of biological control include the parasitism of hyphae, the production of secondary metabolites, the production of siderophores, and the production of extracellular enzymes such as cellulases, amylase, and chitinase. Extracellular enzymes (e.g., β-1,3-glucanases and chitinases) are responsible for the mycoparasitism exerted by certain strains of *Streptomyces* and the suppression of plant diseases (Singh and Gaur 2016). Moreover, they can produce other agro-active compounds with relevant antimicrobial potential: terpenoids, vitamins, pigments, etc. (Franco-Correa et al. 2010).

Aldesuquy et al. (1998) were among the first to report that endophytic actinomycetes could also improve the growth, vigor, and yield of wheat crops, in different environments and ecological conditions (Marques et al. 2010; Zhang et al. 2012). For instance, *Streptomyces* isolated from rotten wheat straw (e.g., *Streptomyces* sp. UU15 and *Streptomyces vinaceusdrappus* UU11) offer enormous potential for plant growth promotion (PGP) and the formation of agro-active compounds (Singh et al. 2019).

Another relevant area of application of actinobacteria is related to their potential to be applied in biostimulation, bioaugmentation, cellular immobilization, and bioremediation of organic and inorganic contaminants techniques. By means of multi-omics analysis, it is possible to know the mechanisms involved in the formation of bioactive and biosurfactant films, the processes associated with the recycling and degradation of complex substances or pesticides and in the recovery of soils contaminated by heavy metals. Through the design of mixed specific crops and the use of microorganism consortium systems (including *Streptomyces*), new biocontrol strategies for the toxicity of contaminants may be devised. However, further studies are still necessary to optimize the role of these microorganisms in removing pollutants from contaminated environments (Timková et al. 2018).

A recent Special Issue on Plant–Microbe Interaction (Schirawski and Perlin 2018) has summarized and discussed the current understanding of plant–microbe interactions, providing a general overview of soil microbes that positively affect plant growth, and the elucidating mechanistic strategies of plant pathogens and microbiomes of seeds and roots. However, a mini-review of the most recent developments focused only on the applications of the genus *Streptomyces* would be of use to academic and industrial researchers, undergraduates and postgraduates alike, working in this field. This chapter aims to deliver a panorama of the extensive and rapidly growing applications of these microorganisms in sustainable agriculture.

13.2 Composting Process Enhancement with *Streptomyces* spp.

In spite of the spontaneous microbiological nature of the composting process, the addition of selected microbial starters can speed up the process and improve its quality. Thus, the inoculation of such selected microorganisms as part of the preparation of multifunctional biofertilizers can be a promising approach to reduce the amounts of synthetic fertilizers used in agriculture.

Thermophilic and highly cellulolytic *Streptomyces* can be isolated either from soil (Ramírez and Coha 2003), solid waste compost (Strom 1985), compost of agricultural wastes (Jang and Chang 2005), or compost-treated soils (Feng et al. 2014). In 2003, Ramírez and Coha (2003) isolated 145 cellulolytic thermophilic actinomycete strains from over 70 compost, hay, dung, and soil samples. They found 10 cellulolytic actinomycete strains with a high yield in cellulases (endoglucanase, β-glucosidase, and exoglucanase activities), and concluded that *Streptomyces* sp. 7CMC10 was the strain that featured highest activity levels (corresponding to 20.14, 5.40, and 2.61 $UI \cdot mg^{-1}$ of protein, respectively).

Cellulases production by a strain isolated from a Brazilian forest soil, *Streptomyces drozdowiczii*, was studied using agro-industrial products (Grigorevski de Lima et al. 2005; Semedo 2004), and that from *S. malaysiensis* was evaluated by Nascimento et al. (2009) using submerged fermentation for a mixture of brewer's spent grain and corn steep liquor. *S. viridobrunneus* strain has been shown to be cellulolytic too, due to its ability to decompose cellulose from agro-industrial residues (Da Vinha et al. 2010). Moreover, Ventorino et al. (2016) highlighted the potential of *S. argenteolus* AE58P from a biotechnology perspective, since its behavior as a biocatalyst-producing bacterium makes it an exciting candidate for lignocellulose conversion and composting processes.

Other cellulolytic streptomycetes for composting purposes are *S. celluloflavus*, *Streptomyces* sp. SirexAA-E, *S. reticuli*, *Streptomyces* sp. Amel2xE9, *Streptomyces* sp. LamerLS-31b, *Streptomyces* sp. DpondAA-B6, *Streptomyces* sp. KhCrAH-340, *Streptomyces* sp. LaPpAH 95, and *Streptomyces* sp. ATexAB-D23 (Hillis et al. 2016). This study showed that the ability of *Streptomyces* spp. to degrade cellulose in a rapid manner would be restricted to two clades of host-associated strains, and that, although plant biomass-degrading genes (CAZy) are common in microorganisms of this genus, crucial enzyme families would be enriched in highly cellulolytic strains. The authors verified the importance of the CebR transcriptional repressor and a highly expressed cellulase (viz. GH6 family cellobiohydrolase) to the cellulolytic phenotype. Their evolutionary analyses identified intricate genomic modifications—which include the acquisition and selective retention of CAZy genes and transcriptional regulators—that would drive plant biomass deconstruction in *Streptomyces* spp. Thus, they suggested that some symbiotic streptomycetes have been selected in host-associated niches because of their increased cellulose degrading activity, which makes those strains the most relevant for composting processes enhancement. The distribution of the cellulolytic ability in *Streptomyces* genus is shown in Fig. 13.1.

According to the line of action proposed by Pugliese et al. (2008), in the following sections, we summarize a selection of *Streptomyces* spp. strains (for the most part coming from compost) for remediation purposes, as plant growth promoters and to control plant pathogens.

13.3 Soils and Water Bioremediation with *Streptomyces* spp.

Phytoremediation is a technology that uses plants to restore soils contaminated with trace elements or pesticides (Cao et al. 2016). This technique depends on the host plant–microorganisms interaction, and the bioremediation with actinobacteria is among the most popular options for the cleaning of contaminated sites (Alvarez et al. 2017). Such actinobacteria biomass used in soil remediation may be produced, for example, from cultures based on sugarcane vinasse (Aparicio et al. 2017).

Streptomyces spp. possess very interesting properties in terms of metabolic diversity, ability to rapidly colonize substrates, formation of mycelia, and production of spores in unfavorable conditions (Alexander 1991). These microorganisms have different metabolic pathways in which toxic compounds are used as an energy source for cell processes through fermentation, respiration, and co-metabolism. So as to survive in toxic environments, they have developed resistance to heavy metals

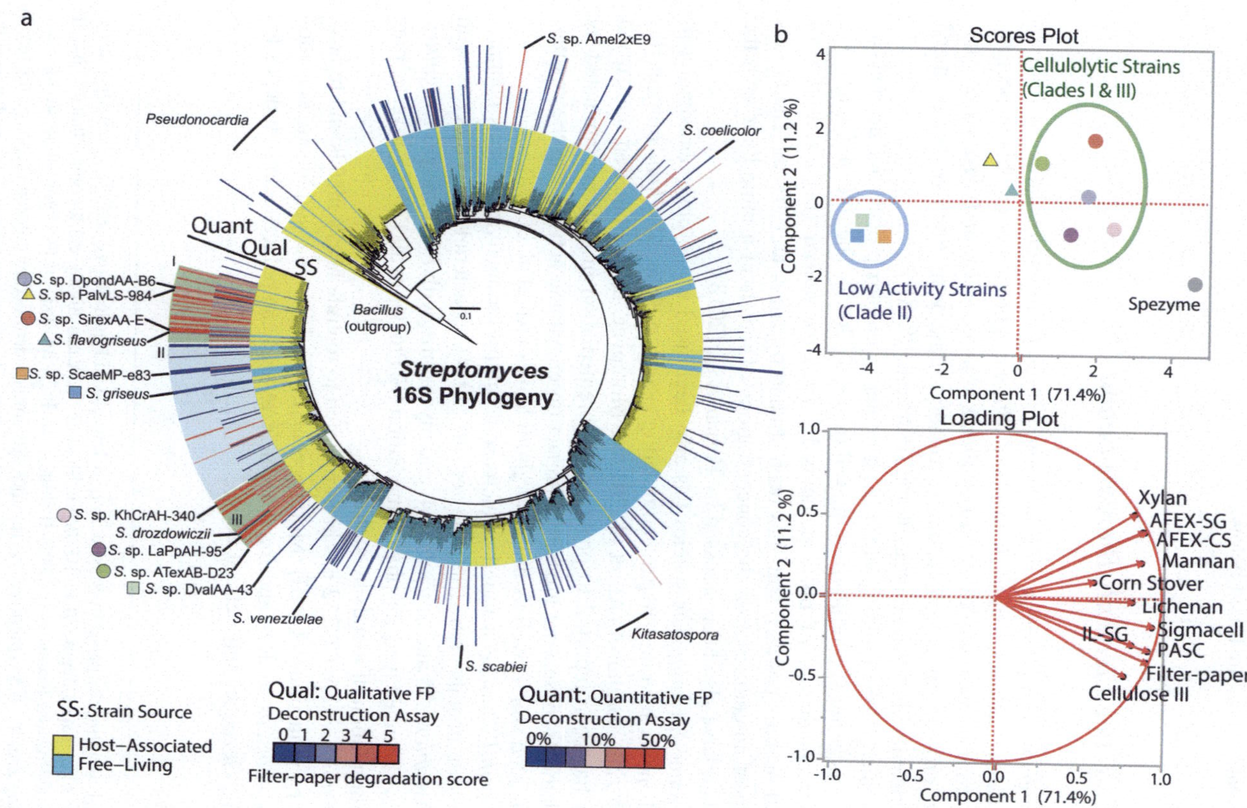

Fig. 13.1 Distribution of cellulolytic ability in *Streptomyces* genus. (**a**) 16S rRNA gene phylogenetic tree of 1141 *Streptomyces* strains from free-living (cyan) and host-associated (yellow) environments. The tree is annotated with qualitative cellulose (filter paper) degradation scores (0: no growth in 3 weeks, 5: filter

and mechanisms aimed at maintaining homeostasis (Brar et al. 2006). These combinations of these mechanisms could lead to an extraordinarily resistant bacterium, as noted by Timková et al. (2018). Some success cases of their application to bioremediation are illustrated below.

13.3.1 Trace Elements Bioremediation

From an ecological perspective, it has been proven that the inoculation of *Robinia pseudoacacia* L. (black locust) rhizosphere with the appropriate rhizobia (especially those from the *Streptomyces, Mesorhizobium*, *Rhodococcus*, and *Variovorax* genera) can provide an environmentally friendly strategy to improve heavy metal contaminated soils by phytoremediation (Fan et al. 2018).

The remediation of mining sites is a particularly complex process and depends on the physicochemical conditions of the soil (Ma et al. 2015). The role of *S. pactum* in the phytoremediation of trace elements by *Brassica juncea* (L.) Czern (brown mustard) in mine polluted soils has been studied by Ali et al. (2017c). The effects of compost amendments on the concentrations of metals in a mine floor with the same plant have also been assessed by Forján et al. (2018).

The same *Streptomyces* species has also been used in the phytoremediation of trace elements by other hyperaccumulating plants such as *Sorghum bicolor* (L.) Moench (i.e., sorghum) (Ali et al. 2017b). Moreover, its combination with wood biochar has been reported to promote phytoremediation in soils contaminated by trace elements, observing a positive impact on enzymatic activities in the smelter-contaminated soil as well as in the sorghum leaves (Ali et al. 2017a).

S. mirabilis has been found to increase sorghum productivity in soils contaminated with metals too (Schütze et al. 2014). Bacterial amendments with *S. acidiscabies* E13 and *S. tendae* F4, and mycorrhiza with *Rhizophagus irregularis* have also been assayed for phytoremediation by Phieler et al. (2015), investigating the accumulation of metals in sorghum.

Streptomyces sp. CG252 has been reported to tolerate heavy metals and remove Cr(VI) by reduction to Cr(III) (Morales et al. 2007). Likewise, soil bioaugmentation by *Streptomyces* sp. R25 and *Bacillus* sp. ZAN-044 can reduce cadmium deposited in plants (Jezequel and Lebeau 2008).

Fig. 13.1 (continued) paper deconstruction in 1 week) and quantitative cellulose degrading activities (% filter paper degraded in 10 days). Shading indicates highly cellulolytic clades I and III (green) and related low-activity clade II (blue). (**b**) PCA analysis of the lignocellulosic biomass-degrading activity of *Streptomyces* secretomes. Strains are identified by colored shapes on the tree in panel A. The scores plot shows similarity of polysaccharide degrading activity, and the loading plot indicates which substrates influence components 1 and 2 of the scores plot. Reprinted from Hillis et al. (2016) under Creative Commons Attribution License, CC BY 4.0

13.3.2 Pesticides Bioremediation

The application of a consortium of *Streptomyces* spp. to eliminate mixtures of pesticides in different soil systems has been the subject of recent attention (Fuentes et al. 2017). The successful elimination of multiple organochlorine pesticides with strains of *Streptomyces* sp. A5 and their influence on the cytotoxicity of the treated systems was recently reported by Fuentes et al. (2018).

Streptomyces spp. have been studied for the controlled elimination of a mixture of organophosphorus pesticides (diazinon and chlorpyrifos insecticides) by Briceño et al. (2017). In a similar fashion, a remarkably high degradation efficiency of *Streptomyces* sp. AH-B strain against quinclorac has been reported by Lang et al. (2018): in liquid medium, it achieved 97.2% removal after 18 days, and in soil, it attained a degradation of 87.5% after 42 days.

Even recalcitrant chlorinated pesticides like lindane may be degraded by this genus: the elimination of high concentrations of lindane in soils has been demonstrated using stable microemulsions and *Streptomyces* sp. M7. In the work by Saez et al. (2017), a soybean oil-based microemulsion allowed to solubilize 66% of the lindane present in the aqueous medium, i.e., a 4.5 times higher amount than when a surfactant was used. The authors recommended the use of microemulsions formed with soybean oil, Tween 80 (polysorbate 80) and 1-pentanol as a soil washing technology and for the ex situ bioremediation of wastewater polluted with lindane or with other hydrophobic organic compounds.

Streptomyces sp. M7, MC1, A5, and *Amycolatopsis tucumanensis* DSM 45259 have also been tested for the simultaneous elimination of lindane and chromium from different contaminated systems (Aparicio et al. 2018a, b). The viability of these strains was confirmed after the bioremediation process, so these studies pose an approximation to what may be carried out at field scale.

13.4 *Streptomyces* spp. as Plant-Growth Promoters and Biofertilizers

In agricultural practices, it should be emphasized that lignocellulosic wastes should be valorized as substrates, and the conversion of agricultural residues with microorganisms by composting processes for the preparation of multifunctional biofertilizers should thus be encouraged, replacing the fertilizers of synthetic origin. Moreover, the inclusion of species with the ability to control plant pathogens should be a priority. During the initial stages of composting, the decomposition of phytotoxic organic substances is carried out due to the presence of organic acids, ammonia, and ethylene oxide (Mehta et al. 2014). The maturity of the compost can be determined by several physical–chemical and biological or microbiological parameters (discussed in Chap. 8 of this book), while the phytotoxicity tests involve the determination of the seed germination index (GI) of the compost extract. If GI > 101, the substrate is considered as a phytostimulant and it is suitable for application as a fertilizer (Rashad et al. 2010).

In the last few years, Actinobacteria—especially the *Streptomyces* genus because of its soil dominant saprophytic nature and strong antimicrobial potential—have attracted attention as efficient plant growth promoters (Franco-Correa et al. 2010) and have inspired their application as biofertilizers to boost plant productivity. The applications in terms of PGP (and also in biocontrol of phytopathogens) of *Streptomyces* spp. have been discussed in a recent review paper by Vurukonda et al. (2018). Another recent review paper by Sathya et al. (2017), focused on grain legumes sustainable production and protection, also explains how actinobacteria can promote plant growth.

The genus *Streptomyces* comprises many species that have different functions in the promotion of plants growth, and whose characterization can be attained through different methods (Table 13.1). In 2016, Passari et al. (2016) proposed *Streptomyces* sp. strain DBT204 isolates (recovered from *Solanum lycopersicum* L., i.e., tomato) as biofertilizers to improve tomato and chili seedlings growth, observing an increase in the productivity of the crop thanks to the phytohormone producing potential and the metabolites with phytohormone mimicking activity of this strain. The strain *S. roseoflavus* NKZ-259 has also been demonstrated to be a promoter of plant growth for pepper and tomato, both in greenhouse and field tests, and simultaneously behaved as a biological agent and biopesticide agent, inhibiting pathogenic fungi (tomato gray mold) (Shi et al. 2018).

Endophytic *S. olivaceoviridis, S. viridis, S. atrovirens, S. rochei*, and *S. rimosus* have been shown to improve seed germination as well as root elongation and growth (El-Tarabily 2008; Khamna et al. 2010). Also in 2016, Tamreihao et al. (2016) reported the biocontrol and PGP activities of an *S. corchorusii* strain (UCR3-16) upon its application to rice as a biofertilizer agent. In 2018, Chaiharn et al. (2018) reported that rice root length was significantly increased by *Streptomyces* isolate KT 6-4-1. According to a study by Wang (2018), conducted in field conditions, *S. griseoplanus* (namely PSA1) in combination with P fertilizer increased soil available P content, enhanced plant growth, and enhanced grain yield of maize by 11%. In a similar fashion, Singh et al. (2016) reported that metabolites of *Streptomyces* sp. significantly enhanced biomass yield (3.58-fold increase versus control plants). In 2019, Singh et al. (2019) reported a seedling assay and antagonism test of *S. rochei* UU07 (a strain that grows well at pH 4.5–9.4 and that shows tolerance to salt), revealing it as an excellent PGP and a potential antagonist for *Rhizoctonia solani*. They confirmed that rotten wheat straw can be a source for the isolation of actinobacteria with PGP traits, which may be used as consortia for the composting of different agriculture waste materials and to increase crop yields.

The synergistic effects of the joint inoculation of the endophyte bacterium *S. griseoflavus* P4 and *Bradyrhizobium japonicum* (*B. diazoefficiens*) SAY3-7 on nitrogen fixation, nutrient uptake, plant growth, nodulation, and seed yield of

Table 13.1 Characterization and functions related to plants growth promotion of different *Streptomyces* spp.

Streptomyces spp.	Sample	Plant	Characterization	Function in PGP	References
S. humidis	Sandy soil with debris of *Diplotaxis tenuifolia*	Cucumber	Quantification of total bacteria: real time-PCR (polymerase chain reaction) *Streptomyces* spp.-specific population patterns: PCR-DGGE (denaturing gradient gel electrophoresis)	Disease suppression in the cucumber–*F. oxysporum* f. sp. *radicis-cucumerinum* system	Klein et al. (2013)
S. kumnigenesis *S. mutabilis* *S. enissocaesilis* *S. djakartensis* *S. nobilis* *S. kunmingenesis* *S. enissocaesilis*	Soil samples collected from the rhizhosphere	Wheat and tomato	Morphological, biochemical, and physiological characterization: melanin pigment, decomposition of oxalic acid and other organic acids, cell wall type, tyrosine hydrolysis by actinomycetes, xanthine and hypoxanthine utilization, starch hydrolysis. Genomic DNA isolation, PCR amplification and sequencing of the 16s rRNA gen	IAA production, phosphate solubilization, siderophores,, HCN production and 1-aminocyclopropane-1-carboxylate deaminase production	Anwar et al. (2016)
Streptomyces sp. strain DBT204	Seeds (IARI, Pusa Campus, New Delhi, India)	Chili and tomato	Morphological: primary identification of endophytic bacteria. Spore chain morphology: field emission gun/scanning electron microscope (FEG-SEM) Phylogenetic analysis: PCR	Growth promotion: significant increase in shoot length, root length and plant weight	Passari et al. (2016)
Streptomyces sp.	Vertisol soil samples (Icrisar, Pachanteru, India)	Seeds of chickpea variety ICCV 2	Identification by 16S rDNA sequencing	Improved soil health and crop growth	Sathya et al. (2016)

(continued)

Table 13.1 (continued)

Streptomyces spp.	Sample	Plant	Characterization	Function in PGP	References
Streptomyces isolate L3 *Streptomyces* isolate KT 6-4-1 *Streptomyces* isolate ST 3	Soil samples	Rice seeds (variety RD-6)	Morphological and cultural characteristics: scanning electron microscope Chemotaxonomic and physiological characteristics: ISP method Secondary identification: 16S rDNA sequence analysis	Long-term root disease control. Phosphate solubilization, through excretion of siderophores Plant growth promotion	Kpomblekou-A and Tabatabai (1994)

soybean crops have been investigated by Htwe et al. (2018) and Htwe and Yamakawa (2016). Analogous studies were conducted for *S. griseoflavus* P4 with various *Bradyrhizobium* strains (Aung and Takeo 2015) and different crops (cereals, horticultural crops, and leguminous plants), finding a beneficial effect of *S. griseoflavus* P4 in all cases.

The PGP behavior of Actinobacteria (particularly of *Streptomyces* spp.) discussed above is accomplished by different mechanisms (Passari et al. 2015), which include the production of secondary metabolites like antibiotics (e.g., chloramphenicol, fluconazole, nalidixic acid, trimethoprim, streptomycin, and rifamycin) (Doumbou et al. 2001), 1-aminocyclopropane-1-carboxylate deaminase (El-Tarabily et al. 2008), phytohormones (e.g., indole acetic acid, kinetin) (Rashad et al. 2015), desferrioxamine and coelichelin siderophores (Khamna et al. 2008); and phosphate solubilization (Mohandas et al. 2013).

Aforementioned secondary metabolites play an important role in regulatory activities and as antagonistic agents, agrobiologicals, and pharmacological agents (Harir et al. 2018). For example, endophytic *S. hygroscopicus* TP_A045 produces metabolites such as pteridic acids A and B that show auxin-like activity and that induce root elongation in *Phaseolus vulgaris* L. (common bean) (Igarashi et al. 2002).

In addition, the genus *Streptomyces* is well known as a generator of extracellular polymeric substances (EPS) that act as siderophores or Fe^{2+} iron transporters, which can be chelated, mobilized, solubilized, and assimilated by active transport mechanisms (Singh et al. 2016). Poaceae (barley and wheat grasses) are able to effectively sequester iron by releasing phytosiderophores through their roots, and *Streptomyces* spp. can increase the availability and absorption of iron. Oats are also capable of assimilating iron by microbial siderophores (Kraemer et al. 2006), and so are the seeds of chickpea (Sathya et al. 2016). With regard to the latter crop, Gopalakrishnan et al. (2015) confirmed grain yield and plant growth enhancement by using broad-spectrum *Streptomyces* spp.

Among the isolates tested in the biological restoration of tailings landfills, *Streptomyces* sp. R05.33 and *Phyllobacterium* R01.34 have been claimed to have the greatest potential to act as PGP rhizobacteria. Zappelini et al. (2018) showed how *Streptomyces* spp. could colonize the roots of *Betula pendula* Roth (silver birch) in a red gypsum landfill in unfavorable conditions, forming spores and favoring the growth of the plant. A similar approach (birch and strains of actinomycetes as rooting agents) was also successful in the restoration of coal mine landfills (Ostash et al. 2014).

Entering fully in specific aspects of composting, Sharma et al. (2017) developed an efficient microorganism (EM) compost or bioorganic fertilizer optimized by dose of EM compost and inocula of various microorganisms (viz. *S. globisporous* C3, *Candida tropicalis* Y6, *Phanerochaete chrysosporium* VV18, and *Lactobacillus* spp.) to improve soil fertility and stimulate plant growth. The physicochemical characteristics of the EM mature compost used in ornamental crops were as follows: total C 26%; total N 1.66%; C/N ratio 15.66; humus 7.55%; available P 0.31%; pH 7.8; EC 0.38 $S{\cdot}m^{-1}$.

Verma et al. (2015) noted that the application of EM compost improved the lycopene content of tomato fruits, provided nutrients, and reduced the cultivation cost. The application of EM to plants would also induce higher levels of calcium compared to untreated plants (Daiss et al. 2007).

Results of the synergistic potential of compost with two *Streptomyces* strains for PGP and biocontrol against *R. solani* on pepper evidenced that this combination was highly efficient for the disease suppression (Wang et al. 2015).

In a recent contribution by Dimitrijevic et al. (2017), a superior fertilization effect of a compost obtained from mixed herbs waste was gained after inoculation with three novel mesophilic strains (viz. *S. spororaveus* CKS2, *S. microflavus* CKS6, and *S. fulvissimus* CKS 7).

13.5 *Streptomyces* as Biocontrol Agents

Traditionally, crop rotation has been an effective strategy to combat pathogens, and soil solarization has been used, for instance, to fight *F. oxysporum* (Benlioglu et al. 2005; Huang et al. 2012). Since disinfection with methyl bromide and other pesticides is currently a prohibited practice (Cebolla et al. 2000), it is necessary to find biological agents that are more efficient and safer.

Interference competition, a relevant strategy in interspecific interactions, refers to the production of secondary metabolites (e.g., enzymes, toxins, biosurfactants, antibiotics, and volatiles) that can suppress microbial opponents (Hibbing et al. 2009). In this regard, actinomycetes, particularly *Streptomyces* spp., are receiving increasing attention as biological agents (Cuesta et al. 2012) to control soil-borne pathogens (Getha and Vikineswary 2002; Gopalakrishnan et al. 2011) and to produce antibiotics. In fact, *Streptomyces* is the largest antibiotic-producing genus against clinical microorganisms and parasites (Castillo et al. 2002; Hwang et al. 2001). In addition, they produce other bioactive compounds of clinical importance such as immunosuppressants (Watve et al. 2001).

In sustainable agriculture, the potential of *Streptomyces* to produce antibiotics and other secondary metabolites can be used to control phytopathogenic fungi and bacteria (Schrey and Tarkka 2008) and even to act as nematicidals (Santos et al. 2016). Moreover, the use of *Streptomyces* as agro-antibiotics (Demain 2009) and as biocontrol agents against numerous fungal and viral pathogens is becoming increasingly popular and is reaching the commercialization stage in many countries (Trejo-Estrada et al. 1998; Macagnan et al. 2008; Gopalakrishnan et al. 2011; McGhee and Sundin 2011; Liu et al. 2014; Peng et al. 2014). Volatile antifungal compounds produced by *Streptomyces* species play an important role in biocontrol and are important because they can act as fungicides, bactericides, nematicides, herbicides, insecticides, molluscicides, etc. (Rey and Dumas 2017).

A selection of very promising *Streptomyces* spp. and their associated secondary metabolites, together with their applications, is presented in Table 13.2. The chemical structures of some of those secondary metabolites are depicted below, in

Table 13.2 Selected *Streptomyces* spp. with their associated secondary metabolites and proposed applications

Streptomyces spp.	Secondary metabolites	Applications	References
S. ambofaciens	Oxytetracycline, spiramycin, albopeptin A, albopeptin B, alpomycin	Antibacterial	Bunet et al. (2008), Pernodet et al. (1993)
S. alboflavus	Oxytetracycline, tetracycline, desertomycin A	Antibacterial	Dam et al. (2014), Ji et al. (2012), Wang et al. (2013a)
S. althioticus	Althiomycin	Antibacterial	Luo et al. (2016), Vining (2014)
S. amphotericinicus	Amphotericin	Antifungal	Cao et al. (2017)
S. anandii	Pentaene G8, gilvocarcin V, gilvocarcin M, gilvocarcin E	Antifungal	Balitz et al. (1981), Singh and Srivastava (1998)
S. atratus	Atramycin A, hydrazidomycins A, hydrazidomycins B, hydrazidomycins C, rufomycins A, rufomycins B	Antifungal	Deutsche Forschungsgemeinschaft. Senatskommission zur Beurteilung von Stoffen in der Landwirtschaft. (2001), Ueberschaar et al. (2011)
S. atrovirens	Indole-3-acetic acid	Plant growth promotion	Abd-Alla et al. (2013)
S. avermitilis	Ivermectin, abamectin	Insecticide and antihelmintic	Burg et al. (1979), Takahashi (2002)
S. bellus	Althiomycin	Antibacterial	BacDive (2019a)
S. cacaoi	Polyoxine	Antifungal	Chen et al. (2009), Funayama and Isono (2014), Goodfellow et al. (1988)
S. cellulosae	Fungichromin (pentamycin)	Antifungal	Harrison et al. (1986), Laskin and Lechevalier (1977), Li et al. (1989)
S. chattanoogensis	Natamycin	Antifungal	Du et al. (2009), Jiang et al. (2013)
S. chrestomyceticus	Lycopene, pyrrolostatin, paromomycin, aminocidin, aminosidine, neomycin E, neomycin F	Antibacterial	Kato et al. (1993), Prakash and Sharma (2014)
S. crystallinus	Hygromycin A (totomycin)	Antibacterial	Afifi et al. (2012)
S. cuspidosporus	Valinomycin	Antibacterial	Maheswari and Chandra (2000)
S. fabae	Antimicrobial activity	Antibacterial	Nguyen and Kim (2015)
S. filipinensis	pentalenolactone I, hygromycin A, filipin	Antifungal	Payero et al. (2015), Uyeda et al. (2014)

(continued)

Table 13.2 (continued)

Streptomyces spp.	Secondary metabolites	Applications	References
S. flavofungini	Bafilomycin K, aminoacylase	Antifungal	Uri and Békési (1958)
S. fradiae	Neomycin, tylosin, fosfomycin	Antibacterial	Janssen et al. (1989), Waksman and Lechevalier (1949)
S. glomeratus	Beromycin, nogalamycin	Antibacterial	Blumauerová et al. (1980)
S. griseiniger	Nigericin	Antibacterial	BacDive (2019b)
S. griseochromogenes	Blasticidin A, blasticidin B, blasticidin C, blasticidin S, pentalenene, cytomycin	Antifungal	Cutler and Cutler (1999)
S. griseoviridis	Etamycin, griseoviridin, bactobolin, prodigiosin R1, rosophilin	Antifungal	Kawasaki et al. (2009), Xie et al. (2012)
S. griseus	Produces 32 different structural types of secondary metabolites of commercial importance	Antibacterial	Graf et al. (2007), Ohnishi et al. (2008), Schatz et al. (1944)
S. halstedii	Bafilomycin B1, bafilomycin C1, deltamycin A2, deltamycin A3, magnamycin B, vicenistatin	Antifungal	Hochstein and Murai (1954)
S. heilongjiangensis	Borrelidin	Antibacterial	Liu et al. (2012)
S. hiroshimensis	Prodigiosin	Antifungal	Magae et al. (1993)
S. hygroscopicus	Geldanamycin, hygromycin B, nigericin, validamycin, cyclothiazomycin	Antifungal and antibacterial	Murakami et al. (1986)
S. kanamyceticus	Kanamycin	Antifungal and antibacterial	BacDive (2019c)
S. kasugaensis	Kasugamycin, thiolutin	Antibacterial	Hotta et al. (1996)
S. koyangensis	4-Phenyl-3-butenoic acid	Antifungal	Lee (2005), Lee et al. (2005)
S. kurssanovii	Chitinase, N-(phenylacetyl)-2-butenediamide, fumaramidmycin	Antifungal	Il'ina et al. (2000), Tikhonov et al. (1998)
S. laurentii	Thiostrepton	Antibacterial	Trejo et al. (1977)
S. lavendofoliae	Fosfazinomycin A, fosfazinomycin B, piperastatin B	Antifungal	Al-Humiany (2011), Murakami et al. (2008)
S. lavendulae	Streptothricin, lavendamycin	Antifungal	August et al. (1996), Sheldon et al. (1999)

(continued)

Table 13.2 (continued)

Streptomyces spp.	Secondary metabolites	Applications	References
S. lydicus	Actithiazic acid, natamycin, lydimycin, streptolydigin, 1-deoxygalactonojirimycin	Antifungal	Atta et al. (2015), Gómez et al. (2012), Yuan and Crawford (1995)
S. narbonensis	Narbomycin, josamycin	Antibacterial	van Balken (1997)
S. netropsis	Netropsin, distamycin A, mycoheptin	Antifungal and antibacterial	Ekzemplyarov (1977), Neilan et al. (2014)
S. omiyaensis	Chloramphenicol, pentalenolactone P	Antibacterial	Alam et al. (2004)
S. phytohabitans	Novonestmycin A, novonestmycin B	Antifungal	Bian et al. (2012), Wan et al. (2014)
S. platensis	Oxytetracycline, platensimycin, migrastatin, isomigrastatin, platencin, dorrigocin A, dorrigocin B	Antibacterial	Peterson et al. (2014), Smanski et al. (2009)
S. prasinus	Prasinomycin, validamycin, prasinon A, prasinon B	Antifungal and antibacterial	Box et al. (1973)
S. rochei	Borrelidin, butyrolactol A, butyrolactol B, uricase, streptothricin	Antifungal	Augustine et al. (2005), Irdani et al. (1996), Kanini et al. (2013)
S. roseosporus	Daptomycin	Antibacterial	Miao (2005)
S. spectabilis	Hangtaimycin, gentamicin, kanamycin, neomycin B, sisomicin, tobramycin, paromomycin, spectinabilin, spectinomycin, aminocyclitol, actinospectacin, prodigiosin, streptovaricin	Antibacterial	Kakinuma et al. (1976), Zuo et al. (2016)
S. sulfonofaciens	Pluracidomycin	Antibacterial	Miyadoh et al. (1983)
S. variegatus	Prodigiosin	Antifungal	Sveshnikova et al. (1983)
S. vitaminophilus	Pyrrolomycin	Antifungal and antibacterial	Mahan et al. (2016)
S. zaomyceticus	Zaomycin, pikromycin, glumamycin, foroxomithine	Antibacterial	Wegler (1981)

Figs. 13.2 and 13.3. Table 13.3 shows different species of *Streptomyces* and their role as biocontrol agents.

Many species of actinomycetes can inhibit various pathogenic fungi (Al-Askar et al. 2014; Hwang et al. 2001; Lim et al. 2000). Some strains of *S. humidus* (e.g., strain S5–55) have been reported to feature antagonism against *Phytophthora capsici* through direct antibiosis (sodium phenylacetate and phenylacetic acid)

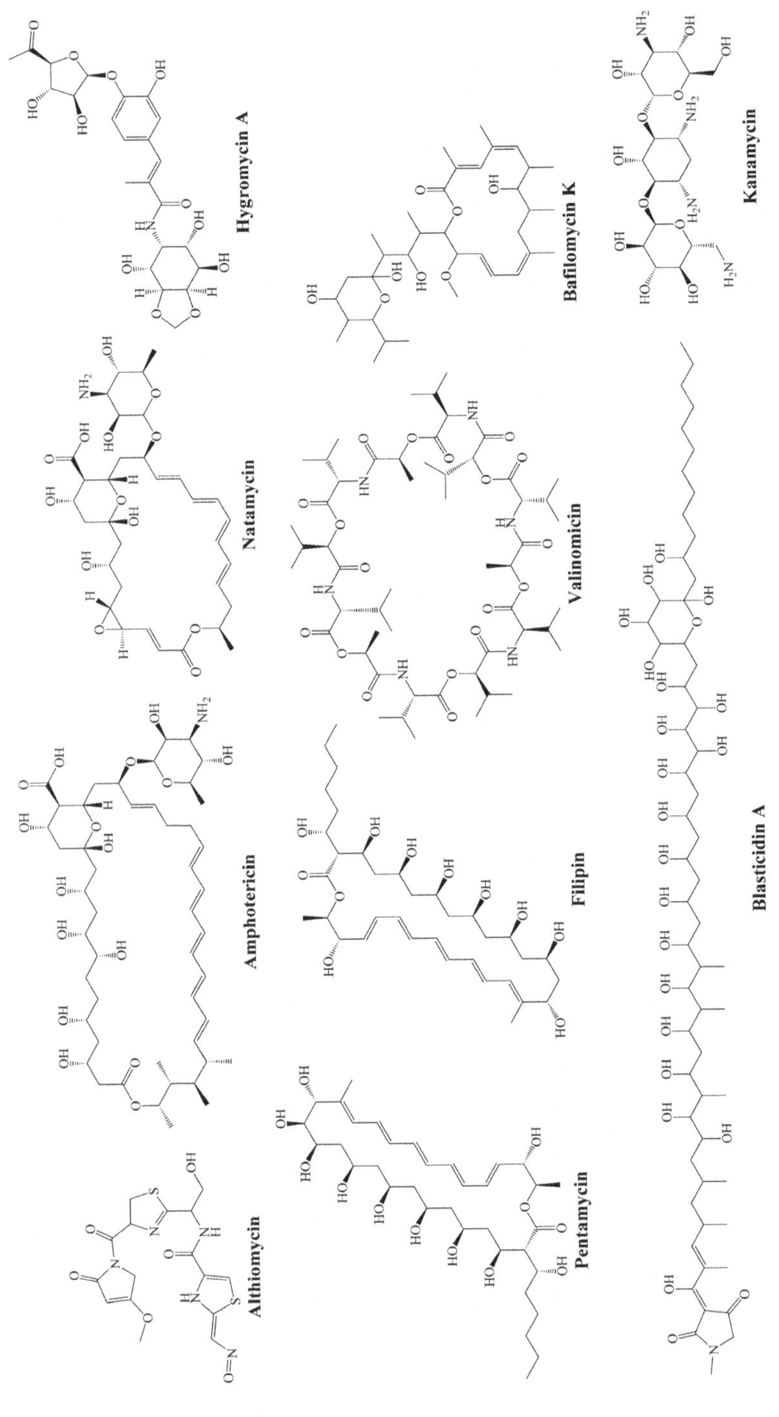

Fig. 13.2 Chemical structures of some of the secondary metabolites produced by the *Streptomyces* spp. in Table 13.2

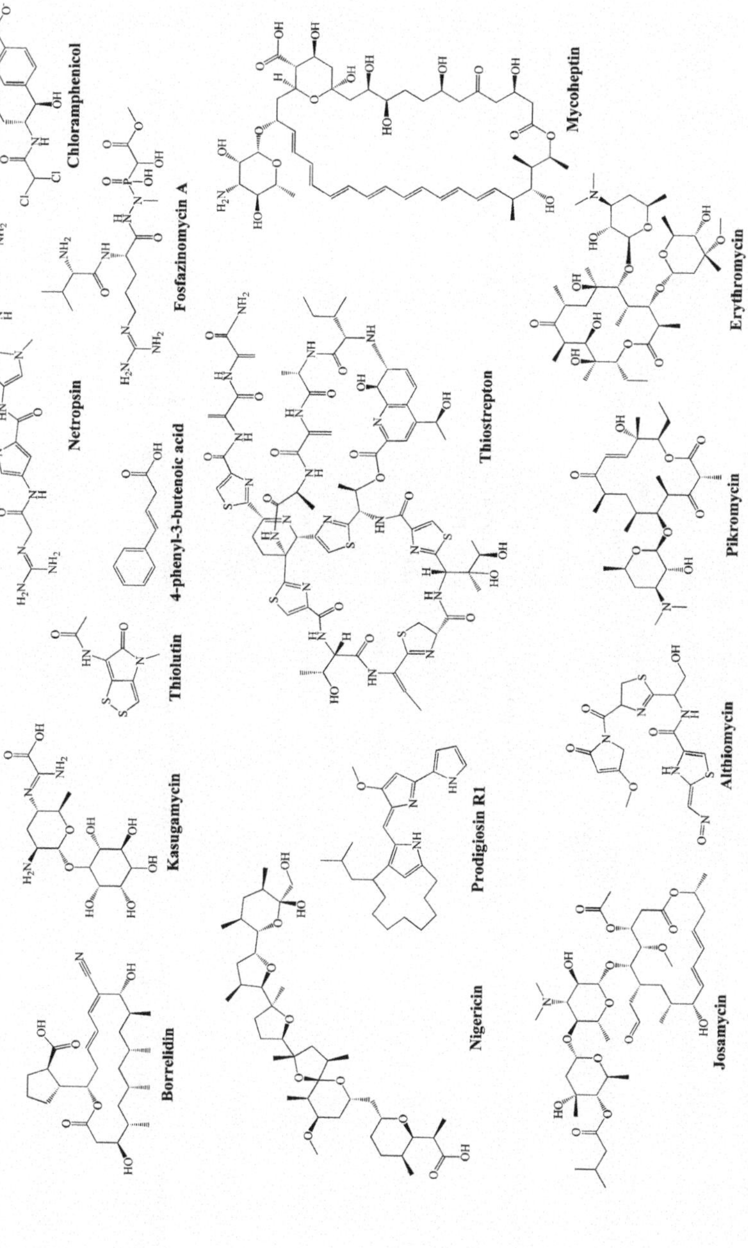

Fig. 13.3 Chemical structures of some of the secondary metabolites produced by the *Streptomyces* spp. in Table 13.2 (continued)

Table 13.3 Functions of different *Streptomyces* spp. against plant pathogens

Streptomyces spp.	Sample	Plant	Pathogen	Disease	Functions	Reference
S. humidis	Sandy soil with debris of *Diplotaxis tenuifolia*	Cucumber	*F. oxysporum* f. sp. *radicis-cucumerinum*	Root rot	Shifts in root microbiome associated with reduction of pathogen root colonization	Klein et al. (2013)
S. roseoflavus strain NKZ-259	Soil samples from the Qilian Mountains (Qinghai province, China)	Tomato and pepper	*Botrytis cinerea* *Fulvia fulva* *Curvularia lunata* *Fusarium oxysporum* *Rhizopus stolonifer* *Ustilaginoidea oryza* *Botryosphaeria ribis* *Bipolaris sorokiniana* *Bipolaris maydis* *Botryosphaeria dothidea* *Phyllosticta ampelicida* *Valsa ceratosperma* *Colletotrichum gloeosporioides* *Alternaria alternata* *Exserohilum turcicum* *Fusarium graminearum* *Pyricularia oryzae* *Rhizoctonia cerealis*	Seedlings rot	Inhibition of mycelial growth Pathogen outbreaks were blocked after NKZ-259 fermentation broth was applied to infected detached leaves Growth promotion of tomato and pepper seedlings	Shi et al. (2018)
S. albospinus CT205	Cucumber seeds (Jiangsu, China)	Cucumber	*Fusarium oxysporum* NJAU-2 *F. oxysporum* f. sp. *cucumerinum*	Hyphal deformation Death incidence	Inhibition of the development of *F. oxysporum* NJAU-2 mycelium Induced resistance against ampicillin, cefotaxime, and amoxicillin Promoted plant growth Secreted hydrolytic enzymes that led to the leakage and	Wang et al. (2016)

					breakdown of pathogenic fungal cells	
S. pratensis LMM15	Leaves of tomatoes (Weinan, Shaanxi province, China)	Tomato	*Botrytis cinerea*	Gray mold	Inhibited hyphal growth Reduced decay incidence of gray mold on tomato fruits.	Lian et al. (2017)
S. vinaceusdrappus S5 MW2	Samples of tomato seeds	Tomato (var. Navratan)	*R. solani*	Root rot Gray mold	Inhibition of *R. solani* growth up to 65% With chitin enhanced PGP parameters	Yandigeri et al. (2015)

(Hwang et al. 2001). Nonetheless, there is still substantial interest in finding novel strains with the ability to produce safer and more effective antifungal agents.

Zhao et al. (2012) evaluated *Streptomyces* spp. for the control of *Didymella bryoniae* (responsible for gummy stem blight) and to promote growth of *Cucumis melo* L. (muskmelon). *S. hygroscopicus* (strain B04), isolated by Shen et al. (2016) from the rhizosphere soil of a healthy strawberry plant, was found to strongly inhibit the growth of phytopathogenic fungi such as *Fusarium oxysporum*. This strain can utilize many carbon sources and can produce extracellular fungal cell wall-degrading enzymes (e.g., β-1,3-glucanase, chitinase, protease, and cellulase). In a similar fashion, *S. albospinus* CT205 can inhibit the growth of *F. oxysporum* NJAU-2 (responsible for Fusarium wilt of cucumber, which leads to serious economic losses), effectively reducing the presence of this pathogen in the rhizosphere of cucumber. The biological characteristics of CT205 were measured by Wang et al. (2016), and this strain was shown to produce β-glucanase, chitinase, and a heat-resistant antagonistic substance.

In 2017, Lian et al. (2017) showed that the strain *S. pratensis* LMM15 has the ability to inhibit *Botrytis cinerea* mycelial growth and to reduce lesion expansion associated with gray mold on detached fruits and leaves. Shi et al. (2018) isolated from soil samples from Qinghai (China) a strain named NKZ-259, identified as *S. roseoflavus*, which displays high antagonistic activity against 6 fungal pathogens: *Botrytis cinerea, Curvularia lunata, Colletotrichum gloeosporioides, Fulvia fulva, Rhizoctonia cerealis*, and *Ustilaginoidea oryzae*. The use of its fermentation broth reduced tomato gray mold incidence by 66.67%, and promoted the growth of pepper and tomato seedlings, resulting in a significant increase in fresh weight, plant height, and root length. Indoleacetic acid (IAA) phytohormone production was involved.

Lytic enzymes such as chitinase and lipase that bacteria from the *Streptomyces* genus produce (Santos et al. 2016) allowed their use as biocontrol agents against nematodes (*Scutellonema bradys* in yam plants). The production of chitinase enzymes caused the destruction of the nematode cuticle, in which chitin is an important constituent (Park et al. 2002). Moreover, substances that are toxic for the phytonematodes (e.g., ammonia) were released in the decomposition process of chitin.

Finally, an important aspect to take into consideration is that antagonistic strains need to be combined with an appropriate substrate in order to improve their biological control efficacy against soil-borne diseases. Solid-state fermentation of agro-industrial residues with *Streptomyces* spp. to produce bioorganic fertilizers is currently a promising strategy for the management of agro-industrial waste. As noted above (see Shen et al. (2016)), the fermentation equipment involved is simple, economical, and suitable for multiple applications in production. In their work, solid shallow-tray fermentation of various combinations of pig manure compost, vermicompost, wheat bran, and rapeseed meal were inoculated with *S. hygroscopicus* strain B04, and the resulting bioorganic fertilizer showed a remarkable activity against fungal pathogens.

13.6 *Streptomyces* spp. in the Formation of Biofilms

Streptomyces have the capacity to form biofilms (Kim et al. 2019), which are defined as "*a bacterial community immersed in a liquid medium, with the ability to adhere to a substrate or surface and interfaces, which are embedded in an extracellular matrix produced by themselves, and which show an altered phenotype in terms of the degree of cell multiplication or the expression of their genes*" (Donlan and Costerton 2002).

In most biofilms formation, unicellular organisms come together, forming a community attached to a solid surface and covered in an exopolysaccharide matrix (Satpathy et al. 2016). The biofilm formation process consists of 5 stages: (1) preconditioning of the adhesion surface by macromolecules, either present in the bulk fluid or intentionally coated on the surface (Simões et al. 2010); (2) reversible cellular adhesion, in which attachment occurs most effortlessly on rougher and more hydrophobic surfaces, and on surfaces coated by conditioning films; (3) clonal expansion of the bacteria takes place and the production of exopolysaccharides begins, which together with proteins and nucleic acids form the matrix of the biofilm during the irreversible adhesion stage (Reisner et al. 2003); (4) maturation, in which biofilms develop into an organized structure that can be flat or mushroom-shaped (Chmielewski and Frank 2003; Klausen et al. 2003); and (5) dispersion, when the environmental conditions are unfavorable and the cells return to their planktonic form (Sauer et al. 2002).

The formation of microbial biofilm is a very complex process (Shi and Zhu 2009), which depends on the joint interaction of factors such as the environment, surface characteristics, and characteristics of the cell itself. To the best of the authors' knowledge, only a limited number of studies on biofilms of streptomycetes have been reported in the literature (de Jong et al. 2009; Khiyami et al. 2005; Kim and Kim 2004; Morales et al. 2007; Winn et al. 2014). From a biotechnology point of view, the most important were *S. setonii* 75Vi2 (ATCC 39116), which degraded microbial inhibitors in diluted corn stover and starch pyrolysis liquors (Khiyami et al. 2005), and *S. griseus*, which was cultivated as a biofilm in a tubular reactor (Winn et al. 2014).

Compost is a medium capable of providing electrochemically active biofilms for the oxidation of organic compounds and the transfer of electrons (Dulon et al. 2006). Parot et al. (2008) investigated the effect of adding acetate to garden compost to promote the development of electrochemically efficient biofilms.

In agricultural soils, the stage prior to the invasion of plant tissues occurs on the surface of roots, leaves, or seeds, through the establishment of sessile populations of phytopathogenic bacteria (Monier and Lindow 2003). The development of biofilms contributes to phytopathogenic bacteria's virulence by way of the blockage of xylem vessels (Mansfield et al. 2012). On the other hand, biofilms can participate in the biological control processes of pathogens through mutualistic relationships between rhizobacteria and plants (Chin-A-Woeng et al. 2000; Espinosa-Urgel et al. 2002). The self-aggregation characteristic of bacteria has implications for the production of inoculants for agriculture (Bogino et al. 2013) and the protection effects of biofilms would reduce the toxicity of the compounds to the cells (Burmølle et al. 2014). In

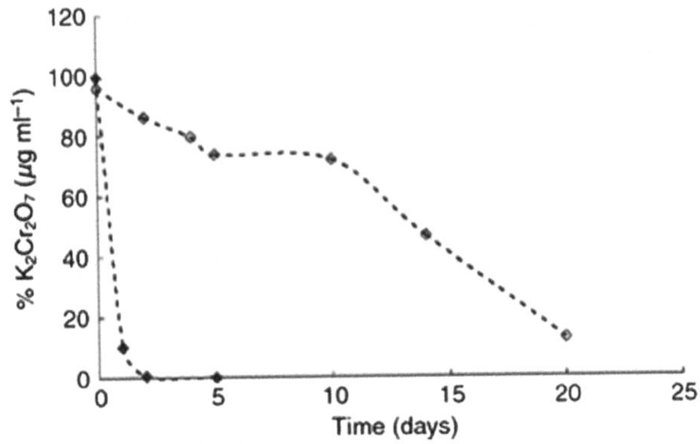

Fig. 13.4 Cr(VI) removal assay in M63 medium. Dashed lines indicate removal by cells forming biofilms on glass beads (filled diamond) or by control planktonic cells (open diamond). Reproduced with permission from Morales et al. (2007). Copyright John Wiley & Sons

particular, the biofilms formed by *Streptomyces* are capable of eliminating $K_2Cr_2O_7$ from contaminated soils. In Fig. 13.4 it can be seen that the biofilms of the strain *Streptomyces* sp. CG252 completely eliminated $K_2Cr_2O_7$ in 3 days, while the planktonic form of this same strain was only able to eliminate 80–90% and needed more than double the time. Therefore, this strain of *Streptomyces* can be a promising candidate for detoxification of sites containing this heavy metal (Morales et al. 2007).

13.7 Conclusions

Compost is a self-heated substrate formed when organic materials are broken down and recycled by successive groups of microorganisms in various composting stages. Several studies have shown that the compost microbial population is highly dominated by actinomycetes and that their presence can potentially serve as an indicator of compost maturity, provided that they participate in suppressing pathogens in the curing stage. Among the microbial biomass that colonizes composts, the versatile *Streptomyces* species produce various lytic enzymes that can break down cellulose and other insoluble organic polymers and produce nutrients that can be used by plants. Because of their capability to form spores and subsist to adverse conditions in the soil, they are also more competitive than other microbes. *Streptomyces* are active producers of antibiotics and compost colonized by these microorganisms contains strains that have plant growth-promoting abilities and that can be used as biofertilizers. Various *Streptomyces* spp. are antagonists of plant pathogens and can thus save the plant from attacks by dangerous fungi and bacteria. As such, they can be regarded as biocontrol agents in several cropping systems.The identification of efficient *Streptomyces* spp. is necessary to stimulate further research on their utility in converting agricultural wastes to organic manure for soil amendment and to attain higher crop productivity.

Acknowledgments The authors would like to gratefully acknowledge the financial support of the European Regional Development Fund and the regional Ministry of Education of Junta de Castilla y León through project VA258P18, and the European Union funding through project LIFE+ AMMONIA TRAPPING (LIFE15-ENV/ES/000284).

References

Abd-Alla MH, El-Sayed E-SA, Rasmey A-HM (2013) Indole-3-acetic acid (IAA) production by *Streptomyces atrovirens* isolated from rhizospheric soil in Egypt. J Biol Earth Sci 3:182–193

Afifi MM, Atta HM, Elshanawan AA, Abdoul-rao UM, El-Adly AM (2012) Biosynthesis of hygromycin-B antibiotic by *Streptomyces crystallinus* AZ151 isolated from Assuit, Egypt. Bacteriol J 2:46–65. https://doi.org/10.3923/bj.2012.46.65

Alam MZ, Manchur MA, Anwar MN (2004) Isolation, purification, characterization of cellulolytic enzymes produced by the isolate *Streptomyces omiyaensis*. Pak J Biol Sci 7:1647–1653. https://doi.org/10.3923/pjbs.2004.1647.1653

Al-Askar AA, Abdulkhair WM, Rashad YM, Hafez EE, Ghoneem KM, Baka ZA (2014) *Streptomyces griseorubens* E44G: a potent antagonist isolated from soil in Saudi Arabia. J Pure Appl Microbiol 8:221–230

Aldesuquy HS, Mansour FA, Abo-Hamed SA (1998) Effect of the culture filtrates of *Streptomyces* on growth and productivity of wheat plants. Folia Microbiol 43:465–470. https://doi.org/10.1007/bf02820792

Alexander M (1991) Introduction to soil microbiology. R.E. Krieger Publishing Company, Malabar, FL

Al-Humiany AU-RAA (2011) Taificidin1 and Taificidin2, two anti-microbial agents isolated from the fermentation broth of *Streptomyces roseodistaticus* TA15 and *Streptomyces lavendofoliae* TA17. Res J Microbiol 6:328–342. https://doi.org/10.3923/jm.2011.328.342

Ali A, Guo D, Mahar A, Ma F, Li R, Shen F, Wang P, Zhang Z (2017a) *Streptomyces pactum* assisted phytoremediation in Zn/Pb smelter contaminated soil of Feng County and its impact on enzymatic activities. Sci Rep 7. https://doi.org/10.1038/srep46087

Ali A, Guo D, Mahar A, Wang P, Ma F, Shen F, Li R, Zhang Z (2017b) Phytoextraction of toxic trace elements by *Sorghum bicolor* inoculated with *Streptomyces pactum* (Act12) in contaminated soils. Ecotoxicol Environ Saf 139:202–209. https://doi.org/10.1016/j.ecoenv.2017.01.036

Ali A, Guo D, Mahar A, Wang Z, Muhammad D, Li R, Wang P, Shen F, Xue Q, Zhang Z (2017c) Role of *Streptomyces pactum* in phytoremediation of trace elements by *Brassica juncea* in mine polluted soils. Ecotoxicol Environ Saf 144:387–395. https://doi.org/10.1016/j.ecoenv.2017.06.046

Alvarez A, Saez JM, Davila Costa JS, Colin VL, Fuentes MS, Cuozzo SA, Benimeli CS, Polti MA, Amoroso MJ (2017) Actinobacteria: current research and perspectives for bioremediation of pesticides and heavy metals. Chemosphere 166:41–62. https://doi.org/10.1016/j.chemosphere.2016.09.070

Anwar S, Ali B, Sajid I (2016) Screening of rhizospheric actinomycetes for various *in-vitro* and *in-vivo* plant growth promoting (PGP) traits and for agroactive compounds. Front Microbiol 7. https://doi.org/10.3389/fmicb.2016.01334

Aparicio JD, Benimeli CS, Almeida CA, Polti MA, Colin VL (2017) Integral use of sugarcane vinasse for biomass production of actinobacteria: potential application in soil remediation. Chemosphere 181:478–484. https://doi.org/10.1016/j.chemosphere.2017.04.107

Aparicio JD, Raimondo EE, Gil RA, Benimeli CS, Polti MA (2018a) Actinobacteria consortium as an efficient biotechnological tool for mixed polluted soil reclamation: experimental factorial design for bioremediation process optimization. J Hazard Mater 342:408–417. https://doi.org/10.1016/j.jhazmat.2017.08.041

Aparicio JD, Saez JM, Raimondo EE, Benimeli CS, Polti MA (2018b) Comparative study of single and mixed cultures of actinobacteria for the bioremediation of co-contaminated matrices. J Environ Chem Eng 6:2310–2318. https://doi.org/10.1016/j.jece.2018.03.030

Araragi M (1979) Comparison of actinomycete flora between tropical and temperate upland farm soils. Soil Sci Plant Nutr 25:245–254. https://doi.org/10.1080/00380768.1979.10433165

Atta HM, El-Sayed AS, El-Desoukey MA, Hassan M, El-Gazar M (2015) Biochemical studies on the Natamycin antibiotic produced by *Streptomyces lydicus*: fermentation, extraction and biological activities. J Saudi Chem Soc 19:360–371. https://doi.org/10.1016/j.jscs.2012.04.001

August PR, Rahn JA, Flickinger MC, Sherman DH (1996) Inducible synthesis of the mitomycin C resistance gene product (MCRA) from *Streptomyces lavendulae*. Gene 175:261–267. https://doi.org/10.1016/0378-1119(96)00172-2

Augustine S, Bhavsar S, Kapadnis B (2005) Production of a growth dependent metabolite active against dermatophytes by *Streptomyces rochei* AK 39. Indian J Med Res 121:164–170

Aung ZH, Takeo Y (2015) Enhanced plant growth and/or nitrogen fixation by leguminous and non-leguminous crops after single or dual inoculation of *Streptomyces griseoflavus* P4 with *Bradyhizobium strains*. Afr J Microbiol Res 9:2337–2344. https://doi.org/10.5897/ajmr2015.7796

BacDive (2019a) *Streptomyces bellus*. https://doi.org/10.13145/bacdive15048.20180622.3

BacDive (2019b) *Streptomyces griseiniger*. https://doi.org/10.13145/bacdive16422.20180622.3

BacDive (2019c) *Streptomyces kanamyceticus*. https://doi.org/10.13145/bacdive15333.20180622.3

Balitz DM, O'Herron FA, Bush J, Vyas DM, Nettleton DE, Grulich RE, Bradner WT, Doyle TW, Arnold E, Clardy J (1981) Antitumor agents from *Streptomyces anandii*: gilvocarcins V, M and E. J Antibiot 34:1544–1555. https://doi.org/10.7164/antibiotics.34.1544

Benlioglu S, Boz O, Yildiz A, Kaskavalci G, Benlioglu K (2005) Alternative soil solarization treatments for the control of soil-borne diseases and weeds of strawberry in the Western Anatolia of Turkey. J Phytopathol 153:423–430. https://doi.org/10.1111/j.1439-0434.2005.00995.x

Bian G-K, Qin S, Yuan B, Zhang Y-J, Xing K, Ju X-Y, Li W-J, Jiang J-H (2012) *Streptomyces phytohabitans* sp. nov., a novel endophytic actinomycete isolated from medicinal plant *Curcuma phaeocaulis*. Antonie Van Leeuwenhoek 102:289–296. https://doi.org/10.1007/s10482-012-9737-8

Blumauerová M, Podojil M, Gauze GF, Maksimova TS, Panoš J, Vaněk Z (1980) Spontaneous variability of *Streptomyces glomeratus*, a producer of the anthracycline antibiotics beromycins. Folia Microbiol 25:207–212. https://doi.org/10.1007/bf02877339

Bogino P, Oliva M, Sorroche F, Giordano W (2013) The role of bacterial biofilms and surface components in plant-bacterial associations. Int J Mol Sci 14:15838–15859. https://doi.org/10.3390/ijms140815838

Box S, Cole M, Yeoman G (1973) Prasinons A and B: potent insecticides from *Streptomyces prasinus*. J Appl Environ Microbiol 26:699–704

Brar SK, Verma M, Surampalli RY, Misra K, Tyagi RD, Meunier N, Blais JF (2006) Bioremediation of hazardous wastes—a review. Pract Period Hazard Toxic Radioact Waste Manag 10:59–72. https://doi.org/10.1061/(asce)1090-025x(2006)10:2(59)

Briceño G, Vergara K, Schalchli H, Palma G, Tortella G, Fuentes MS, Diez MC (2017) Organophosphorus pesticide mixture removal from environmental matrices by a soil *Streptomyces* mixed culture. Environ Sci Pollut Res 25:21296–21307. https://doi.org/10.1007/s11356-017-9790-y

Bunet R, Mendes MV, Rouhier N, Pang X, Hotel L, Leblond P, Aigle B (2008) Regulation of the synthesis of the angucyclinone antibiotic alpomycin in *Streptomyces ambofaciens* by the autoregulator receptor AlpZ and its specific ligand. J Bacteriol 190:3293–3305. https://doi.org/10.1128/jb.01989-07

Burg RW, Miller BM, Baker EE, Birnbaum J, Currie SA, Hartman R, Kong YL, Monaghan RL, Olson G, Putter I, Tunac JB, Wallick H, Stapley EO, Oiwa R, Omura S (1979) Avermectins, new family of potent anthelmintic agents: producing organism and fermentation. Antimicrob Agents Chemother 15:361–367. https://doi.org/10.1128/aac.15.3.361

Burmølle M, Ren D, Bjarnsholt T, Sørensen SJ (2014) Interactions in multispecies biofilms: do they actually matter? Trends Microbiol 22:84–91. https://doi.org/10.1016/j.tim.2013.12.004

Cao S, Wang W, Wang F, Zhang J, Wang Z, Yang S, Xue Q (2016) Drought-tolerant *Streptomyces pactum* Act12 assist phytoremediation of cadmium-contaminated soil by *Amaranthus hypochondriacus*: great potential application in arid/semi-arid areas. Environ Sci Pollut Res 23:14898–14907. https://doi.org/10.1007/s11356-016-6636-y

Cao T, Mu S, Lu C, Zhao S, Li D, Yan K, Xiang W, Liu C (2017) *Streptomyces amphotericinicus* sp. nov., an amphotericin-producing actinomycete isolated from the head of an ant (*Camponotus japonicus* Mayr). Int J Syst Evol Microbiol 67:4967–4973. https://doi.org/10.1099/ijsem.0.002382

Castillo UF, Jensen JB, Condron MAM, Yaver D, Robison R, Stevens D, Porter H, Hess WM, Teplow DB, Ford EJ, Strobel GA, Albert H (2002) Munumbicins, wide-spectrum antibiotics produced by *Streptomyces* NRRL 30562 endophytic on *Kennedia nigricans*. Microbiology 148:2675–2685. https://doi.org/10.1099/00221287-148-9-2675

Cebolla V, Busto J, Ferrer A, Miguel A, Maroto V (2000) Methyl bromide alternatives on horticultural crops. Acta Hortic:237–242. https://doi.org/10.17660/ActaHortic.2000.532.32

Chaiharn M, Pathom-Aree W, Sujada N, Lumyong S (2018) Characterization of phosphate solubilizing *Streptomyces* as a biofertilizer. Chiang Mai J Sci 45:701–716

Chen W, Huang T, He X, Meng Q, You D, Bai L, Li J, Wu M, Li R, Xie Z, Zhou H, Zhou X, Tan H, Deng Z (2009) Characterization of the polyoxin biosynthetic gene cluster from *Streptomyces cacaoi* and engineered production of polyoxin H. J Biol Chem 284:10627–10638. https://doi.org/10.1074/jbc.M807534200

Chin-A-Woeng TFC, Bloemberg GV, Mulders IHM, Dekkers LC, Lugtenberg BJJ (2000) Root colonization by phenazine-1-carboxamide-producing bacterium *Pseudomonas chlororaphis* PCL1391 is essential for biocontrol of tomato foot and root rot. Mol Plant Microbe Interact 13:1340–1345. https://doi.org/10.1094/mpmi.2000.13.12.1340

Chmielewski RAN, Frank JF (2003) Biofilm formation and control in food processing facilities. Compr Rev Food Sci Food Saf 2:22–32. https://doi.org/10.1111/j.1541-4337.2003.tb00012.x

Cuesta G, García-de-la-Fuente R, Abad M, Fornes F (2012) Isolation and identification of actinomycetes from a compost-amended soil with potential as biocontrol agents. J Environ Manag 95: S280–S284. https://doi.org/10.1016/j.jenvman.2010.11.023

Cutler HG, Cutler SJ (1999) Biologically active natural products: agrochemicals. CRC Press, Boca Raton, FL

Da Vinha FNM, Gravina-Oliveira MP, Franco MN, Macrae A, da Silva Bon EP, Nascimento RP, Coelho RRR (2010) Cellulase production by *Streptomyces viridobrunneus* SCPE-09 using lignocellulosic biomass as inducer substrate. Appl Biochem Biotechnol 164:256–267. https://doi.org/10.1007/s12010-010-9132-8

Daiss N, Lobo MG, Socorro AR, Brückner U, Heller J, Gonzalez M (2007) The effect of three organic pre-harvest treatments on Swiss chard (*Beta vulgaris* L. var. *cycla*) quality. Eur Food Res Technol 226:345–353. https://doi.org/10.1007/s00217-006-0543-2

Dam HG, Zhang B, Cai G, Wang H, Li D, Yang X, An X, Zheng X, Tian Y, Zheng W, Zheng T (2014) *Streptomyces alboflavus* RPS and its novel and high algicidal activity against harmful algal bloom species *Phaeocystis globosa*. PLoS One 9. https://doi.org/10.1371/journal.pone.0092907

de Jong W, Wösten HAB, Dijkhuizen L, Claessen D (2009) Attachment of *Streptomyces coelicolor* is mediated by amyloidal fimbriae that are anchored to the cell surface via cellulose. Mol Microbiol 73:1128–1140. https://doi.org/10.1111/j.1365-2958.2009.06838.x

Demain AL (2009) Antibiotics: natural products essential to human health. Med Res Rev 29:821–842. https://doi.org/10.1002/med.20154

Deutsche Forschungsgemeinschaft. Senatskommission zur Beurteilung von Stoffen in der Landwirtschaft (2001) Potentially harmful organisms and substances in feedstuffs and animal faeces. Wiley-VCH; Deutsche Forschungsgemeinschaft, Weinheim, p 5

Diaz LF, De Bertoldi M, Bidlingmaier W (2007) Compost science and technology. Waste management series. Elsevier, Amsterdam, p 8

Dimitrijevic S, Radanovic D, Antic-Mladenovic S, Milutinovic M, Rajilic-Stojanovic M, Dimitrijevic-Brankovic S (2017) Enhanced fertilization effect of a compost obtained from mixed herbs waste inoculated with novel strains of mesophilic bacteria. Hemijska industrija 71:503–513. https://doi.org/10.2298/hemind170327013d

Donlan RM, Costerton JW (2002) Biofilms: survival mechanisms of clinically relevant microorganisms. Clin Microbiol Rev 15:167–193. https://doi.org/10.1128/cmr.15.2.167-193.2002

Doumbou CL, Hamby Salove MK, Crawford DL, Beaulieu C (2001) Actinomycetes, promising tools to control plant diseases and to promote plant growth. Phytoprotection 82. https://doi.org/10.7202/706219ar

Du Y-L, Chen S-F, Cheng L-Y, Shen X-L, Tian Y, Li Y-Q (2009) Identification of a novel *Streptomyces chattanoogensis* L10 and enhancing its natamycin production by overexpressing positive regulator ScnRII. J Microbiol 47:506–513. https://doi.org/10.1007/s12275-009-0014-0

Dulon S, Parot S, Delia M-L, Bergel A (2006) Electroactive biofilms: new means for electrochemistry. J Appl Electrochem 37:173–179. https://doi.org/10.1007/s10800-006-9250-8

Ekzemplyarov ON (1977) Mycoheptin, an antifungal antibiotic. Pharm Chem J 11:285–289. https://doi.org/10.1007/bf00779104

El-Tarabily KA (2008) Promotion of tomato (*Lycopersicon esculentum* Mill.) plant growth by rhizosphere competent 1-aminocyclopropane-1-carboxylic acid deaminase-producing streptomycete actinomycetes. Plant Soil 308:161–174. https://doi.org/10.1007/s11104-008-9616-2

El-Tarabily KA, Nassar AH, Sivasithamparam K (2008) Promotion of growth of bean (*Phaseolus vulgaris* L.) in a calcareous soil by a phosphate-solubilizing, rhizosphere-competent isolate of *Micromonospora endolithica*. Appl Soil Ecol 39:161–171. https://doi.org/10.1016/j.apsoil.2007.12.005

Espinosa-Urgel M, Kolter R, Ramos J-L (2002) Root colonization by *Pseudomonas putida*: love at first sight. Microbiology 148:341–343. https://doi.org/10.1099/00221287-148-2-341

Fan M, Xiao X, Guo Y, Zhang J, Wang E, Chen W, Lin Y, Wei G (2018) Enhanced phytoremediation of *Robinia pseudoacacia* in heavy metal-contaminated soils with rhizobia and the associated bacterial community structure and function. Chemosphere 197:729–740. https://doi.org/10.1016/j.chemosphere.2018.01.102

Feng H, Sun Y, Zhi Y, Mao L, Luo Y, Wei X, Zhou P (2014) Lignocellulose degradation by the isolate of *Streptomyces griseorubens* JSD-1. Funct Integr Genomics 15:163–173. https://doi.org/10.1007/s10142-014-0425-9

Forján R, Rodríguez-Vila A, Cerqueira B, Covelo EF (2018) Effects of compost and technosol amendments on metal concentrations in a mine soil planted with *Brassica juncea* L. Environ Sci Pollut Res 25:19713–19727. https://doi.org/10.1007/s11356-018-2173-1

Franco-Correa M, Quintana A, Duque C, Suarez C, Rodríguez MX, Barea J-M (2010) Evaluation of actinomycete strains for key traits related with plant growth promotion and mycorrhiza helping activities. Appl Soil Ecol 45:209–217. https://doi.org/10.1016/j.apsoil.2010.04.007

Fuentes MS, Raimondo EE, Amoroso MJ, Benimeli CS (2017) Removal of a mixture of pesticides by a *Streptomyces* consortium: influence of different soil systems. Chemosphere 173:359–367. https://doi.org/10.1016/j.chemosphere.2017.01.044

Fuentes MS, Sineli PE, Pons S, de Moreno de LeBlanc A, Benimeli CS, Hill RT, Cuozzo SA (2018) Study of the removal of a pesticides mixture by a *Streptomyces* strain and their effect on the cytotoxicity of treated systems. J Environ Chem Eng 6:6836–6843. https://doi.org/10.1016/j.jece.2018.10.023

Funayama S, Isono K (2014) Biosynthesis of the polyoxins, nucleoside peptide antibiotics: formation of 5-carboxyuracil nucleosides by *Streptomyces cacaoi*. Agric Biol Chem 40:1039–1044. https://doi.org/10.1080/00021369.1976.10862152

Getha K, Vikineswary S (2002) Antagonistic effects of *Streptomyces violaceusniger* strain G10 on *Fusarium oxysporum* f.sp. cubense race 4: indirect evidence for the role of antibiosis in the antagonistic process. J Ind Microbiol Biotechnol 28:303–310. https://doi.org/10.1038/sj/jim/7000247

Gómez C, Olano C, Palomino-Schätzlein M, Pineda-Lucena A, Carbajo RJ, Braña AF, Méndez C, Salas JA (2012) Novel compounds produced by *Streptomyces lydicus* NRRL 2433 engineered mutants altered in the biosynthesis of streptolydigin. J Antibiot 65:341–348. https://doi.org/10.1038/ja.2012.37

Goodfellow M, Williams ST, Mordarski M (1988) Actinomycetes in biotechnology. Academic Press, San Diego

Gopalakrishnan S, Pande S, Sharma M, Humayun P, Kiran BK, Sandeep D, Vidya MS, Deepthi K, Rupela O (2011) Evaluation of actinomycete isolates obtained from herbal vermicompost for the biological control of Fusarium wilt of chickpea. Crop Protect 30:1070–1078. https://doi.org/10.1016/j.cropro.2011.03.006

Gopalakrishnan S, Srinivas V, Alekhya G, Prakash B, Kudapa H, Rathore A, Varshney RK (2015) The extent of grain yield and plant growth enhancement by plant growth-promoting broad-spectrum *Streptomyces* sp. in chickpea. SpringerPlus 4. https://doi.org/10.1186/s40064-015-0811-3

Graf E, Schneider K, Nicholson G, Ströbele M, Jones AL, Goodfellow M, Beil W, RD S¨s, Fiedler H-P (2007) Elloxazinones A and B, new aminophenoxazinones from *Streptomyces griseus* Acta 2871. J Antibiot 60:277–284. https://doi.org/10.1038/ja.2007.35

Grigorevski de Lima AL, Pires do Nascimento R, da Silva Bon EP, Coelho RRR (2005) *Streptomyces drozdowiczii* cellulase production using agro-industrial by-products and its potential use in the detergent and textile industries. Enzym Microb Technol 37:272–277. https://doi.org/10.1016/j.enzmictec.2005.03.016

Harir M, Bendif H, Bellahcene M, Fortas RPZ (2018) *Streptomyces* secondary metabolites. In: Enany S (ed) Basic biology and applications of actinobacteria. InTechOpen, pp 99–121. https://doi.org/10.5772/intechopen.79890

Harrison PH, Noguchi H, Vederas JC (1986) Biosynthesis of polyene antibiotics: intact incorporation of carbon-13 labeled octanoate into fungichromin by *Streptomyces cellulosae*. J Am Chem Soc 108:3833–3834. https://doi.org/10.1021/ja00273a050

Hibbing ME, Fuqua C, Parsek MR, Peterson SB (2009) Bacterial competition: surviving and thriving in the microbial jungle. Nat Rev Microbiol 8:15–25. https://doi.org/10.1038/nrmicro2259

Hillis DM, Book AJ, Lewin GR, McDonald BR, Takasuka TE, Wendt-Pienkowski E, Doering DT, Suh S, Raffa KF, Fox BG, Currie CR (2016) Evolution of high cellulolytic activity in symbiotic *Streptomyces* through selection of expanded gene content and coordinated gene expression. PLoS Biol 14. https://doi.org/10.1371/journal.pbio.1002475

Hochstein FA, Murai K (1954) Magnamycin B, a second antibiotic from *Streptomyces halstedii*. J Am Chem Soc 76:5080–5083. https://doi.org/10.1021/ja01649a020

Hotta K, Ogata T, Ishikawa JUN, Okanishi M, Mizuno S, MoRiOka M, Naganawa H, Okami Y (1996) Mechanismm of multiple aminoglycoside resistance of kasugamycin-producing *Streptomyces kasugaensis* MB273: involvement of two types of acetyltransferases in resistance to astromicin group antibiotics. J Antibiot 49:682–688. https://doi.org/10.7164/antibiotics.49.682

Htwe AZ, Yamakawa T (2016) Low-density co-inoculation with *Bradyrhizobium japonicum* SAY3-7 and *Streptomyces griseoflavus* P4 promotes plant growth and nitrogen fixation in soybean cultivars. Am J Plant Sci 7:1652–1661. https://doi.org/10.4236/ajps.2016.712156

Htwe AZ, Moh SM, Moe K, Yamakawa T (2018) Effects of co-inoculation of *Bradyrhizobium japonicum* SAY3-7 and *Streptomyces griseoflavus* P4 on plant growth, nodulation, nitrogen fixation, nutrient uptake, and yield of soybean in a field condition. Soil Sci Plant Nutr 64:222–229. https://doi.org/10.1080/00380768.2017.1421436

Huang X, Zhang N, Yong X, Yang X, Shen Q (2012) Biocontrol of *Rhizoctonia solani* damping-off disease in cucumber with *Bacillus pumilus* SQR-N43. Microbiol Res 167:135–143. https://doi.org/10.1016/j.micres.2011.06.002

Hwang BK, Lim SW, Kim BS, Lee JY, Moon SS (2001) Isolation and *in vivo* and *in vitro* antifungal activity of phenylacetic acid and sodium phenylacetate from *Streptomyces humidus*. Appl Environ Microbiol 67:3739–3745. https://doi.org/10.1128/aem.67.8.3739-3745.2001

Igarashi Y, Iida T, Sasaki T, Saito N, Yoshida R, Furumai T (2002) Isolation of actinomycetes from live plants and evaluation of antiphytopathogenic activity of their metabolites. Actinomycetologica 16:9–13. https://doi.org/10.3209/saj.16_9

Il'ina AV, Tatarinova N, Tikhonov VE, Varlamov VP (2000) Extracellular proteinases and chitinases, produced by a *Streptomyces kurssanovii* culture. Prikl Biokhim Mikrobiol 36:173–177

Irdani T, Perito B, Mastromei G (1996) Characterization of a *Streptomyces rochei* endoglucanasea. Ann N Y Acad Sci 782:173–181. https://doi.org/10.1111/j.1749-6632.1996.tb40558.x

Jang H-D, Chang K-S (2005) Thermostable cellulases from *Streptomyces* sp.: scale-up production in a 50-l fermenter. Biotechnol Lett 27:239–242. https://doi.org/10.1007/s10529-004-8356-5

Janssen GR, Ward JM, Bibb MJ (1989) Unusual transcriptional and translational features of the aminoglycoside phosphotransferase gene (aph) from *Streptomyces fradiae*. Genes Dev 3:415–429. https://doi.org/10.1101/gad.3.3.415

Jezequel K, Lebeau T (2008) Soil bioaugmentation by free and immobilized bacteria to reduce potentially phytoavailable cadmium. Bioresour Technol 99:690–698. https://doi.org/10.1016/j.biortech.2007.02.002

Ji Z, Qiao G, Wei S, Fan L, Wu W (2012) Isolation and characterization of two novel antibacterial cyclic hexapeptides from *Streptomyces alboflavus* 313. Chem Biodivers 9:1567–1578. https://doi.org/10.1002/cbdv.201100364

Jiang H, Wang Y-Y, Ran X-X, Fan W-M, Jiang X-H, Guan W-J, Li Y-Q (2013) Improvement of natamycin production by engineering of phosphopantetheinyl transferases in *Streptomyces chattanoogensis* L10. Appl Environ Microbiol 79:3346–3354. https://doi.org/10.1128/aem.00099-13

Kakinuma K, Hanson CA, Rinehart KL (1976) Spectinabilin, a new nitro-containing metabolite isolated from *Streptomyces spectabilis*. Tetrahedron 32:217–222. https://doi.org/10.1016/0040-4020(76)87004-4

Kanini GS, Katsifas EA, Savvides AL, Karagouni AD (2013) *Streptomyces rochei* ACTA1551, an indigenous Greek isolate studied as a potential biocontrol agent against *Fusarium oxysporum* f. sp. lycopersici. Biomed Res Int 2013:1–10. https://doi.org/10.1155/2013/387230

Kato S, Shindo K, Kawai H, Odagawa A, Matsuoka M, Mochizuki J (1993) Pyrrolostatin, a novel lipid peroxidation inhibitor from *Streptomyces chrestomyceticus*. Taxonomy, fermentation, isolation, structure elucidation and biological properties. J Antibiot 46:892–899. https://doi.org/10.7164/antibiotics.46.892

Kawasaki T, Sakurai F, S-y N, Hayakawa Y (2009) Prodigiosin biosynthesis gene cluster in the roseophilin producer *Streptomyces griseoviridis*. J Antibiot 62:271–276. https://doi.org/10.1038/ja.2009.27

Khamna S, Yokota A, Lumyong S (2008) Actinomycetes isolated from medicinal plant rhizosphere soils: diversity and screening of antifungal compounds, indole-3-acetic acid and siderophore production. World J Microbiol Biotechnol 25:649–655. https://doi.org/10.1007/s11274-008-9933-x

Khamna S, Yokota A, Peberdy JF, Lumyong S (2010) Indole-3-acetic acid production by *Streptomyces* sp. isolated from some Thai medicinal plant rhizosphere soils. EurAsian J Biosci:23–32. https://doi.org/10.5053/ejobios.2010.4.0.4

Khiyami MA, Pometto AL, Brown RC (2005) Detoxification of corn Stover and corn starch pyrolysis liquors by *Pseudomonas putida* and *Streptomyces setonii* suspended cells and plastic compost support biofilms. J Agric Food Chem 53:2978–2987. https://doi.org/10.1021/jf048224e

Kim YM, Kim JH (2004) Formation and dispersion of mycelial pellets of *Streptomyces coelicolor* A3(2). J Microbiol 42:64–67

Kim Y, Kim H, Beuchat LR, Ryu J-H (2019) Inhibition of *Listeria monocytogenes* using biofilms of non-pathogenic soil bacteria (*Streptomyces* spp.) on stainless steel under desiccated condition. Food Microbiol 79:61–65. https://doi.org/10.1016/j.fm.2018.11.007

Klausen M, Heydorn A, Ragas P, Lambertsen L, Aaes-Jørgensen A, Molin S, Tolker-Nielsen T (2003) Biofilm formation by *Pseudomonas aeruginosa* wild type, flagella and type IV pili mutants. Mol Microbiol 48:1511–1524. https://doi.org/10.1046/j.1365-2958.2003.03525.x

Klein E, Ofek M, Katan J, Minz D, Gamliel A (2013) Soil suppressiveness to Fusarium disease: shifts in root microbiome associated with reduction of pathogen root colonization. Phytopathology 103:23–33. https://doi.org/10.1094/phyto-12-11-0349

Kpomblekou-A K, Tabatabai MA (1994) Effect of organic acids on release of phosphorus from phosphate rocks. Soil Sci 158:442–453. https://doi.org/10.1097/00010694-199415860-00006

Kraemer SM, Crowley DE, Kretzschmar R (2006) Geochemical aspects of phytosiderophore-promoted iron acquisition by plants. Adv Agron 91:1–46. https://doi.org/10.1016/s0065-2113(06)91001-3

Lang Z, Qi D, Dong J, Ren L, Zhu Q, Huang W, Liu Y, Lu D (2018) Isolation and characterization of a quinclorac-degrading Actinobacteria *Streptomyces* sp. strain AH-B and its implication on microecology in contaminated soil. Chemosphere 199:210–217. https://doi.org/10.1016/j.chemosphere.2018.01.133

Laskin AI, Lechevalier HA (1977) CRC handbook of microbiology, 2nd edn. CRC Press, Boca Raton, FL

Lee JY (2005) *Streptomyces koyangensis* sp. nov., a novel actinomycete that produces 4-phenyl-3-butenoic acid. Int J Syst Evol Microbiol 55:257–262. https://doi.org/10.1099/ijs.0.63168-0

Lee JY, Lee JY, Moon SS, Hwang BK (2005) Isolation and antifungal activity of 4-phenyl-3-butenoic acid from *Streptomyces koyangensis* strain VK-A60. J Agric Food Chem 53:7696–7700. https://doi.org/10.1021/jf050957r

Li ZHE, Rawlings BJ, Harrison PH, Vederas JC (1989) Production of new polyene antibiotics by *Streptomyces cellulosae* after addition of ethyl(Z)-16-phenylhexadec-9-enoate. J Antibiot 42:577–584. https://doi.org/10.7164/antibiotics.42.577

Lian Q, Zhang J, Gan L, Ma Q, Zong Z, Wang Y (2017) The biocontrol efficacy of *Streptomyces pratensis* LMM15 on *Botrytis cinerea* in tomato. Biomed Res Int 2017:1–11. https://doi.org/10.1155/2017/9486794

Lim S-W, Kim J-D, Kim B-S, Hwang B-K (2000) Isolation and numerical identification of *Streptomyces humidus* strain S5-55 antagonistic to plant pathogenic fungi. Plant Pathol J 16:189–199

Liu C, Wang X, Yan Y, Wang J, Zhang B, Zhang J, Xiang W (2012) *Streptomyces heilongjiangensis* sp. nov., a novel actinomycete that produces borrelidin isolated from the root surface of soybean [*Glycine max* (L.) Merr]. Int J Syst Evol Microbiol 63:1030–1036. https://doi.org/10.1099/ijs.0.041483-0

Liu Y, Ryu H, Ge B, Pan G, Sun L, Park K, Zhang K (2014) Improvement of Wuyiencin biosynthesis in *Streptomyces wuyiensis* CK-15 by identification of a key regulator, WysR. J Microbiol Biotechnol 24:1644–1653. https://doi.org/10.4014/jmb.1405.05017

Locatelli FM, Goo K-S, Ulanova D (2016) Effects of trace metal ions on secondary metabolism and the morphological development of streptomycetes. Metallomics 8:469–480. https://doi.org/10.1039/c5mt00324e

Luo L, Cai J, Wang C, Lin J, Du X, Zhou A, Xiang M (2016) Purification and characterization of an alkaliphilic endo-xylanase from *Streptomyces althioticus* LMZM and utilization in the pulp paper industry. J Chem Technol Biotechnol 91:1093–1098. https://doi.org/10.1002/jctb.4690

Ma S-C, Zhang H-B, Ma S-T, Wang R, Wang G-X, Shao Y, Li C-X (2015) Effects of mine wastewater irrigation on activities of soil enzymes and physiological properties, heavy metal uptake and grain yield in winter wheat. Ecotoxicol Environ Saf 113:483–490. https://doi.org/10.1016/j.ecoenv.2014.12.031

Macagnan D, Romeiro RS, Pomella AWV, deSouza JT (2008) Production of lytic enzymes and siderophores, and inhibition of germination of basidiospores of *Moniliophthora (ex Crinipellis) perniciosa* by phylloplane actinomycetes. Biol Control 47:309–314. https://doi.org/10.1016/j.biocontrol.2008.08.016

Magae J, Kataoka T, Yamashita M, Tsuji RF, Lee M-H, Yamasaki M, Nagai K (1993) Suppression of killer T cell induction by prodigiosin 25-C. In: Animal cell technology: basic & applied aspects. pp 149–154. doi:https://doi.org/10.1007/978-94-011-2044-9_21

Mahan KM, Klingeman DM, Hettich RL, Parry RJ, Graham DE (2016) Draft genome sequence of *Streptomyces vitaminophilus* ATCC 31673, a producer of pyrrolomycin antibiotics, some of which contain a nitro group. Genome Announc 4. https://doi.org/10.1128/genomeA.01582-15

Maheswari MU, Chandra TS (2000) Production and potential applications of a xylanase from a new strain of *Streptomyces cuspidosporus*. World J Microbiol Biotechnol 16:257–263. https://doi.org/10.1023/a:1008945931108

Mansfield J, Genin S, Magori S, Citovsky V, Sriariyanum M, Ronald P, Dow MAX, Verdier V, Beer SV, Machado MA, Toth IAN, Salmond G, Foster GD (2012) Top 10 plant pathogenic bacteria in molecular plant pathology. Mol Plant Pathol 13:614–629. https://doi.org/10.1111/j.1364-3703.2012.00804.x

Marques APGC, Pires C, Moreira H, Rangel AOSS, Castro PML (2010) Assessment of the plant growth promotion abilities of six bacterial isolates using *Zea mays* as indicator plant. Soil Biol Biochem 42:1229–1235. https://doi.org/10.1016/j.soilbio.2010.04.014

McGhee GC, Sundin GW (2011) Evaluation of kasugamycin for fire blight management, effect on nontarget bacteria, and assessment of kasugamycin resistance potential in *Erwinia amylovora*. Phytopathology 101:192–204. https://doi.org/10.1094/phyto-04-10-0128

Mehta CM, Palni U, Franke-Whittle IH, Sharma AK (2014) Compost: its role, mechanism and impact on reducing soil-borne plant diseases. Waste Manag 34:607–622. https://doi.org/10.1016/j.wasman.2013.11.012

Miao V (2005) Daptomycin biosynthesis in *Streptomyces roseosporus*: cloning and analysis of the gene cluster and revision of peptide stereochemistry. Microbiology 151:1507–1523. https://doi.org/10.1099/mic.0.27757-0

Miyadoh S, Shomura T, Ito T, Niida T (1983) *Streptomyces sulfonofaciens* sp. nov. Int J Syst Bacteriol 33:321–324. https://doi.org/10.1099/00207713-33-2-321

Mohandas S, Poovarasan S, Panneerselvam P, Saritha B, Upreti KK, Kamal R, Sita T (2013) Guava (*Psidium guajava* L.) rhizosphere *Glomus mosseae* spores harbor actinomycetes with growth promoting and antifungal attributes. Sci Hortic 150:371–376. https://doi.org/10.1016/j.scienta.2012.11.019

Monier JM, Lindow SE (2003) Differential survival of solitary and aggregated bacterial cells promotes aggregate formation on leaf surfaces. Proc Natl Acad Sci 100:15977–15982. https://doi.org/10.1073/pnas.2436560100

Morales DK, Ocampo W, Zambrano MM (2007) Efficient removal of hexavalent chromium by a tolerant *Streptomyces* sp. affected by the toxic effect of metal exposure. J Appl Microbiol 103:2704–2712. https://doi.org/10.1111/j.1365-2672.2007.03510.x

Murakami T, Anzai H, Imai S, Satoh A, Nagaoka K, Thompson CJ (1986) The bialaphos biosynthetic genes of *Streptomyces hygroscopicus*: molecular cloning and characterization of the gene cluster. MGG Mol Gen Genet 205:42–53. https://doi.org/10.1007/bf02428031

Murakami S, Harada S, Takahashi Y, Naganawa H, Takeuchi T, Aoyagi T (2008) Piperastatin B: A new selective serine carboxypeptidase inhibitor from *Streptomyces lavendofoliae* MJ908-WF13. J Enzym Inhib 11:51–66. https://doi.org/10.3109/14756369609038222

Nascimento RP, Junior NA, Pereira Jr N, Bon EPS, Coelho RRR (2009) Brewer's spent grain and corn steep liquor as substrates for cellulolytic enzymes production by *Streptomyces malaysiensis*. Lett Appl Microbiol 48:529–535. https://doi.org/10.1111/j.1472-765X.2009.02575.x

Neilan B, Hao C, Huang S, Deng Z, Zhao C, Yu Y (2014) Mining of the pyrrolamide antibiotics analogs in *Streptomyces netropsis* reveals the amidohydrolase-dependent "iterative strategy" underlying the pyrrole polymerization. PLoS One 9. https://doi.org/10.1371/journal.pone.0099077

Nguyen TM, Kim J (2015) Description of *Streptomyces fabae* sp. nov., a producer of antibiotics against microbial pathogens, isolated from soybean (*Glycine max*) rhizosphere soil. Int J Syst Evol Microbiol 65:4151–4156. https://doi.org/10.1099/ijsem.0.000551

Ohnishi Y, Ishikawa J, Hara H, Suzuki H, Ikenoya M, Ikeda H, Yamashita A, Hattori M, Horinouchi S (2008) Genome sequence of the streptomycin-producing microorganism *Streptomyces griseus* IFO 13350. J Bacteriol 190:4050–4060. https://doi.org/10.1128/jb.00204-08

Ostash B, Gren T, Hrubskyy Y, Tistechok S, Beshley S, Baranov V, Fedorenko V (2014) Cultivable actinomycetes from rhizosphere of birch (*Betula pendula*) growing on a coal mine dump in Silets, Ukraine. J Basic Microbiol 54:851–857. https://doi.org/10.1002/jobm.201200551

Park JO, El-Tarabily KA, Ghisalberti EL, Sivasithamparam K (2002) Pathogenesis of *Streptoverticillium albireticuli* on *Caenorhabditis elegans* and its antagonism to soil-borne fungal pathogens. Lett Appl Microbiol 35:361–365. https://doi.org/10.1046/j.1472-765x.2002.01194.x

Parot S, Délia M-L, Bergel A (2008) Forming electrochemically active biofilms from garden compost under chronoamperometry. Bioresour Technol 99:4809–4816. https://doi.org/10.1016/j.biortech.2007.09.047

Passari AK, Mishra VK, Saikia R, Gupta VK, Singh BP (2015) Isolation, abundance and phylogenetic affiliation of endophytic actinomycetes associated with medicinal plants and screening for their in vitro antimicrobial biosynthetic potential. Front Microbiol 6. https://doi.org/10.3389/fmicb.2015.00273

Passari AK, Chandra P, Zothanpuia MVK, Leo VV, Gupta VK, Kumar B, Singh BP (2016) Detection of biosynthetic gene and phytohormone production by endophytic actinobacteria associated with *Solanum lycopersicum* and their plant-growth-promoting effect. Res Microbiol 167:692–705. https://doi.org/10.1016/j.resmic.2016.07.001

Payero TD, Vicente CM, Rumbero Á, Barreales EG, Santos-Aberturas J, de Pedro A, Aparicio JF (2015) Functional analysis of filipin tailoring genes from *Streptomyces filipinensis* reveals alternative routes in filipin III biosynthesis and yields bioactive derivatives. Microb Cell Factories 14. https://doi.org/10.1186/s12934-015-0307-4

Peng D, Li S, Wang J, Chen C, Zhou M (2014) Integrated biological and chemical control of rice sheath blight by *Bacillus subtilis* NJ-18 and jinggangmycin. Pest Manag Sci 70:258–263. https://doi.org/10.1002/ps.3551

Pernodet JL, Alegre MT, Blondelet-Rouault MH, Guerineau M (1993) Resistance to spiramycin in *Streptomyces ambofaciens*, the producer organism, involves at least two different mechanisms. J Gen Microbiol 139:1003–1011. https://doi.org/10.1099/00221287-139-5-1003

Peterson RM, Huang T, Rudolf Jeffrey D, Smanski Michael J, Shen B (2014) Mechanisms of self-resistance in the platensimycin- and platencin-producing *Streptomyces platensis* MA7327 and MA7339 strains. Chem Biol 21:389–397. https://doi.org/10.1016/j.chembiol.2014.01.005

Phieler R, Merten D, Roth M, Büchel G, Kothe E (2015) Phytoremediation using microbially mediated metal accumulation in *Sorghum bicolor*. Environ Sci Pollut Res 22:19408–19416. https://doi.org/10.1007/s11356-015-4471-1

Prakash D, Sharma G (2014) Phytochemicals of nutraceutical importance. CABI/CAB International, Wallingford

Pugliese M, Liu B, Gullino ML, Garibaldi A (2008) Selection of antagonists from compost to control soil-borne pathogens. J Plant Dis Prot 115:220–228. https://doi.org/10.1007/BF03356267

Ramírez P, Coha JM (2003) Enzymatic degradation of cellulose for thermophilic actinomycete: isolation, characterization and cellulolytic activity determination. Rev Peru Biol 10:67–77

Rashad FM, Saleh WD, Moselhy MA (2010) Bioconversion of rice straw and certain agro-industrial wastes to amendments for organic farming systems: 1. Composting, quality, stability and maturity indices. Bioresour Technol 101:5952–5960. https://doi.org/10.1016/j.biortech.2010.02.103

Rashad FM, Fathy HM, El-Zayat AS, Elghonaimy AM (2015) Isolation and characterization of multifunctional *Streptomyces* species with antimicrobial, nematicidal and phytohormone activities from marine environments in Egypt. Microbiol Res 175:34–47. https://doi.org/10.1016/j.micres.2015.03.002

Reisner A, Haagensen JAJ, Schembri MA, Zechner EL, Molin S (2003) Development and maturation of *Escherichia coli* K-12 biofilms. Mol Microbiol 48:933–946. https://doi.org/10.1046/j.1365-2958.2003.03490.x

Rey T, Dumas B (2017) Plenty is no plague: *Streptomyces* symbiosis with crops. Trends Plant Sci 22:30–37. https://doi.org/10.1016/j.tplants.2016.10.008

Saez JM, Casillas García V, Benimeli CS (2017) Improvement of lindane removal by *Streptomyces* sp. M7 by using stable microemulsions. Ecotoxicol Environ Saf 144:351–359. https://doi.org/10.1016/j.ecoenv.2017.06.026

Sanglier JJ, Haag H, Huck TA, Fehr T (1993) Novel bioactive compounds from Actinomycetes: a short review (1988–1992). Res Microbiol 144:633–642. https://doi.org/10.1016/0923-2508(93)90066-b

Santos JFD, Sousa CDS, Soares ACF, Lima FDS, Barbosa DHSG (2016) Actinobacteria and organic fertilizers for management of the nematode *Scutellonema bradys* in yam plants. Rev Caatinga 29:548–558. https://doi.org/10.1590/1983-21252016v29n304rc

Sathya A, Vijayabharathi R, Srinivas V, Gopalakrishnan S (2016) Plant growth-promoting actinobacteria on chickpea seed mineral density: an upcoming complementary tool for sustainable biofortification strategy. 3 Biotech 6. https://doi.org/10.1007/s13205-016-0458-y

Sathya A, Vijayabharathi R, Gopalakrishnan S (2017) Plant growth-promoting actinobacteria: a new strategy for enhancing sustainable production and protection of grain legumes. 3 Biotech 7. https://doi.org/10.1007/s13205-017-0736-3

Satpathy S, Sen SK, Pattanaik S, Raut S (2016) Review on bacterial biofilm: an universal cause of contamination. Biocatal Agric Biotechnol 7:56–66. https://doi.org/10.1016/j.bcab.2016.05.002

Sauer K, Camper AK, Ehrlich GD, Costerton JW, Davies DG (2002) *Pseudomonas aeruginosa* displays multiple phenotypes during development as a biofilm. J Bacteriol 184:1140–1154. https://doi.org/10.1128/jb.184.4.1140-1154.2002

Schatz A, Bugle E, Waksman SA (1944) Streptomycin, a substance exhibiting antibiotic activity against Gram-positive and Gram-negative bacteria. Exp Biol Med 55:66–69. https://doi.org/10.3181/00379727-55-14461

Schirawski J, Perlin M (2018) Plant–microbe interaction 2017—the good, the bad and the diverse. Int J Mol Sci 19. https://doi.org/10.3390/ijms19051374

Schrey SD, Tarkka MT (2008) Friends and foes: streptomycetes as modulators of plant disease and symbiosis. Antonie Van Leeuwenhoek 94:11–19. https://doi.org/10.1007/s10482-008-9241-3

Schütze E, Klose M, Merten D, Nietzsche S, Senftleben D, Roth M, Kothe E (2014) Growth of streptomycetes in soil and their impact on bioremediation. J Hazard Mater 267:128–135. https://doi.org/10.1016/j.jhazmat.2013.12.055

Semedo LTAS (2004) *Streptomyces drozdowiczii* sp. nov., a novel cellulolytic streptomycete from soil in Brazil. Int J Syst Evol Microbiol 54:1323–1328. https://doi.org/10.1099/ijs.0.02844-0

Sharma A, Saha TN, Arora A, Shah R, Nain L (2017) Efficient microorganism compost benefits plant growth and improves soil health in calendula and marigold. Horticult Plant J 3:67–72. https://doi.org/10.1016/j.hpj.2017.07.003

Sheldon PJ, Mao Y, He M, Sherman DH (1999) Mitomycin resistance in *Streptomyces lavendulae* includes a novel drug-binding-protein-dependent export system. J Bacteriol 181:2507–2512

Shen T, Wang C, Yang H, Deng Z, Wang S, Shen B, Shen Q (2016) Identification, solid-state fermentation and biocontrol effects of *Streptomyces hygroscopicus* B04 on strawberry root rot. Appl Soil Ecol 103:36–43. https://doi.org/10.1016/j.apsoil.2016.02.016

Shi X, Zhu X (2009) Biofilm formation and food safety in food industries. Trends Food Sci Technol 20:407–413. https://doi.org/10.1016/j.tifs.2009.01.054

Shi L, Nwet TT, Ge B, Zhao W, Liu B, Cui H, Zhang K (2018) Antifungal and plant growth-promoting activities of *Streptomyces roseoflavus* strain NKZ-259. Biol Control 125:57–64. https://doi.org/10.1016/j.biocontrol.2018.06.012

Simões M, Simões LC, Vieira MJ (2010) A review of current and emergent biofilm control strategies. LWT Food Sci Technol 43:573–583. https://doi.org/10.1016/j.lwt.2009.12.008

Singh SP, Gaur R (2016) Evaluation of antagonistic and plant growth promoting activities of chitinolytic endophytic actinomycetes associated with medicinal plants against *Sclerotium rolfsii* in chickpea. J Appl Microbiol 121:506–518. https://doi.org/10.1111/jam.13176

Singh O, Srivastava S (1998) Dictionary of entomology, plant pathology and nematology. Concept Publishing Company, New Delhi

Singh A, Gupta R, Srivastava M, Gupta MM, Pandey R (2016) Microbial secondary metabolites ameliorate growth, in planta contents and lignification in *Withania somnifera* (L.) Dunal. Physiol Mol Biol Plants 22:253–260. https://doi.org/10.1007/s12298-016-0359-x

Singh RP, Manchanda G, Maurya IK, Maheshwari NK, Tiwari PK, Rai AR (2019) *Streptomyces* from rotten wheat straw endowed the high plant growth potential traits and agro-active compounds. Biocatal Agric Biotechnol 17:507–513. https://doi.org/10.1016/j.bcab.2019.01.014

Smanski MJ, Peterson RM, Rajski SR, Shen B (2009) Engineered *Streptomyces platensis* strains that overproduce antibiotics platensimycin and platencin. Antimicrob Agents Chemother 53:1299–1304. https://doi.org/10.1128/aac.01358-08

Strom PF (1985) Identification of thermophilic bacteria in solid-waste composting. Appl Environ Microbiol 50:906–913

Sveshnikova M, Timuk O, Borisova V, Fedorova G (1983) New species of actinomycete *Streptomyces variegatus* sp. nov. synthesizing an antibiotic of the alpha-hydroxyketopentaene group. Antibiotiki 28:723–728

Takahashi Y (2002) *Streptomyces avermectinius* sp. nov., an avermectin-producing strain. Int J Syst Evol Microbiol 52:2163–2168. https://doi.org/10.1099/ijs.0.02237-0

Tamreihao K, Ningthoujam DS, Nimaichand S, Singh ES, Reena P, Singh SH, Nongthomba U (2016) Biocontrol and plant growth promoting activities of a *Streptomyces corchorusii* strain UCR3-16 and preparation of powder formulation for application as biofertilizer agents for rice plant. Microbiol Res 192:260–270. https://doi.org/10.1016/j.micres.2016.08.005

Tikhonov VE, Radigina LA, Yamskov IA, Gulyaeva ND, Ilyina AV, Anisimova MV, Varlamov VP, Tatarinova NY (1998) Affinity purification of major chitinases produced by *Streptomyces kurssanovii*. Enzym Microb Technol 22:82–85. https://doi.org/10.1016/s0141-0229(97)00099-9

Timková I, Sedláková-Kaduková J, Pristaš P (2018) Biosorption and bioaccumulation abilities of actinomycetes/streptomycetes isolated from metal contaminated sites. Separations 5. https://doi.org/10.3390/separations5040054

Trejo WH, Dean LD, Pluscec J, Meyers E, Brown WE (1977) *Streptomyces laurentii*, a new species producing thiostrepton. J Antibiot 30:639–643. https://doi.org/10.7164/antibiotics.30.639

Trejo-Estrada SR, Paszczynski A, Crawford DL (1998) Antibiotics and enzymes produced by the biocontrol agent *Streptomyces violaceusniger* YCED-9. J Ind Microbiol Biotechnol 21:81–90. https://doi.org/10.1038/sj.jim.2900549

Ueberschaar N, Ndejouong BLST, Ding L, Maier A, Fiebig H-H, Hertweck C (2011) Hydrazidomycins, cytotoxic alkylhydrazides from *Streptomyces atratus*. Bioorg Med Chem Lett 21:5839–5841. https://doi.org/10.1016/j.bmcl.2011.07.108

Uri J, Békési I (1958) Flavofungin, a new crystalline antifungal antibiotic: origin and biological properties. Nature 181:908–908. https://doi.org/10.1038/181908a0

Uyeda M, Mizukami M, Yokomizo K, Suzuki K (2014) Pentalenolactone I and hygromycin A, immunosuppressants produced by *Streptomyces filipinensis* and *Streptomyces hygroscopicus*. Biosci Biotechnol Biochem 65:1252–1254. https://doi.org/10.1271/bbb.65.1252

van Balken JAM (1997) Biotechnological innovations in chemical synthesis. Biotechnology by open learning. Butterworth Heinemann, Oxford

Ventorino V, Ionata E, Birolo L, Montella S, Marcolongo L, de Chiaro A, Espresso F, Faraco V, Pepe O (2016) Lignocellulose-adapted endo-cellulase producing *Streptomyces* strains for bioconversion of cellulose-based materials. Front Microbiol 7. https://doi.org/10.3389/fmicb.2016.02061

Verma S, Sharma A, Kumar R, Kaur C, Arora A, Shah R, Nain L (2015) Improvement of antioxidant and defense properties of tomato (var. Pusa Rohini) by application of bioaugmented compost. Saudi J Biol Sci 22:256–264. https://doi.org/10.1016/j.sjbs.2014.11.003

Vining LC (2014) Chapter 5 - Peptides. In: Tsao GT (ed) Annual reports on fermentation processes, vol 4, 2nd edn. Elsevier, New York, pp 123–163. https://doi.org/10.1016/B978-0-12-040304-2.50012-6

Virolle M-J, Passari AK, Mishra VK, Gupta VK, Yadav MK, Saikia R, Singh BP (2015) *In vitro* and *in vivo* plant growth promoting activities and DNA fingerprinting of antagonistic endophytic actinomycetes associates with medicinal plants. PLoS One 10. https://doi.org/10.1371/journal.pone.0139468

Vurukonda SSKP, Giovanardi D, Stefani E (2018) Plant growth promoting and biocontrol activity of *Streptomyces* spp. as endophytes. Int J Mol Sci 19. https://doi.org/10.3390/ijms19040952

Waksman SA, Lechevalier HA (1949) Neomycin, a new antibiotic active against streptomycin-resistant bacteria, including tuberculosis organisms. Science 109:305–307. https://doi.org/10.1126/science.109.2830.305

Wan Z, Fang W, Shi L, Wang K, Zhang Y, Zhang Z, Wu Z, Yang Z, Gu Y (2014) Novonestmycins A and B, two new 32-membered bioactive macrolides from *Streptomyces phytohabitans* HBERC-20821. J Antibiot 68:185–190. https://doi.org/10.1038/ja.2014.123

Wang C (2018) Phosphorus-release dynamics by phosphate solubilizing actinomycetes and its enhancement of growth and yields in maize. Int J Agric Biol 437–444. doi:https://doi.org/10.17957/ijab/15.0554

Wang C, Wang Z, Qiao X, Li Z, Li F, Chen M, Wang Y, Huang Y, Cui H (2013a) Antifungal activity of volatile organic compounds from *Streptomyces alboflavus* TD-1. FEMS Microbiol Lett 341:45–51. https://doi.org/10.1111/1574-6968.12088

Wang X-J, Zhang J, Wang J-D, Qian P-T, Liu C-X, Xiang W-S (2013b) Novel cyclopentenone derivatives produced by a rare actinobacterial strain *Actinoalloteichus nanshanensis* sp. nov. NEAU 119. Nat Prod Res 27:1863–1869. https://doi.org/10.1080/14786419.2013.771349

Wang H, Han L, Feng J, Zhang X (2015) Evaluation of two *Streptomyces* spp. and compost for growth promotion and biocontrol potential against *Rhizoctonia solani* on pepper. Biocontrol Sci Technol 25:852–857. https://doi.org/10.1080/09583157.2015.1015485

Wang S, Liang Y, Shen T, Yang H, Shen B (2016) Biological characteristics of *Streptomyces albospinus* CT205 and its biocontrol potential against cucumber Fusarium wilt. Biocontrol Sci Technol 26:951–963. https://doi.org/10.1080/09583157.2016.1172203

Watve M, Tickoo R, Jog M, Bhole B (2001) How many antibiotics are produced by the genus *Streptomyces*? Arch Microbiol 176:386–390. https://doi.org/10.1007/s002030100345

Wegler R (1981) Insektizide · Bakterizide · Oomyceten-Fungizide / Biochemische und biologische Methoden · Naturstoffe / Insecticides · Bactericides · OOMYCETE Fungicides / Biochemical and Biological Methods · Natural Products, vol 6. Springer, Berlin. doi:https://doi.org/10.1007/978-3-642-67778-6

Winn M, Casey E, Habimana O, Murphy CD (2014) Characteristics of *Streptomyces griseus* biofilms in continuous flow tubular reactors. FEMS Microbiol Lett 352:157–164. https://doi.org/10.1111/1574-6968.12378

Xie Y, Wang B, Liu J, Zhou J, Ma J, Huang H, Ju J (2012) Identification of the biosynthetic gene cluster and regulatory cascade for the synergistic antibacterial antibiotics griseoviridin and viridogrisein in *Streptomyces griseoviridis*. ChemBioChem 13:2745–2757. https://doi.org/10.1002/cbic.201200584

Yandigeri MS, Malviya N, Solanki MK, Shrivastava P, Sivakumar G (2015) Chitinolytic *Streptomyces vinaceusdrappus* S5MW2 isolated from Chilika lake, India enhances plant growth and biocontrol efficacy through chitin supplementation against *Rhizoctonia solani*. World J Microbiol Biotechnol 31:1217–1225. https://doi.org/10.1007/s11274-015-1870-x

Yu J-H, Seipke RF, Barke J, Brearley C, Hill L, Yu DW, Goss RJM, Hutchings MI (2011) A single *Streptomyces* symbiont makes multiple antifungals to support the fungus farming ant *Acromyrmex octospinosus*. PLoS One 6. https://doi.org/10.1371/journal.pone.0022028

Yuan WM, Crawford DL (1995) Characterization of *Streptomyces lydicus* WYEC108 as a potential biocontrol agent against fungal root and seed rots. Appl Environ Microbiol 61:3119–3128

Zappelini C, Alvarez-Lopez V, Capelli N, Guyeux C, Chalot M (2018) *Streptomyces* dominate the soil under *Betula* trees that have naturally colonized a red gypsum landfill. Front Microbiol 9. https://doi.org/10.3389/fmicb.2018.01772

Zhang J, Liu J, Meng L, Ma Z, Tang X, Cao Y, Sun L (2012) Isolation and characterization of plant growth-promoting rhizobacteria from wheat roots by wheat germ agglutinin labeled with fluorescein isothiocyanate. J Microbiol 50:191–198. https://doi.org/10.1007/s12275-012-1472-3

Zhao J, Xue Q-H, Shen G-H, Xue L, Duan J-L, Wang D-S (2012) Evaluation of *Streptomyces* spp. for biocontrol of gummy stem blight (*Didymella bryoniae*) and growth promotion of *Cucumis melo* L. Biocontrol Sci Tech 22:23–37. https://doi.org/10.1080/09583157.2011.636481

Zuo L, Jiang B, Jiang Z, Zhao W, Li S, Liu H, Hong B, Yu L, Zuo L, Wu L (2016) Hangtaimycin, a peptide secondary metabolite discovered from *Streptomyces spectabilis* CPCC 200148 by chemical screening. J Antibiot 69:835–838. https://doi.org/10.1038/ja.2016.29

The manufacturer's authorised representative in the EU is Springer Nature Customer Service Centre GmbH, Europaplatz 3, 69115 Heidelberg, Germany. If you have any concerns regarding our products, please contact ProductSafety@springernature.com

Printed and bound by CPI Group (UK) Ltd, Croydon, CR0 4YY
22/09/2026
02220948-0015